Remote Sensing Technologies for Monitoring and Prediction of Disasters

Remote Sensing Technologies for Monitoring and Prediction of Disasters

Editor: Aston Butler

Callisto Reference
www.callistoreference.com

Callisto Reference,
118-35 Queens Blvd., Suite 400,
Forest Hills, NY 11375, USA

Visit us on the World Wide Web at:
www.callistoreference.com

ISBN: 978-1-64116-828-1 (Hardback)

Cataloging-in-Publication Data

Remote sensing technologies for monitoring and prediction of disasters / edited by Aston Butler.
p. cm.
Includes bibliographical references and index.
ISBN 978-1-64116-828-1
1. Natural disasters--Remote sensing. 2. Natural disasters--Geographic information systems.
3. Environmental monitoring--Remote sensing. 4. Environmental monitoring--Geographic information systems. 5. Emergency management. I. Butler, Aston.

GB5014 .R46 2023
363.34--dc23

Table of Contents

Preface

This book has been an outcome of determined endeavour from a group of educationists in the field. The primary objective was to involve a broad spectrum of professionals from diverse cultural background involved in the field for developing new researches. The book not only targets students but also scholars pursuing higher research for further enhancement of the theoretical and practical applications of the subject.

Disasters refer to severe disruptions that affect the functioning of a community, to an extent which is beyond its coping capacity using its own resources. Technical, natural, and man-made risks, along with a number of other factors make a community vulnerable and exposed to disasters. Remote sensing is a technique of gathering data about the Earth through the use of satellites and drones. Remote sensing data is helpful in monitoring, predicting, and managing disasters. Satellites contain sensors that are capable of taking measurements at regular intervals and at different wavelengths. All the data collected by these sensors can be used for disaster monitoring. It can also be utilized to estimate the rate of recovery in disaster-affected regions, based on signs such as reconstruction, vegetation regrowth, and debris removal. The topics covered in this extensive book deal with the application of remote sensing technologies for monitoring and prediction of disasters. It will provide comprehensive knowledge to the readers.

It was an honour to edit such a profound book and also a challenging task to compile and examine all the relevant data for accuracy and originality. I wish to acknowledge the efforts of the contributors for submitting such brilliant and diverse chapters in the field and for endlessly working for the completion of the book. Last, but not the least; I thank my family for being a constant source of support in all my research endeavours.

Editor

1

ST-CORAbico: A Spatiotemporal Object-Based Bias Correction Method for Storm Prediction Detected by Satellite

Miguel Laverde-Barajas [1,2,3,4,*,†], Gerald A. Corzo [1], Ate Poortinga [3,5], Farrukh Chishtie [3,5], Chinaporn Meechaiya [3,4], Susantha Jayasinghe [3,4], Peeranan Towashiraporn [3,4], Amanda Markert [6,7], David Saah [3,4,8], Lam Hung Son [9], Sothea Khem [9], Surajate Boonya-Aroonnet [10], Winai Chaowiwat [10], Remko Uijlenhoet [11,‡] and Dimitri P. Solomatine [1,2]

1 IHE-Delft Institute for Water Education, Westvest 7, 2611 AX Delft, The Netherlands; g.corzo@un-ihe.org (G.A.C.); d.solomatine@un-ihe.org (D.P.S.)
2 Department of Water Management, Faculty of Civil Engineering and Geosciences, Delft University of Technology, 2628 CN Delft, The Netherlands
3 SERVIR-Mekong, SM Tower, 24th Floor, 979/69 Paholyothin Road, Samsen Nai Phayathai, Bangkok 10400, Thailand; apoortinga@sig-gis.com (A.P.); fchishtie@sig-gis.com (F.C.); chinaporn.m@adpc.net (C.M.); susantha@adpc.net (S.J.); peeranan@adpc.net (P.T.); dsaah@sig-gis.com (D.S.)
4 Asian Disaster Preparedness Center, SM Tower, 24th Floor, 979/69 Paholyothin Road, Samsen Nai Phayathai, Bangkok 10400, Thailand
5 Spatial Informatics Group, LLC, 2529 Yolanda Ct., Pleasanton, CA 94566, USA
6 Earth System Science Center, The University of Alabama in Huntsville, 320 Sparkman Drive, Huntsville, AL 35805, USA; amanda.m.weigel@nasa.gov
7 SERVIR Science Coordination Office, NASA Marshall Space Flight Center, 320 Sparkman Drive, Huntsville, AL 35805, USA
8 Geospatial Analysis Lab, University of San Francisco, 2130 Fulton Street, San Francisco, CA 94117, USA
9 Mekong River Commission, Flood Management and Mitigation Programme, 576 National Road No. 2, Sangkat Chak Angre Krom, Khan Menachey, Phnom Penh 12353, Cambodia; son@mrcmekong.org (L.H.S.); khem@mrcmekong.org (K.S.)
10 Hydroinformatic Institute, 901 Ngam Wong Wan Road, Lat Yao, Chatuchak, Bangkok 10900, Thailand; surajate@hii.or.th (S.B.-A.); winai@hii.or.th (W.C.)
11 Hydrology and Quantitative Water Management Group, Department of Environmental Sciences, Wageningen University, 6700 AA Wageningen, The Netherlands; remko.uijlenhoet@wur.nl
* Correspondence: m.laverde@un-ihe.org
† SM Tower, 24th Floor, 979/69 Paholyothin Road, Samsen Nai Phayathai, Bangkok 10400, Thailand.
‡ Current address: Delft University of Technology, 2628 CN Delft, The Netherlands.

Abstract: Advances in near real-time rainstorm prediction using remote sensing have offered important opportunities for effective disaster management. However, this information is subject to several sources of systematic errors that need to be corrected. Temporal and spatial characteristics of both satellite and in-situ data can be combined to enhance the quality of storm estimates. In this study, we present a spatiotemporal object-based method to bias correct two sources of systematic error in satellites: displacement and volume. The method, Spatiotemporal Contiguous Object-based Rainfall Analysis for Bias Correction (ST-CORAbico), uses the spatiotemporal rainfall analysis ST-CORA incorporated with a multivariate kernel density storm segmentation for describing the main storm event characteristics (duration, spatial extension, volume, maximum intensity, centroid). Displacement and volume are corrected by adjusting the spatiotemporal structure and the intensity distribution, respectively. ST-CORAbico was applied to correct the early version of the Integrated Multi-satellite Retrievals for the Global Precipitation Mission (GPM-IMERG) over the Lower Mekong basin in Thailand during the monsoon season from 2014 to 2017. The performance of ST-CORABico is compared against the Distribution Transformation (DT) and Gamma Quantile Mapping (GQM) probabilistic methods. A total of 120 storm events identified over the study area were classified into short and long-lived storms by using a k-means cluster analysis method. Examples for both storm event types describe the error reduction due to location and magnitude by ST-CORAbico. The results showed that the displacement and magnitude correction made by ST-CORAbico considerably reduced RMSE and bias of GPM-IMERG. In both storm event types, this method showed a lower impact on the spatial correlation of the storm event. In comparison with DT and GQM, ST-CORAbico showed a superior performance, outperforming both approaches. This spatiotemporal bias correction method offers a new approach to enhance the accuracy of satellite-derived information for near real-time estimation of storm events.

Keywords: bias correction; satellite-based precipitation; spatiotemporal analysis; object-based method; storm events

---

1. Introduction

Rainfall is a key component in the hydrological cycle and the primary source of freshwater in many regions. However, climatic extremes, such as floods and landslides that are caused by heavy rainfall events, pose a great threat to communities causing loss of life and damage to properties [1,2]. Flood monitoring and water management applications require a high accuracy representation of rainfall in extreme conditions [3]. This is of particular importance in tropical monsoonal climates, such as the Mekong basin region, where convective storm events are localised in space and time.

Traditionally, rain gauges and ground-based weather radar networks have provided the most reliable precipitation data at the catchment scale [4]. However, in many areas around the world, these data are either scarce or not available. An irregular distribution of these ground-based observation stations makes it difficult to discern spatiotemporal features of convective storms and their associated rainfall fields. Therefore, satellite-based remote sensing measurements have become an important source of rainfall data e.g., [5–7]. These satellite-based measurements enable monitoring of storm events at a quasi-global scale in near real-time. In comparison to other remote systems, such as ground-based weather radar, satellite sensor observations provide a wider coverage, being able to acquire data on storm systems in regions where topographic variations limit or obstruct weather radars [8].

Satellite-based precipitation estimates are derived by combining visible to long-wave infrared (VIS/IR) sensors from the Geosynchronous Earth Orbit (GEO) satellite with Passive Microwave (PMW) sensors from the Low Earth Orbit (LEO) satellites. VIS/IR sensors are relevant to measure albedo and cloud top temperature with a high temporal and spatial resolution [9,10]. On the other hand, PMW sensors can penetrate clouds for measuring thermal emissions, which are attenuated by raindrops with a 3-h interval [11,12]. Currently available and commonly used Satellite-based Precipitation Products (SPP) include the Precipitation Estimation from Remotely Sensed Information using Artificial Neural Networks (PERSIANN) [13], the NOAA Climate Prediction Center MORPHing technique (CMORPH [14]), the Multi-satellite Precipitation Analysis from the Tropical Rainfall Measurement Mission (TMPA) [15,16], and the Integrated Multi-satellitE Retrievals from the Global Precipitation Measurement (GPM-IMERG) [17]. Sun et al. [18] details a comprehensive review of the main global available satellite-based precipitation datasets.

Despite advances in the field of remote sensing, SPP information is subject to several systematic and random errors that require correction e.g., [19–21]. A wide range of bias correction methodologies has been developed to improve the performance of SPP, leveraging ground-based observations. Several examples include linear scaling, local intensity scaling, the power and distribution transformation methods, and Gamma Quantile mapping e.g., [8,22,23]. These methods all adjust SPP as function of rainfall intensity values, ignoring important systematic errors, such as those that are caused by displacement and timing.

In the field of weather forecasting, displacement error in storm prediction has been taken using spatial verification methods into account [24–26]. These "nontraditional" methods do not rely on point-to-point matches between the observed and estimated fields for avoiding double penalties (e.g., rainfall estimated but not observed and vice versa) that are commonly found in traditional approaches. Methods can be broadly grouped into neighbour or fuzzy [27,28], scale separation [29–31], object-based e.g., [24–26], and field transformation [32,33]. The first two categories can be described as spatial filtering methods, in which the verification statistics are evaluated at coarser resolutions to provide information about the scale of the performance. Object-based and field transformation are considered as displacement verification methods when estimated rainfall fields, defined as an object, are spatially manipulated (displacement, rotation, scaling, etc.) to try to fit the observed value.

Several studies have used spatial verification methods to analyse and correct systematic error of SPP based on the characteristics of matched storm objects, such as location, rotation, intensity, and shape [34–36]. For instance, Demaria et al. [37] used the object-based method, Contiguous Rainfall Analysis (CRA, Ebert and McBride [24]), to correct the location error of CMORPH, PERSIANN, and the TMPA datasets over the Plata basin. Recently, Le Coz et al. [38] used the field transformation method, called Feature Calibration and Alignment technique (FCA), to correct the error due to location in the GPM-IMERG late version over Sub-Saharan Africa. These methods have been useful for correcting the displacement errors when the grid resolution is high and the storm event is small, while preserving the higher spatial variability of SPP storm. However, these methodologies are constrained by the two-dimensional analysis of the storm event.

The spatiotemporal analysis can provide a much deeper analysis on aspects of the entire life-cycle of the storm event, including time span, speed, evolution, among others. In the literature, error analysis using spatiotemporal approaches has been useful to evaluate the performance of several spatial rainfall products. For example, Ref. [39,40] used the Object-Based Diagnostic Evaluation time-domain (MODE-TD) that was proposed by Bullock [41] to evaluate the convection-allowing forecast from the Weather Forecast Model over the United States. Recently, Laverde-Barajas et al. [42] used the Spatiotemporal Contiguous Rainfall Analysis (ST-CORA; Laverde-Barajas et al. [43]) in the Southeast region of Brazil for analysing the error composition of the CMORPH SPP and evaluated the individual hydrological response of two systematic error sources: location and magnitude. This study demonstrated the importance of spatial and temporal storm characteristics to analyse the main systematic error sources in SPP.

Spatiotemporal storm analysis incorporated into bias correction methods is key to reduce several sources of systematic error in SPP. In this study, we present a spatiotemporal object-based bias correction method to reduce several systematic errors in storm events estimated by satellite. The method, called Spatiotemporal Contiguous Object-based Rainfall Analysis for Bias Correction (ST-CORAbico), uses the main storm characteristics of satellite and observed events detected by the ST-CORA method to remove errors due to displacement in space and time and volume. This method is evaluated over the lower Mekong Basin in Thailand to correct several storm event types in the Integrated Multi-satellitE Retrievals for GPM (GPM-IMERG) early version during the monsoon season from 2014 to 2017. The performance of ST-CORAbico is compared against two widely used probabilistic methods—Distribution Transformation and Gamma Quantile Mapping. This manuscript is organised, as follows: Section 2 describes the study and the rainfall data-sets; Section 3 details the methodology of ST-CORAbico; Sections 4 and 5 contain the results and associated discussion; and finally, Section 6 presents the conclusions and future work.

2. Study Area and Data

2.1. Study Area and Period

The study was conducted in Isan, in the northeastern region of Thailand (Figure 1). This region is part of the Mekong-river basin, with an average discharge of 475 km^3/year. The Mekong river basin covers a total drainage area of 795.000 km^2 and it is characteristically divided into the upper and lower basins. The Lower Mekong river in Thailand is located in the Lower Mekong basin, covering 23% of the total drainage area (184.000 km^2) and representing 36% of the country (Figure 1).

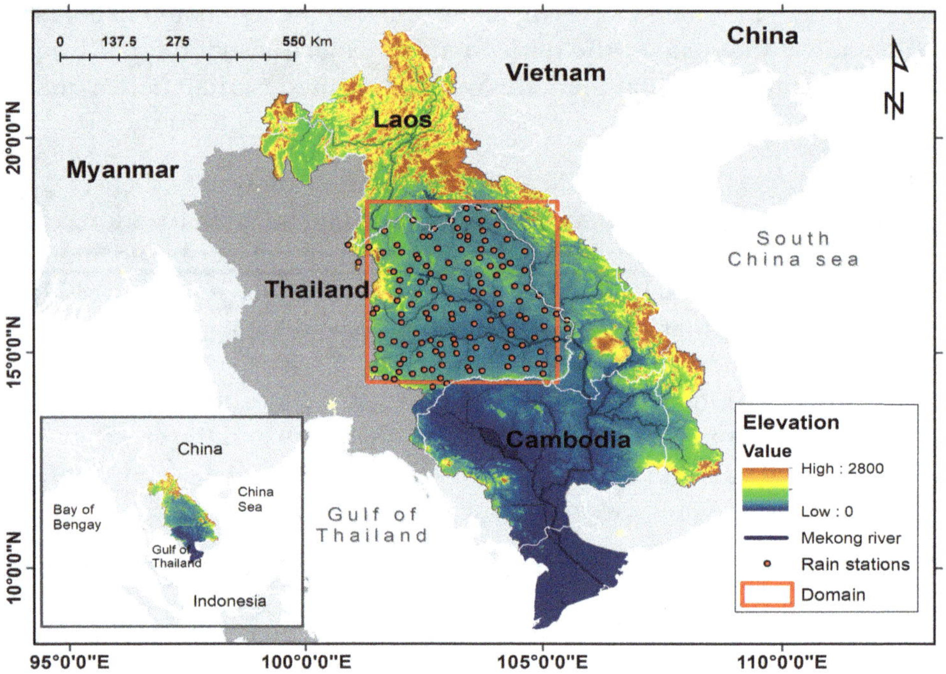

Figure 1. Digital elevation of the Lower Mekong Basin in Thailand. Red rectangle corresponds to the study area in the Isan region. Red dots represent the hourly rain gauge stations from www.thaiwater.net. Blue line represents the Lower Mekong river network.

Heavy rainfall often occurs in the Lower Mekong river basin during the Asian monsoon period between June to October. In this period, the covariability of the Indian summer monsoon and East Asian summer monsoon impacts the Lower Mekong basin, causing high convective storm events that are the main trigger for flash floods and landslides [44,45]. We focused on the wet months for the years 2014 to 2017, as these disasters are mostly occurring during the wet season.

2.2. Satellite-based Precipitation Data

ST-CORAbico was used to correct storm events detected by the Integrated Multi-satellitE Retrievals for GPM (GPM-IMERG). GPM-IMERG is a high-resolution global precipitation product produced by [17] from the NASA/JAXA Global Precipitation Measurement (GPM) mission. The product combines information from multiple infrared, passive-microwave, and satellite-radar sensors to provide rainfall estimations at 0.1-degree spatial resolution every half-hour. GPM-IMERG computes Early, Late, and Final runs. The first two runs are near real-time versions of IMERG and are available at six hours and 18 hours latency, respectively. In the Early version, rainfall estimations are propagated forward while the Late has both forward and backward propagation allowing the incorporation of climatological gauge data. The final version is obtained three months after the measurements. In this run GPM-IMERG ingests the monthly rainfall analysis from the Global Precipitation Climatology Centre (GPCC; Schneider et al. [46]). This version is used for scientific purposes, as it is considered to be the most reliable version [17]. In this study, we evaluated the early run of GPM-IMERG at a half-hourly temporal resolution. This version was selected, given it is the lowest latency data product available, a crucial aspect for operational applications.

2.3. In-Situ Ground-Based Rainfall Observations

Hourly rainfall data during the monsoon season from 2014 to 2017 were obtained from a dense network of 138 rain gauge stations from the Thailand Integrated Water Resource Management System, operated by the Hydro-Informatics Institute (HII) in Thailand (Figure 1). Data were quality controlled by mirroring the density distribution functions of neighbouring stations to remove outlier values considered as noise. These observations were then further interpolated using the Ordinary Kriging interpolator from Golden Software [47] at 0.1 degrees for each hour to match the spatial and temporal resolution of the GPM-IMERG data. This method was selected due to the moderated topographic conditions of the study area and density of the rain gauge measurements. Raingauge data were interpolated using an exponential variogram with a sill of 1.14 (mm^2/h^2), a range of 4.4 km and a nugget of 0.66 (mm^2/h^2). It must be noted that interpolation methods for rainfall data are subject to uncertainty e.g., [48,49]; however, in our case, a dense and optimal rain gauge distribution can reduce the level of uncertainty from the interpolation method [50].

3. Methodology

ST-CORAbico was developed in order to analyse the spatiotemporal characteristics of storm events and bias correct the main sources of systematic error in satellites. Figure 2 shows the methodology of ST-CORAbico. In this section, we describe the elements for storm analysis and bias correction in ST-CORAbico.

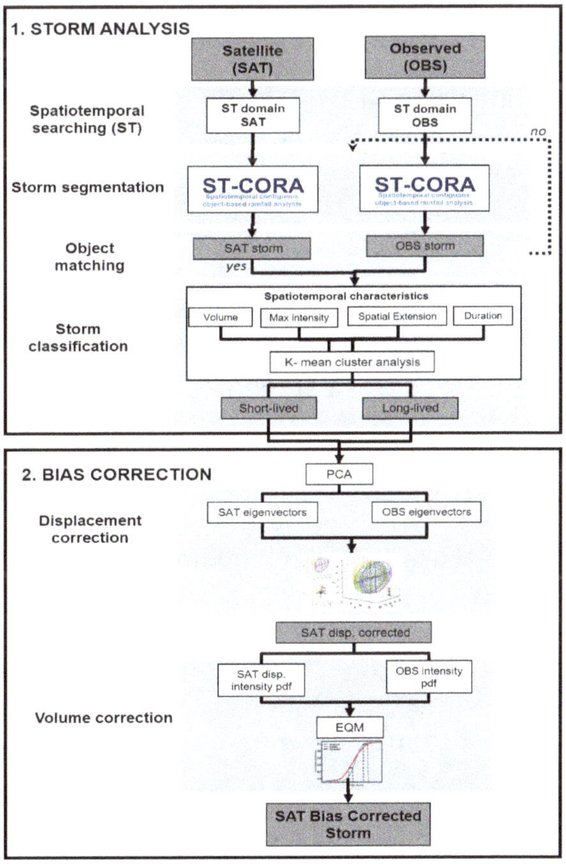

Figure 2. Diagram of the Spatiotemporal Contiguous Object-based Rainfall Analysis for bias correction (ST-CORAbico) method. Grey boxes represent the input and output products while white boxes describe the methodological process for storm analysis and bias correction components.

3.1. Storm Analysis

In the storm analysis, ST-CORAbico uses ST-CORA to analyse the spatiotemporal characteristics of the storm events observed and detected by satellites. This process requires the definition of the spatial and temporal domain in order to reduce the computational time of ST-CORA. We applied a spatiotemporal searching algorithm to predetermine the region of analysis in ST-CORA. This algorithm uses the spatial searching algorithm concept that was proposed by Guttman [51] to index areas with rainfall information in both datasets. The indexing is made in a two-dimensional space compressing the latitude and longitude dimensions using a maximum intensity value as a reference. Once the spatiotemporal domain is defined, we use ST-CORA in the observed and SPP dataset to identify storms in the rainfall data. In this study, ST-CORA incorporates a multivariate kernel density function for storm segmentation.

3.1.1. Storm Segmentation Using the Spatiotemporal Object-Based Rainfall Analysis with Multivariate Kernel Density Segmentation

ST-CORA was applied to analyse the spatiotemporal characteristics of storm events at the catchment scale (duration, spatial extent, magnitude, and centroid). This method enables the feature extraction of different storm event types, classified based on hydrometeorological criteria. ST-CORA uses a multidimensional connected labelling component algorithm to associate connected voxels in space and time (a volume generalisation of pixels) into a disjoint object labelled with a unique classifier. This operation is built upon binary information that was created by voxels, considered to be 'effective rainfall'. Effective rainfall voxels $S_{[x,y,t]}$ are defined according to rainfall voxels $R_{x,y,t}$ above the rainfall intensity threshold IT, as:

$$S_{[x,y,t]} := \begin{cases} 1, & \text{if } R_{x,y,t} \geq \text{IT}. \\ 0, & \text{otherwise}. \end{cases} \tag{1}$$

where, IT is defined by the user and $S_{[x,y,t]}$ is defined in terms of 1 = "true" or 0 ="false". In this study, we used IT = 1mm/h to define effective rainfall [52]. Once binary voxels are created, the connected labelling component algorithm scans all voxels in a neighbour system (from top to bottom and left to right), assigning preliminary labels to $S_{[x,y,t]}$, as follows:

$$c(S_{[x,y,t]}) = \{N_{[x,y,t]} \in \alpha s : S_{CR} = S_N\} \tag{2}$$

where, $c(S_{[x,y,t]})$ is a preliminary label, S_{CR}, S_N are properties of the voxel $S_{[x,y,t]}$ and its neighbours $N_{[x,y,t]}$, respectively, while αs is the neighbour system in space and time. The labelling process $c(S_{[x,y,t]})$ is repeated to resolve equivalence classes of the spatiotemporal object.

Bethel et al. [53] found that object segmentation, while using image thresholding, such as the connected component labelling method, has limitations for edge detection in data with unknown topology. In the original ST-CORA, a size-filtering algorithm and morphological closing method are incorporated in order to remove both small noisy objects and a false merging effect, respectively. However, this process is based on a binary object not taking into account the intensity value of voxels. To overcome this limitation, we have incorporated a Multivariate Kernel Density Estimation (KDE) approach to segment rainfall objects when considering their four dimensions. This method assumes a non-parametric probability density distribution technique for d-dimensional data. Notably, KDE has been widely used in many fields for image detection and object tracking, e.g., [54–58]. Multivariate kernel density is estimated at point x from a random sample $X_1, X_2, ...X_n$ from a density function, f,

$$\widehat{f_K}(x) = \frac{1}{n} \sum_{i=1}^{n} K_h (x - x_i) \tag{3}$$

where K corresponds to the kernel function and h is the bandwidth matrix. Choosing the bandwidth matrix can be restricted to a class of positive diagonal matrices [59]. In the literature, there are several bandwidth selection methods for kernel density estimation [59,60]. For this approach, we use the normal reference rule-of-thumb proposed by Henderson and Parmeter [61]. This method estimates the bandwidth while assuming that the density distribution function follows a Gaussian distribution.

The process of edge detection using KDE is based on the Edge Detection by Density method that was developed by Pereira et al. [55]. This process evaluates the multivariate density distribution of the density of a four-dimensional (4D) rainfall object (Figure 3), and segments the object based on the density threshold, u. This threshold identifies the storm edges that are lower than a probability percentage. This parameter is calculated by analysing the relationship between threshold delineation and the connected intensity value. We found that the 25th distribution percentile for u threshold showed good results for storm segmentation over the Lower Mekong Basin, especially for intense storm events, which are characteristic of monsoon environments.

Rainfall Objects are considered to be storm based on the Critical Mass Threshold (CMT), which is defined as the minimum volume of rainfall (km^3) necessary to be considered as an extreme event [62]. The value of CMT is calculated locally based on the sensitivity between the spatial extent and the total object volume [37,63]. In this analysis, we also incorporated the sensitivity of CMT to the maximum intensity of the storm in order to evaluate the response of intense storm events in the study area. Based on the sensitivity analysis of those parameters, we selected a CMT of 0.01 km^3 for storm events with a maximum intensity greater than 10 mm/h. In the study area, these events correspond to rainfall objects bigger than 2000 km^2.

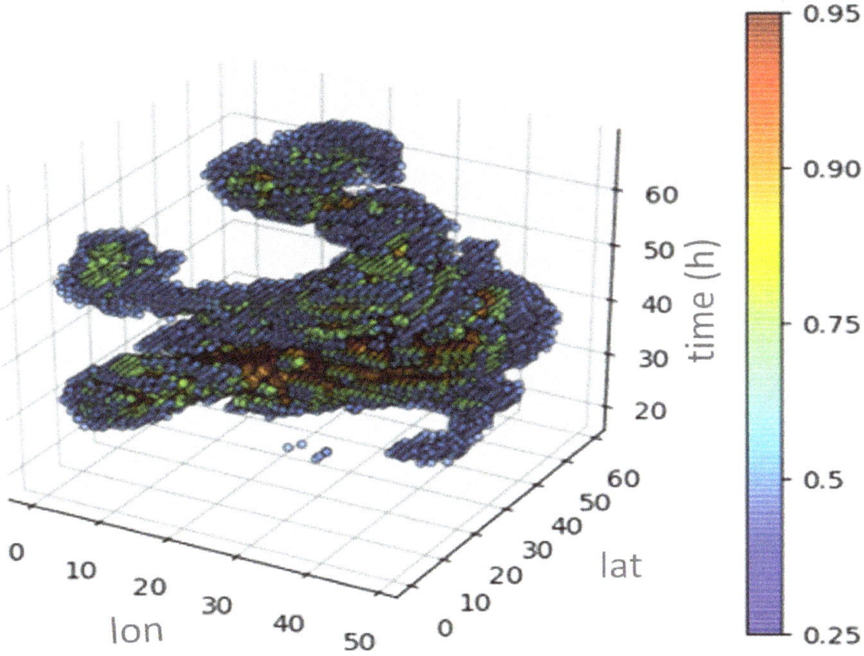

Figure 3. Multivariable kernel density of a storm object in space and time. Example for the storm event 2014-07.

3.1.2. Matching Process

As a result of ST-CORA, multiple storm events are identified in both observed and satellite data sets. However, it is necessary to determine the observed and estimated storm matches. We used the Intersection-over-Union measure (IoU) in order to evaluate the level of similarity between predicted and observed data. IoU is defined as the ratio between the size of the intersection and the union of both objects. This is represented by the following equation:

$$IoU = \frac{TP}{FP + TP + FN} \tag{4}$$

where, TP represents true positives and FP and FN are false positives and negatives, respectively. The selection of the intersection rate value determines the level of matching between objects. If the values are too low, multiple objects will be indexed with the same object. On the other hand, high values indicate that the object does not have any match. We found the IoU value to be 30 percent in the selected study area, which is a good balance for matching the observed and satellite storm events.

3.1.3. Storm Classification

Once all of the storm events are identified, ST-CORA classifies storm events into two types: small convective systems with a short duration (short-lived) and long duration systems extended over large areas (long-lived) [64,65]. We used an unsupervised K-means cluster analysis method to classify short- and long-lived storm events based on the four main storm characteristics (duration, spatial extent, maximum intensity, and total volume). This method divides n observations into k clusters in which each observation is a member of the cluster that minimises the objective function J, as follows:

$$J = \sum_{j=1}^{k} \sum_{i=1}^{n} \|X_{(x,y,z,c)_i^j} - C_j\|^2 \tag{5}$$

where, X is the storm with dimensions x, y, z, c corresponding to the storm characteristics duration, spatial extent, maximum intensity and total volume, respectively. C is the centroid of the cluster k and the absolute number represents the minimum Euclidean distance to C.

3.2. Bias Correction

Bias correction is the second component of the ST-CORAbico method. This component is based on the systematic error source extraction for SPP that was proposed by Laverde-Barajas et al. [42]. Based on the error decomposition for storm estimation defined by Ebert and McBride [24], satellite error is composed of systematic and aleatory errors due to displacement, volume, and pattern, as:

$$E_{total} = E_{displacement} + E_{volume} + E_{pattern} \tag{6}$$

where displacement and volume represent the systematic errors and pattern is the aleatory error calculated as follows:

$$E_{displacement} = E_{total} - E_{shifted} \tag{7}$$

$$E_{volume} = E_{total} - E_{magnitude} \tag{8}$$

$$E_{pattern} = E_{shifted} - E_{volume} \tag{9}$$

In Equations (7) and (8), location is the main source of error due to displacement, while the magnitude is the corresponding source of error for volume. Using the error subtraction from Laverde-Barajas et al. [42], ST-CORAbico corrects displacement and volume error using the following process:

Displacement Correction

Displacement correction corresponds to removing the shifting effect of the estimated storm. In this step, the Principal Component Analysis method (PCA, Johnson and Hebert [66]) is used in order to obtain the weighted centroid and orthogonal variables (eigenvectors, eigenspace) of the SPP storm and reference data (Figure 4a). Once the geometric properties of the objects are obtained, the weighted centroid is matched and the object is rotated accordingly to fit the eigenvectors of the reference storm data.

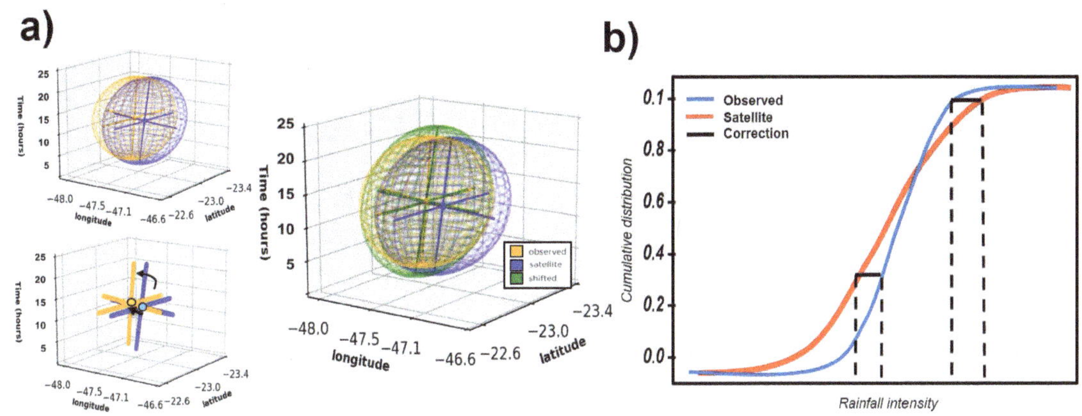

Figure 4. ST-CORAbico systematic error subtraction. (**a**) In error correction, storm centroid and eigenvectors derived from principal component of GPM-IMERG are fitted to the observed event; and, (**b**) magnitude subtraction using both satellite and observed empirical distribution functions, with respect to intensity.

Volume Correction

Volume correction corresponds to the subtraction of the magnitude source of error of the SPP. Using the statistical Empirical Quantile Method (EQM) [67], magnitude error is subtracted by adjusting all moments of the empirical cumulative distribution functions $(ecdfs)$ of the SPP in terms of intensity, with respect to the reference data (Figure 4b). EQM builds the $ecdf$ for the observed $(ecdf_{obs})$ and the satellite $(ecdf_{sat})$ while using the intensity storm distribution I_s, as:

$$EQM = ecdf_{obs}^{-1}(ecdf_{sat}(I_s)) \tag{10}$$

3.3. Evaluation of ST-CORAbico

The evaluation was done by comparing the bias-corrected results with two widely used probabilistic bias correction methods— the Distribution Transformation (DT) method and the Gamma Quantile Mapping (GQM). The DT method was originally developed for the statistical downscaling of climate model data [68]. The method corrects the mean and difference in variation of the SPP by matching the satellite and the observed distribution based on Equation (11):

$$DT = (SAT(t) - \mu_{sat})DT_\tau + \tau_{sat} \times DT_\mu \tag{11}$$

where, μ and τ are the mean and standard deviation of the observed and satellite, respectively. DT_μ and DT_τ are the mean and standard deviation ratio between the observed and satellite data at time t.

The Gamma Quantile Mapping method uses the same methodology as the Empirical Quantile mapping method (10), based on the assumption that both observed OBS and satellite SAT intensity follows a gamma distribution [69]. DT and GQM are implemented for each time-step in order to correct the storm event. The bias correction performance is evaluated based on three widely used error metrics: the Root Mean Square Error (Equation (12)) for evaluating the magnitude error, the bias level (Equation (13)) to evaluate the systematic bias, and the correlation coefficient (Equation (14)) in order to analyse the linear correlation between the observed and the bias-corrected storm event.

$$RMSE = \sqrt{\frac{1}{N}\sum_{i=1}^{N}(OBS_i - SAT_i)^2} \tag{12}$$

$$Bias = \frac{\sum_{i=1}^{N}(SAT_i - OBS_i)}{\sum_{i=1}^{N}(OBS_i)} \tag{13}$$

$$r = \frac{\sum_{i=1}^{N}(SAT_i - \overline{SAT})(OBS_i - \overline{OBS})}{\sqrt{\sum_{i=1}^{N}(SAT_i - \overline{SAT})^2}\sqrt{\sum_{i=1}^{N}(OBS_i - \overline{OBS})^2}} \tag{14}$$

where, OBS represents the rainfall values of the reference rain gauge data and SAT are the satellite and the bias-corrected storm obtained with each method.

4. Results

4.1. Storm Analysis

We identified 120 storm events observed and estimated by GPM-IMERG at an hourly scale for the 2014–2017 monsoon seasons. Figure 5 shows the scatter plot of the main storm characteristics (total volume, duration, spatial extent, and maximum intensity) and classification between short- and long-lived storm events using the k-means cluster analysis. For all events, 68 storms (56%) were classified as short-lived storms while 52 (44%) of storms were classified as long-lived events. Short-lived events had a duration that ranged between three and 17 h, with a maximum spatial extent of 42 thousand km^2. Long-lived events had a duration ranging between 18 and 31 h and covered between 54 and 110 thousand km^2. In terms of total volume and maximum intensity, short-lived events have a total volume of up to 0.15 km^3 with low and intense storms ranging from 3 to 82 mm/h. On the other hand, long-lived storms are comprised of medium and high-intensity events with a total volume ranging from 0.27 to 0.65 km^3. Table 1 describes the observed storm characteristics for short and long-lived event types.

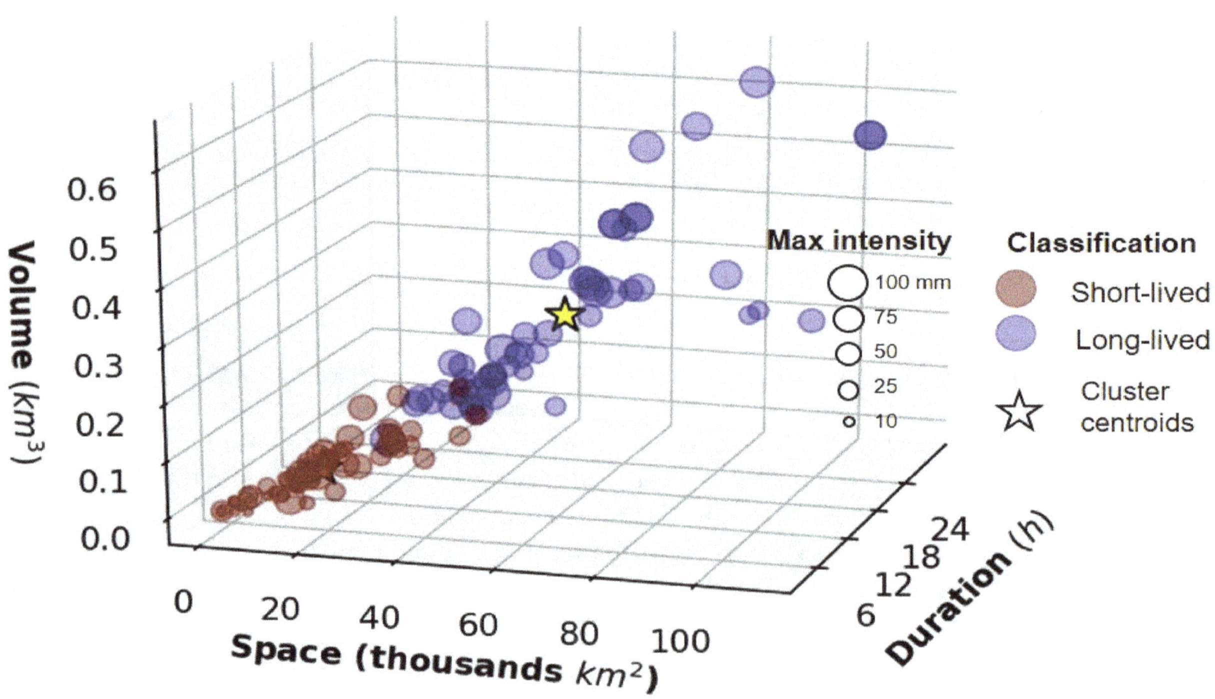

Figure 5. Short- and long-lived cluster analysis classification for observed events during monsoon season 2014–2017.

Table 1. Storm characteristics for short- and long-lived event types.

Storm Type	Statistics	Duration (h)	Spatial Extent (km^2)	Maximum Intensity (mm/h)	Total Volume (km^3)
Short-lived storm	mean	9	15,097	33.0	0.04
	min	3	1900	3.6	0.01
	max	17	42,300	82.0	0.15
long-lived storm	mean	18	54,400	71.4	0.27
	min	10	24,300	31.6	0.07
	max	31	110,600	100.0	0.64

4.2. Results for Bias Correction

We selected a short-lived and a long-lived storm in order to describe the workflow for displacement and volume correction made by ST-CORAbico. Figures 6 and 7 present the bias correction steps for each storm event type. Panel (a) shows the spatial distribution of the observed and satellite events as well as the bias-corrected satellite storm events that were obtained from the correction of location and magnitude errors in ST-CORAbico. Panel (b) describes the displacement and volume corrections. Panel (c) presents the four-dimensional (4D) spatiotemporal evolution of the observed and satellite as well as the bias-corrected storm (time in the z-axis). Panel (d) shows the bias and RMSE statistics as well the scatter and correlation between the observed storm and original and bias-corrected satellite events.

Both examples (Figures 6 and 7) show the importance of bias correction. In both the short and long-lived event scenarios, GPM-IMERG had a longer duration with a larger footprint. However, the long-lived event presented a better spatial agreement than the short-lived event. In terms of magnitude, GPM-IMERG considerably overestimated the total volume and rainfall intensity of the storm. Overall, the performance of GPM-IMERG shows a positive bias and high RMSE, mostly being caused by an excess of rainfall. The correlation coefficients for short- and long-lived event scenarios were 0.7 and 0.5, respectively.

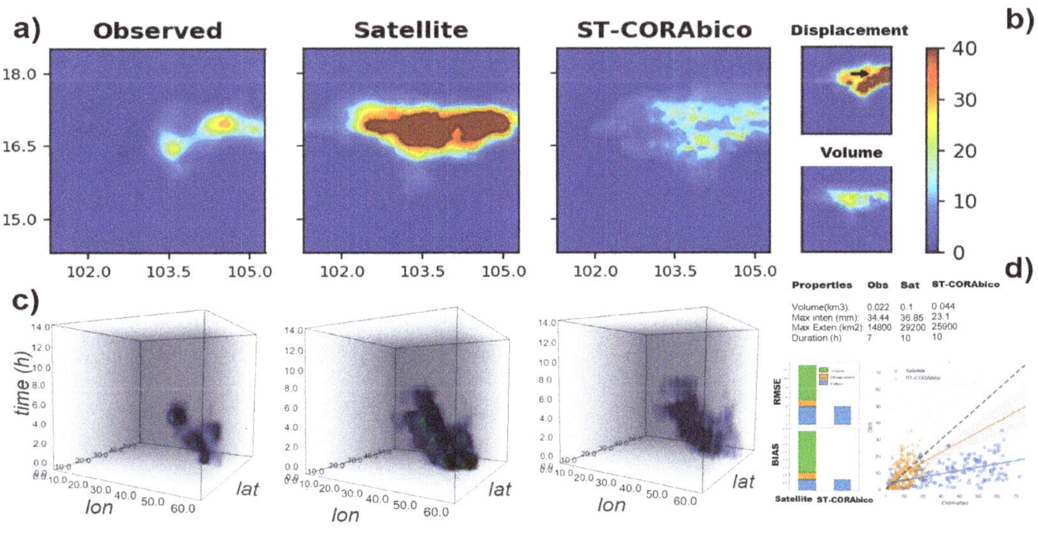

Figure 6. Performance of ST-CORAbico for a short-lived storm event (2014-08-27). (**a**) total events for observed, satellite and ST-CORAbico; (**b**) volume, displacement correction maps; (**c**) four-dimensional (4D) spatiotemporal evolution (lat, lon, time, intensity); and, (**d**) bias, RMSE statistics, and scatter and correlation between observed and estimated rainfall values.

The corrections in displacement and volume made by ST-CORAbico displayed notable changes in the satellite storm structure. In both scenarios, RMSE and bias were mostly reduced by correction due to volume, contributing 40 to 60% of the RMSE reduction and around 70% of the total bias reduction for both events. Displacement correction had an important impact on the reorientation of the satellite storm. The individual correction contributed to 5% of the RMSE correction and 10% reduction of the total bias for the short-lived event. In the case of the long-lived scenario, displacement correction contributed almost 15% of the RMSE reduction and 20% of the total bias reduction. In terms of the correlation coefficient, displacement and volume correction that were made by ST-CORAbico showed a marginal impact on the spatial correlation for the short-lived events. For the long-lived scenario, this did not impact the spatial correlation.

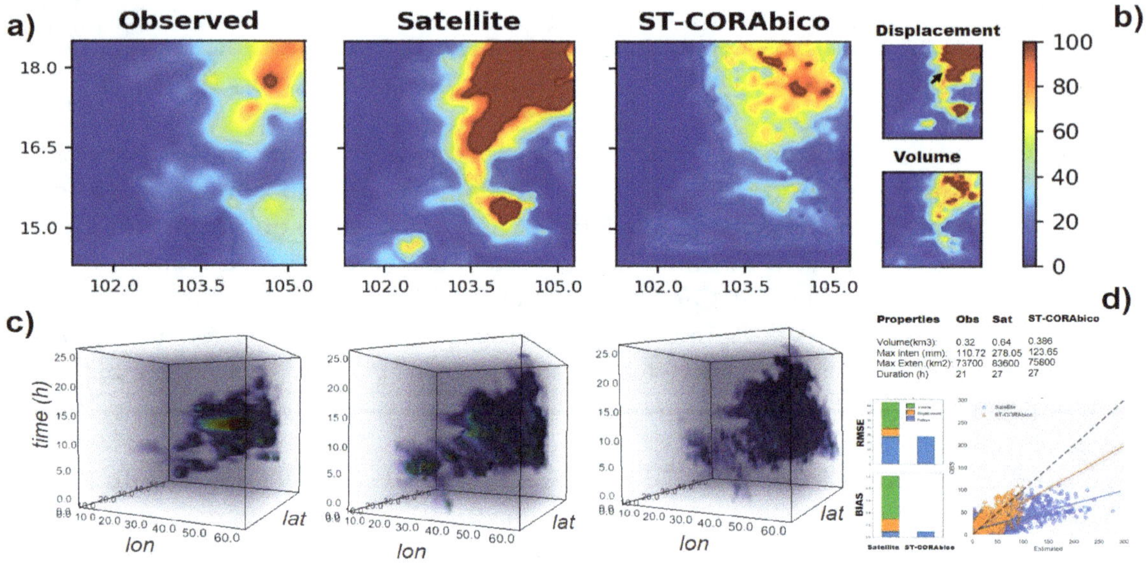

Figure 7. Performance of ST-CORAbico for a long-lived storm event (2014-07-21). (**a**) total events for observed, satellite and ST-CORAbico; (**b**) volume, displacement correction maps; (**c**) 4D spatiotemporal evolution (lat, long, time, intensity); and, (**d**) bias, RMSE statistics and scatter and correlation between observed and estimated rainfall values.

Figure 8 presents the performance of ST-CORAbico for short- and long-lived storm events. This figure describes the density distribution of RMSE (a–b), bias (c–d), and correlation coefficient (e–f) of the short- and long-lived storms estimated by GPM-IMERG, ST-CORAbico, and the individual corrections due to displacement and volume. It was found that ST-CORAbico has a smaller error distribution in RMSE and bias for short- and long-lived storm events when compared with the original GPM-IMERG. This error reduction is mostly caused by the correction due to volume. Displacement correction was an important factor in reducing the bias, especially for long-lived storm events. The results from the correlation coefficient showed that ST-CORAbico had a marginal effect on the spatial correlation of the storm event. Overall, it was found that ST-CORAbico considerably reduced the systematic error of GPM-IMERG.

4.3. Model Comparison

ST-CORAbico was compared with the Distribution Transformation method (DT) and the Gamma Quantile Mapping (GQM) method. Using the short- and long-lived storm scenarios that are presented above, Figure 9 presents the spatial differences and linear correlation between the total observed storms and the bias-corrected events obtained by ST-CORAbico, DT, and GQM. The results for both storm event scenarios showed that ST-CORAbico had the lowest spatial difference among the evaluated methods. For the short-lived storm scenario, ST-CORAbico displayed the highest correlation coefficient (*r*: 0.41) and the lowest RMSE and bias (RMSE: 4.05 mm; bias: 0.74) when compared with DT (*r*: 0.40; RMSE: 5.4 mm; bias: 1.17); and, GQM (*r*: 0.39 RMSE: 6.09mm and bias: 1.5). In the case of the long-lived storm, ST-CORAbico and DT showed a notable error reduction in contrast to the GQM method that showed the biggest differences. For this storm scenario, ST-CORAbico had the best performance (*r*: 0.71 RMSE: 18.02 mm; bias: 0.09), followed by DT (*r*: 0.68, RMSE: 23.0 mm; bias: 0.32), and finally GQM (*r*: 0.62, RMSE: 43.77 mm; bias: 0.97).

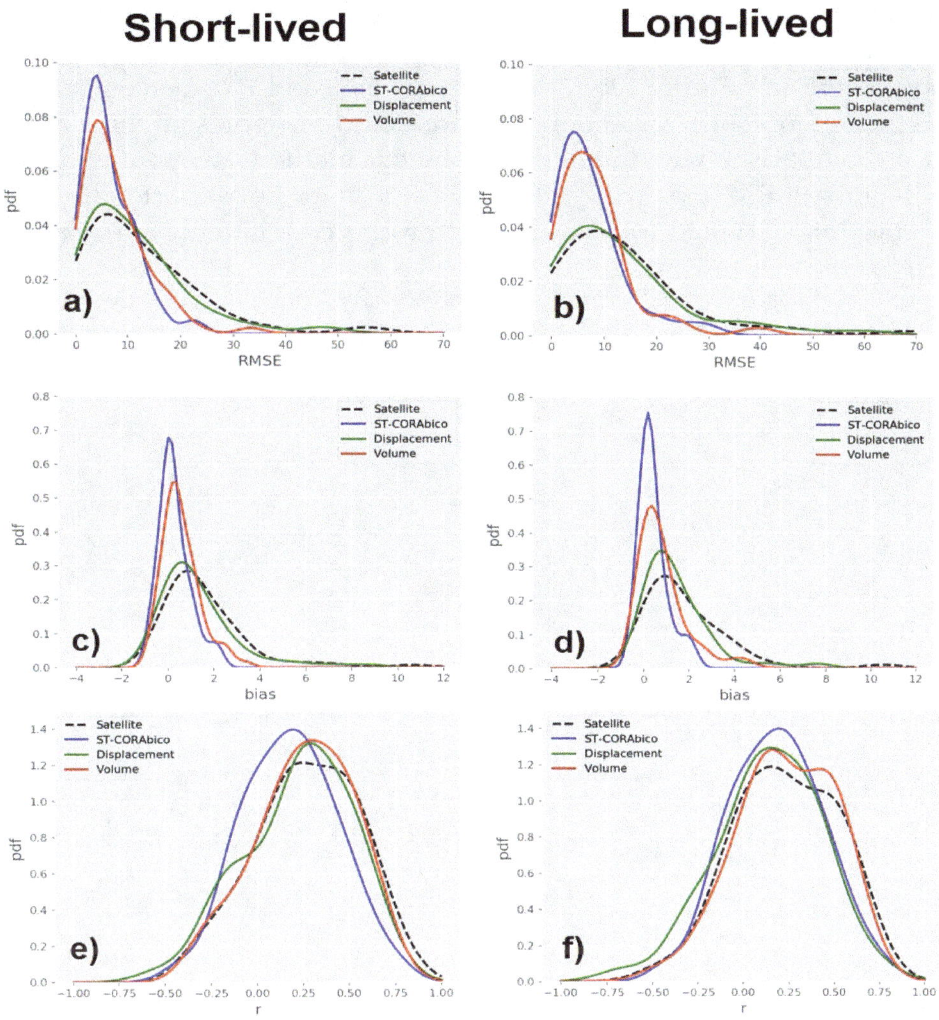

Figure 8. Satellite and bias corrected error distribution for short and long-lived events during monsoon seasons 2014–2017: (**a**,**b**) RMSE; (**c**,**d**) bias; and, (**e**,**f**) correlation coefficient.

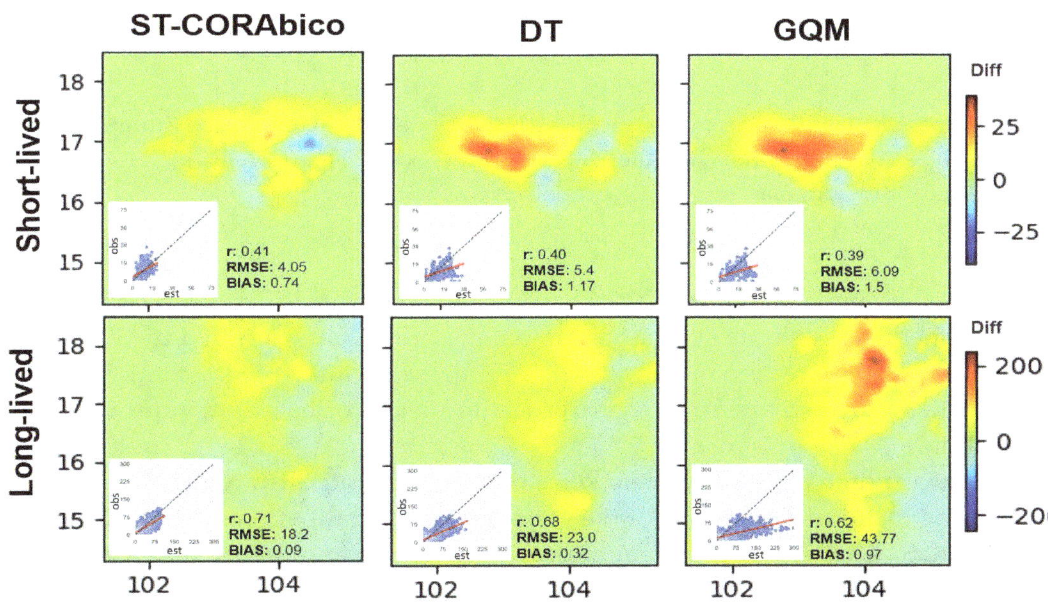

Figure 9. Comparison between ST-CORAbico vs Distribution Transformation (DT) and Gamma Quantile Mapping (GQM).

Figure 10 presents the comparison between ST-CORAbico, DT, and GQM for short- and long-lived storm events. The boxplots show the distributions of the RMSE (Figure 10a,b), the bias (Figure 10c,d), and the correlation coefficient (Figure 10d,f) between the 25% and 75% percentiles for the original GPM-IMERG and the different bias correction methods. The dots represent the individual error for each storm event. In comparison with the two probabilistic methods, we found that ST-CORAbico consistently had the lowest RMSE as well as the lowest bias for both short- and long-lived storm events. ST-CORAbico and DT had a lower impact on the correlation coefficient, especially for short-lived events.

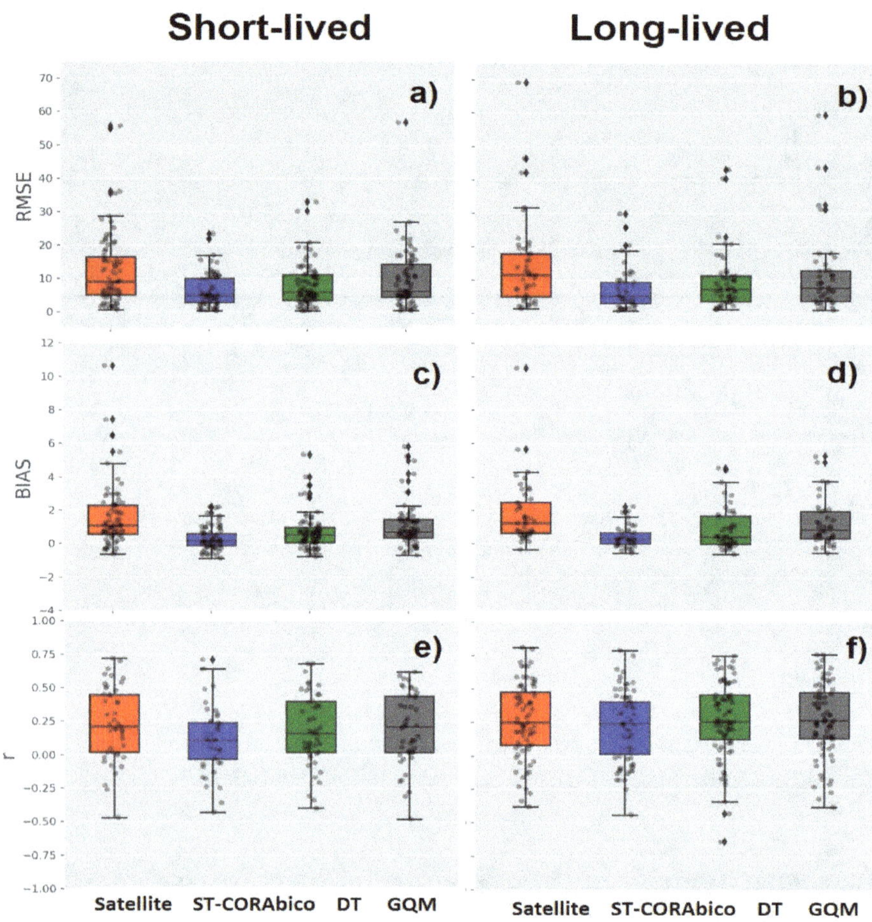

Figure 10. Comparison between the satellite GPM-IMERG (red), ST-CORAbico (blue), Distribution Transformation (green) and Gamma Quantile Mapping (grey) error dispersion during monsoon seasons 2014–2017. (**a**,**b**) RMSE; (**c**,**d**) bias; and, (**e**,**f**) correlation coefficient.

5. Discussion

ST-CORAbico is a spatiotemporal object-based bias correction method that was designed to reduce the displacement and volume systematic errors of storm events detected by SPP. In comparison to spatial object-based bias correction methods e.g., [37,38], the inclusion of the temporal component of the storm event reduced additional error effects due to timing and orientation, improving the efficiency of the bias correction.

This research incorporated a multivariate kernel distribution algorithm into ST-CORA to segment the storm event using the four dimensions of the storm event. In comparison to binary segmentation in the previous version, ST-CORA with KDE segmentation was able to delineate intense storm events by removing unreal storm configurations as a consequence of false merging and false separation of storms due to the multidimensional connected labelling component algorithm. Based on the analysis of KDE threshold delineation and the connected intensity, we found that storm events that were segmented by the 25th percentile of the distribution showed a good result for segmenting intense storms with strong connection. However, further improvement is required.

The implementation of ST-CORAbico described the individual error correction due to displacement and volume. Results in the Lower Mekong basin indicated that volume errors were the main error correction, primarily resulting from the high overestimation of GPM-IMERG. These results agreed with multiple findings regarding hourly GPM-IMERG in monsoonal areas [70,71]. Overall, volume and displacement errors effectively contributed to the reduction of bias and RMSE, demonstrating the importance of reducing both of these systematic errors in satellite correction.

We acknowledge certain limitations of the study. Firstly, the uncertainty arising from the spatial interpolation method that was used for rain gauge values was not fully addressed in this research. Volume and especially displacement corrections in ST-CORAbico can be affected by the type of interpolation methods used to represent the spatiotemporal distribution of the observed storm. A dense rain gauge network can reduce the level of uncertainty; however, it is important to evaluate the impact of the type of interpolation method on the performance of ST-CORAbico, as mentioned above. Another limitation arises from the sensitivity of IoU percentage to match observed and estimated storm events. Higher levels do not always correspond to similar events, which affects the bias correction. This process required an in-depth sensitivity analysis of IoU in order to reduce the automatic storm matching. Additional analysis is required in order to identify why there is a strong correlation between observed and predicted storms in a spatiotemporal environment. In this study, we validated the performance of ST-CORAbico by comparing its performance against two widely used probabilistic methods. However, error metrics were calculated using the observed values, as there is no independent validation dataset available. Further implementations should consider an independent dataset to validate the error correction of the ST-CORAbico method.

This study was conducted in collaboration with the SERVIR-Mekong project and the Mekong River Commissions (MRC). SERVIR-Mekong is harnessing space and geospatial technologies to help decision-makers and key civil society groups to integrate geospatial information into their decision-making, planning, and communication. The application of this methodology can be used for various scientific purposes, including flood risk and water management. More specifically, the methodology enhances the input rainfall data, which are a crucial component of flood and drought early warning systems, landslide monitoring, as well as other water-related decision support systems. Future work will include the integration of machine learning technologies for near real-time bias correction of rainfall data when field data are scarce. In this regard, machine learning models will be trained and optimised using legacy field data and deployed on a near real-time basis.

6. Conclusions

We proposed a new spatiotemporal bias correction method for storm prediction detected by satellites. The method, called Spatiotemporal Contiguous Object-based Rainfall Analysis for bias correction (ST-CORAbico), analyses the main spatiotemporal characteristics of the observed and estimated storm events to correct systematic error sources due to displacement and volume. This methodology has two main elements: storm analysis for the segmentation and classification of storm event; and bias correction for correcting error due displacement and volume. In the storm analysis, we applied the ST-CORA method with a multivariate kernel segmentation in order to identify the spatiotemporal structure of the storm event. This method was applied over the Lower Mekong basin in Thailand to correct the GPM-IMERG Early version during the monsoon seasons from 2014 to 2017. The performance of ST-CORAbico was evaluated against the Distribution Transformation and the Gamma Quantile Mapping methods based on the reduction of RMSE, bias, and correlation coefficient. The results were divided by classifying the storm events into short- and long-lived storm events while using the k-means cluster analysis method.

We classified 68 storms (56%) as short-lived storms and 52 (44%) as long-lived events. The results of both storm event types showed that ST-CORAbico reduced the RMSE and bias of GPM-IMERG.

Volume correction was the major error source due to the overestimation present in GPM-IMERG. Location error was most important in the reduction of the bias. ST-CORAbico displayed a marginal impact on the spatial structure of the satellite-derived rainfall, showing the original structure of the rainfall data.

The comparison of ST-CORAbico with the Distribution Transformation and the Gamma Quantile Mapping methods showed that ST-CORAbico had the lowest RMSE as well as the lowest bias in both short and long-lived events. In terms of the correlation coefficient, ST-CORAbico and DT had a lower impact on the correlation coefficient, especially for short-lived events.

ST-CORAbico improves the accuracy of satellite-derived near real-time information on storm events. It can be used in various flood monitoring and water management applications. Our future studies will also incorporate machine learning methods and related technologies in order to correct storm events in real-time, in situations where field observation data are scarce.

Author Contributions: Conceptualization, M.L.-B., G.A.C. and A.P.; Data curation, M.L.-B. and S.J.; Funding acquisition, P.T., L.S., S.K. and W.C.; Investigation, M.L., G.A.C., A.P., F.C., C.M., D.P.S. and S.B.-A.; Methodology, M.L.-B.; Project administration, C.M. and P.T.; Resources, M.L.-B., S.J., P.T., L.H.S., S.K., S.B.-A. and W.C.; Supervision, G.A.C., A.P., F.C., C.M., P.T., A.M., D.P.S., W.C., R.U. and D.P.S.; Visualization, C.M., S.B.-A. and W.C.; Writing—original draft, M.L.-B., A.P., F.C. and A.M.; Writing—review and editing, M.L.-B., G.A.C., A.P., F.C., C.M., S.J., P.T., A.M., D.S., L.H.S., K.S., S.B.-A., R.U. and D.P.S. All authors have read and agreed to the published version of the manuscript.

Acknowledgments: The authors would like to acknowledge the Asian Disaster Preparedness Center, in Thailand and the USAID-NASA SERVIR-Mekong program for the support provided. We would like to thank Hannah Priestley for her meticulous style corrections. GPM-IMERG information was extracted from SERVIR-Mekong's Virtual Gauge and Stream Gauge Data service tool (VRSGS). Additionally, we would like to thanks the Hydroinformatics Institute in Thailand for the provision of hourly rain gauge data and also the space agencies responsible for the satellite data used in this research.

References

1. Bui, Y.T.; Orange, D.; Visser, S.; Hoanh, C.T.; Laissus, M.; Poortinga, A.; Tran, D.T.; Stroosnijder, L. Lumped surface and sub-surface runoff for erosion modeling within a small hilly watershed in northern Vietnam. *Hydrol. Process.* **2014**, *28*, 2961–2974. [CrossRef]
2. Markert, K.N.; Schmidt, C.M.; Griffin, R.E.; Flores, A.I.; Poortinga, A.; Saah, D.S.; Muench, R.E.; Clinton, N.E.; Chishtie, F.; Kityuttachai, K.; et al. Historical and operational monitoring of surface sediments in the lower mekong basin using Landsat and Google Earth Engine cloud computing. *Remote Sens.* **2018**, *10*, 909. [CrossRef]
3. Poortinga, A.; Bastiaanssen, W.; Simons, G.; Saah, D.; Senay, G.; Fenn, M.; Bean, B.; Kadyszewski, J. A self-calibrating runoff and streamflow remote sensing model for ungauged basins using open-access earth observation data. *Remote Sens.* **2017**, *9*, 86. [CrossRef]
4. Price, K.; Purucker, S.T.; Kraemer, S.R.; Babendreier, J.E.; Knightes, C.D. Comparison of radar and gauge precipitation data in watershed models across varying spatial and temporal scales. *Hydrol. Process.* **2014**, *28*, 3505–3520. [CrossRef]
5. AghaKouchak, A.; Nakhjiri, N. A near real-time satellite-based global drought climate data record. *Environ. Res. Lett.* **2012**, *7*, 044037. [CrossRef]
6. Azarderakhsh, M.; Rossow, W.B.; Papa, F.; Norouzi, H.; Khanbilvardi, R. Diagnosing water variations within the Amazon basin using satellite data. *J. Geophys. Res. Atmos.* **2011**, *116*. [CrossRef]
7. Pan, M.; Li, H.; Wood, E. Assessing the skill of satellite-based precipitation estimates in hydrologic applications. *Water Resour. Res.* **2010**, *46*. [CrossRef]
8. Tesfagiorgis, K.; Mahani, S.; Krakauer, N.; Khanbilvardi, R. Bias correction of satellite rainfall estimates using a radar-gauge product—A case study in Oklahoma (USA). *Hydrol. Earth Syst. Sci.* **2011** *15*, 2631–2647. [CrossRef]
9. Griffith, C.G.; Woodley, W.L.; Grube, P.G.; Martin, D.W.; Stout, J.; Sikdar, D.N. Rain estimation from geosynchronous satellite imagery—Visible and infrared studies. *Mon. Weather Rev.* **1978**, *106*, 1153–1171. [CrossRef]
10. Vicente, G.A.; Scofield, R.A.; Menzel, W.P. The operational GOES infrared rainfall estimation technique. *Bull. Am. Meteorol. Soc.* **1998**, *79*, 1883–1898. [CrossRef]

11. Marzano, F.S.; Palmacci, M.; Cimini, D.; Giuliani, G.; Turk, F.J. Multivariate statistical integration of satellite infrared and microwave radiometric measurements for rainfall retrieval at the geostationary scale. *IEEE Trans. Geosci. Remote Sens.* **2004**, *42*, 1018–1032. [CrossRef]
12. Tapiador, F. A physically based satellite rainfall estimation method using fluid dynamics modelling. *Int. J. Remote Sens.* **2008**, *29*, 5851–5862. [CrossRef]
13. Sorooshian, S.; Hsu, K.L.; Gao, X.; Gupta, H.V.; Imam, B.; Braithwaite, D. Evaluation of PERSIANN system satellite-based estimates of tropical rainfall. *Bull. Am. Meteorol. Soc.* **2000**, *81*, 2035–2046. [CrossRef]
14. Joyce, R.J.; Janowiak, J.E.; Arkin, P.A.; Xie, P. CMORPH: A method that produces global precipitation estimates from passive microwave and infrared data at high spatial and temporal resolution. *J. Hydrometeorol.* **2004**, *5*, 487–503. [CrossRef]
15. Huffman, G.J.; Bolvin, D.T.; Nelkin, E.J.; Wolff, D.B.; Adler, R.F.; Gu, G.; Hong, Y.; Bowman, K.P.; Stocker, E.F. The TRMM multisatellite precipitation analysis (TMPA): Quasi-global, multiyear, combined-sensor precipitation estimates at fine scales. *J. Hydrometeorol.* **2007**, *8*, 38–55. [CrossRef]
16. Huffman, G.J.; Adler, R.F.; Bolvin, D.T.; Nelkin, E.J. The TRMM multi-satellite precipitation analysis (TMPA). In *Satellite Rainfall Applications for Surface Hydrology*; Springer: Berlin/Heidelberg, Germany, 2010; pp. 3–22.
17. Huffman, G.J.; Bolvin, D.T.; Nelkin, E.J. Integrated Multi-satellitE Retrievals for GPM (IMERG) technical documentation. *NASA/GSFC Code* **2015**, *612*, 2019.
18. Sun, Q.; Miao, C.; Duan, Q.; Ashouri, H.; Sorooshian, S.; Hsu, K.L. A review of global precipitation data sets: Data sources, estimation, and intercomparisons. *Rev. Geophys.* **2018**, *56*, 79–107. [CrossRef]
19. Thiemig, V.; Rojas, R.; Zambrano-Bigiarini, M.; Levizzani, V.; De Roo, A. Validation of satellite-based precipitation products over sparsely gauged African river basins. *J. Hydrometeorol.* **2012**, *13*, 1760–1783. [CrossRef]
20. Guo, H.; Chen, S.; Bao, A.; Hu, J.; Gebregiorgis, A.S.; Xue, X.; Zhang, X. Inter-comparison of high-resolution satellite precipitation products over Central Asia. *Remote Sens.* **2015**, *7*, 7181–7211. [CrossRef]
21. Kimani, M.W.; Hoedjes, J.C.B.; Su, Z. An Assessment of satellite-derived rainfall products relative to ground observations over East Africa. *Remote Sens.* **2017**, *9*, 430. [CrossRef]
22. Vila, D.A.; De Goncalves, L.G.G.; Toll, D.L.; Rozante, J.R. Statistical evaluation of combined daily gauge observations and rainfall satellite estimates over continental South America. *J. Hydrometeorol.* **2009**, *10*, 533–543. [CrossRef]
23. Habib, E.; Haile, A.T.; Sazib, N.; Zhang, Y.; Rientjes, T. Effect of bias correction of satellite-rainfall estimates on runoff simulations at the source of the Upper Blue Nile. *Remote Sens.* **2014**, *6*, 6688–6708. [CrossRef]
24. Ebert, E.; McBride, J. Verification of precipitation in weather systems: Determination of systematic errors. *J. Hydrol.* **2000**, *239*, 179–202. [CrossRef]
25. Davis, C.A.; Brown, B.G.; Bullock, R.; Halley-Gotway, J. The method for object-based diagnostic evaluation (MODE) applied to numerical forecasts from the 2005 NSSL/SPC Spring Program. *Weather Forecast.* **2009**, *24*, 1252–1267. [CrossRef]
26. Wernli, H.; Paulat, M.; Hagen, M.; Frei, C. SAL—A novel quality measure for the verification of quantitative precipitation forecasts. *Mon. Weather Rev.* **2008**, *136*, 4470–4487. [CrossRef]
27. Roberts, N. An investigation of the ability of a storm scale configuration of the Met Office NWP model to predict floodproducing rainfall. *UK Met Off. Tech. Rep.* **2005**, *455*, 80.
28. Marsigli, C.; Montani, A.; Paccagnella, T. Verification of the COSMOLEPS new suite in terms of precipitation distribution. *COSMO Newsl.* **2006**, *6*, 134–141.
29. Casati, B.; Ross, G.; Stephenson, D. A new intensity-scale approach for the verification of spatial precipitation forecasts. *Meteorol. Appl. A J. Forecast. Pract. Appl. Train. Tech. Model.* **2004**, *11*, 141–154. [CrossRef]
30. Casati, B. New developments of the intensity-scale technique within the Spatial Verification Methods Intercomparison Project. *Weather Forecast.* **2010**, *25*, 113–143. [CrossRef]
31. Mittermaier, M.P. Using an intensity-scale technique to assess the added benefit of high-resolution model precipitation forecasts. *Atmos. Sci. Lett.* **2006**, *7*, 36–42. [CrossRef]
32. Keil, C.; Craig, G.C. A displacement-based error measure applied in a regional ensemble forecasting system. *Mon. Weather Rev.* **2007**, *135*, 3248–3259. [CrossRef]
33. Keil, C.; Craig, G.C. A displacement and amplitude score employing an optical flow technique. *Weather Forecast.* **2009**, *24*, 1297–1308. [CrossRef]
34. Skok, G.; Tribbia, J.; Rakovec, J.; Brown, B. Object-based analysis of satellite-derived precipitation systems over the low-and midlatitude Pacific Ocean. *Mon. Weather Rev.* **2009**, *137*, 3196–3218. [CrossRef]

35. Li, J.; Hsu, K.; AghaKouchak, A.; Sorooshian, S. An object-based approach for verification of precipitation estimation. *Int. J. Remote Sens.* **2015**, *36*, 513–529. [CrossRef]
36. Li, J.; Hsu, K.L.; AghaKouchak, A.; Sorooshian, S. Object-based assessment of satellite precipitation products. *Remote Sens.* **2016**, *8*, 547. [CrossRef]
37. Demaria, E.; Rodriguez, D.; Ebert, E.; Salio, P.; Su, F.; Valdes, J.B. Evaluation of mesoscale convective systems in South America using multiple satellite products and an object-based approach. *J. Geophys. Res. Atmos.* **2011**, *116*. [CrossRef]
38. Le Coz, C.; Heemink, A.; Verlaan, M.; Veldhuis, M.C.; van de Giesen, N. Correcting position error in precipitation data using image morphing. *Remote Sens.* **2019**, *11*, 2557. [CrossRef]
39. Clark, A.J.; Kain, J.S.; Marsh, P.T.; Correia, J., Jr.; Xue, M.; Kong, F. Forecasting tornado pathlengths using a three-dimensional object identification algorithm applied to convection-allowing forecasts. *Weather Forecast.* **2012**, *27*, 1090–1113. [CrossRef]
40. Clark, A.J.; Bullock, R.G.; Jensen, T.L.; Xue, M.; Kong, F. Application of object-based time-domain diagnostics for tracking precipitation systems in convection-allowing models. *Weather Forecast.* **2014**, *29*, 517–542. [CrossRef]
41. Development and implementation of MODE time domain object-based verification In Proceedings of the 24th Conference Weather and Forecasting, Seattle, WA, USA, 24–27 January 2011.
42. Laverde-Barajas, M.; Corzo Perez, G.; Chishtie, F.; Poortinga, A.; Uijlenhoet, R.; Solomatine, D. Decomposing satellite-based rainfall errors in flood estimation: Hydrological responses using a spatiotemporal object-based verification method. *J. Hydrol.* **2020**, 125554. [CrossRef]
43. Laverde-Barajas, M.; Corzo, G.; Bhattacharya, B.; Uijlenhoet, R.; Solomatine, D.P. Spatiotemporal Analysis of Extreme Rainfall Events Using an Object-Based Approach. In *Spatiotemporal Analysis of Extreme Hydrological Events*; Corzo, G., Varouchakis, E.A., Eds.; Elsevier: Amsterdam, The Netherlands, 2019; pp. 95–112. [CrossRef]
44. Delgado, J.; Merz, B.; Apel, H. A climate-flood link for the lower Mekong River. *Hydrol. Earth Syst. Sci.* **2012**, *16*, 1533–1541. [CrossRef]
45. Yang, R.; Zhang, W.K.; Gui, S.; Tao, Y.; Cao, J. Rainy season precipitation variation in the Mekong River basin and its relationship to the Indian and East Asian summer monsoons. *Clim. Dyn.* **2019**, *52*, 5691–5708. [CrossRef]
46. Schneider, U.; Fuchs, T.; Meyer-Christoffer, A.; Rudolf, B. *Global Precipitation Analysis Products of the GPCC*; Global Precipitation Climatology Centre (GPCC), DWD, Internet Publication; Deutscher Wetterdienst: Offenbach, Germany, 2008; Volume 112.
47. Golden Software, I. *Surfer 12 Users' Guide: Powerful Contouring, Gridding, and Surface Mapping*, 12th ed.; Golden Software, Inc.: Golden, CO, USA, 2014.
48. Ly, S.; Sohier, C.; Charles, C.; Degré, A. Effect of raingage density, position and interpolation on rainfall-discharge modelling. In *EGU2012 Geophysical Research Abstracts*; European Geophysical Society: Katlenburg-Lindau, Germany, 2012; p. 2592.
49. Li, J.; Heap, A.D. Spatial interpolation methods applied in the environmental sciences: A review. *Environ. Model. Softw.* **2014**, *53*, 173–189. [CrossRef]
50. Chen, Y.C.; Wei, C.; Yeh, H.C. Rainfall network design using kriging and entropy. *Hydrol. Process. Int. J.* **2008**, *22*, 340–346. [CrossRef]
51. Guttman, A. R-Trees: A Dynamic Index Structure for Spatial Searching. In Proceedings of the 1984 ACM SIGMOD International Conference on Management of Data, Association for Computing Machinery, New York, NY, USA, 18–21 June 1984; pp. 47–57. Available online: http://www-db.deis.unibo.it/courses/SI-LS/papers/Gut84.pdf (accessed on 30 September 2020).
52. Ebert, E.E.; Gallus, W.A., Jr. Toward better understanding of the contiguous rain area (CRA) method for spatial forecast verification. *Weather Forecast.* **2009**, *24*, 1401–1415. [CrossRef]
53. Bethel, E.W.; Childs, H.; Hansen, C. *High Performance Visualization: Enabling Extreme-Scale Scientific Insight*; CRC Press: Boca Raton, FL, USA, 2012.
54. Chen, T.B.; Lu, H.H.S.; Lee, Y.S.; Lan, H.J. Segmentation of cDNA microarray images by kernel density estimation. *J. Biomed. Inf.* **2008**, *41*, 1021–1027. [CrossRef]
55. Pereira, O.; Torres, E.; Garcés, Y.; Rodríguez, R. Edge detection based on kernel density estimation. In Proceedings of the International Conference on Image Processing, Computer Vision, and Pattern Recognition (IPCV), The Steering Committee of The World Congress in Computer Science, Computer, Engineering and Applied Computing WorldComp, Las Vegas, NV, USA, 17–20 July 2017; pp. 123–128.

56. Wang, L.; Lu, J.; Li, X.; Huan, Z.; Liang, J.; Chen, S. Learning arbitrary-shape object detector from bounding-box annotation by searching region-graph. *Pattern Recognit. Lett.* **2017**, *87*, 171–176. [CrossRef]
57. Zivkovic, Z.; Van Der Heijden, F. Efficient adaptive density estimation per image pixel for the task of background subtraction. *Pattern Recognit. Lett.* **2006**, *27*, 773–780. [CrossRef]
58. Berjón, D.; Cuevas, C.; Morán, F.; García, N. Real-time nonparametric background subtraction with tracking-based foreground update. *Pattern Recognit.* **2018**, *74*, 156–170. [CrossRef]
59. Hyndman, R.L.; Zhang, X.; King, M.L. Bandwidth selection for multivariate kernel density estimation using MCMC. In *Econometric Society 2004 Australasian Meetings*; Econometric Society: Melbourne, Australia, 7–9 July 2004.
60. Zhang, X.; King, M.L.; Hyndman, R.J. A Bayesian approach to bandwidth selection for multivariate kernel density estimation. *Comput. Stat. Data Anal.* **2006**, *50*, 3009–3031. [CrossRef]
61. Henderson, D.J.; Parmeter, C.F. Normal reference bandwidths for the general order, multivariate kernel density derivative estimator. *Stat. Probab. Lett.* **2012**, *82*, 2198–2205. [CrossRef]
62. Grams, J.S.; Gallus, W.A., Jr.; Koch, S.E.; Wharton, L.S.; Loughe, A.; Ebert, E.E. The use of a modified Ebert–McBride technique to evaluate mesoscale model QPF as a function of convective system morphology during IHOP 2002. *Weather Forecast.* **2006**, *21*, 288–306. [CrossRef]
63. Steiner, M.; Houze, R.A., Jr.; Yuter, S.E. Climatological characterization of three-dimensional storm structure from operational radar and rain gauge data. *J. Appl. Meteorol.* **1995**, *34*, 1978–2007. [CrossRef]
64. Molini, L.; Parodi, A.; Siccardi, F. Dealing with uncertainty: An analysis of the severe weather events over Italy in 2006. *Nat. Hazards Earth Syst. Sci* **2009**, *9*, 1775–1786. [CrossRef]
65. Molini, L.; Parodi, A.; Rebora, N.; Craig, G. Classifying severe rainfall events over Italy by hydrometeorological and dynamical criteria. *Q. J. R. Meteorol. Soc.* **2011**, *137*, 148–154. [CrossRef]
66. Johnson, A.E.; Hebert, M. Using spin images for efficient object recognition in cluttered 3D scenes. *IEEE Trans. Pattern Anal. Mach. Intell.* **1999**, *21*, 433–449. [CrossRef]
67. Themeßl, M.J.; Gobiet, A.; Heinrich, G. Empirical-statistical downscaling and error correction of regional climate models and its impact on the climate change signal. *Clim. Chang.* **2012**, *112*, 449–468. [CrossRef]
68. Brown, B.G.; Bullock, Y.R.; Davis, C.A.; Gotway, J.H.; Chapman, M.B.; Takacs, A.; Gillel, E.; Manning, K.; Mahoney, J.L. New verification approaches for convective weather forecasts. In Proceedings of the 11th Conference on Aviation, Range, and Aerospace, Hyannis, MA, USA, 4–8 October 2004; pp. 3–8.
69. Piani, C.; Haerter, J.; Coppola, E. Statistical bias correction for daily precipitation in regional climate models over Europe. *Theor. Appl. Climatol.* **2010**, *99*, 187–192. [CrossRef]
70. Oliveira, R.; Maggioni, V.; Vila, D.; Morales, C. Characteristics and diurnal cycle of GPM rainfall estimates over the central Amazon region. *Remote Sens.* **2016**, *8*, 544. [CrossRef]
71. Tang, G.; Ma, Y.; Long, D.; Zhong, L.; Hong, Y. Evaluation of GPM Day-1 IMERG and TMPA Version-7 legacy products over Mainland China at multiple spatiotemporal scales. *J. Hydrol.* **2016**, *533*, 152–167. [CrossRef]

2

Potential of GPM IMERG Precipitation Estimates to Monitor Natural Disaster Triggers in Urban Areas

Augusto Getirana [1,2,*], Dalia Kirschbaum [1], Felipe Mandarino [3], Marta Ottoni [4], Sana Khan [1,5] and Kristi Arsenault [1,2]

1 Hydrological Sciences Laboratory, NASA Goddard Space Flight Center, Greenbelt, MD 20771, USA; dalia.b.kirschbaum@nasa.gov (D.K.); sana.khan@nasa.gov (S.K.); kristi.r.arsenault@nasa.gov (K.A.)
2 Science Applications International Corporation, Greenbelt, MD 20771, USA
3 Instituto Pereira Passos, Rio de Janeiro 22221-070, Brazil; felipe.mandarino@rio.rj.gov.br
4 Geological Survey of Brazil, Rio de Janeiro 22290-240, Brazil; marta.ottoni@cprm.gov.br
5 Earth System Science Interdisciplinary Center, University of Maryland, College Park, MD 20740, USA
* Correspondence: augusto.getirana@nasa.gov

Abstract: Extreme rainfall can be a catastrophic trigger for natural disaster events at urban scales. However, there remains large uncertainties as to how satellite precipitation can identify these triggers at a city scale. The objective of this study is to evaluate the potential of satellite-based rainfall estimates to monitor natural disaster triggers in urban areas. Rainfall estimates from the Global Precipitation Measurement (GPM) mission are evaluated over the city of Rio de Janeiro, Brazil, where urban floods and landslides occur periodically as a result of extreme rainfall events. Two rainfall products derived from the Integrated Multi-satellite Retrievals for GPM (IMERG), the IMERG Early and IMERG Final products, are integrated into the Noah Multi-Parameterization (Noah-MP) land surface model in order to simulate the spatial and temporal dynamics of two key hydrometeorological disaster triggers across the city over the wet seasons during 2001–2019. Here, total runoff (TR) and rootzone soil moisture (RZSM) are considered as flood and landslide triggers, respectively. Ground-based observations at 33 pluviometric stations are interpolated, and the resulting rainfall fields are used in an in-situ precipitation-based simulation, considered as the reference for evaluating the IMERG-driven simulations. The evaluation is performed during the wet seasons (November-April), when average rainfall over the city is 4.4 mm/day. Results show that IMERG products show low spatial variability at the city scale, generally overestimate rainfall rates by 12–35%, and impacts on TR and RZSM vary spatially mostly as a function of land cover and soil types. Results based on statistical and categorical metrics show that IMERG skill in detecting extreme events is moderate, with IMERG Final performing slightly better for most metrics. By analyzing two recent storms, we observe that IMERG detects mostly hourly extreme events, but underestimates rainfall rates, resulting in underestimated TR and RZSM. An evaluation of normalized time series using percentiles shows that both satellite products have significantly improved skill in detecting extreme events when compared to the evaluation using absolute values, indicating that IMERG precipitation could be potentially used as a predictor for natural disasters in urban areas.

Keywords: IMERG; extreme events; precipitation; natural disaster; urban flood; landslide; soil moisture; total runoff

1. Introduction

There is a notion that disasters are outcomes of inappropriate adjustments between people and environment [1]. An example of inappropriate adjustments is the fast-urban growth observed in the past decades, putting humans in increased natural hazard risks related to environmental degradation that are intrinsic to urban development. Many large cities in developing countries lack urban planning and management, resulting in inappropriate land use, often composed of informal and irregular settlements in high hazard risk areas (e.g., along steep hills and riverbanks) [2]. Impacts of disasters in such cities are explained by the retroactive relationships between natural hazards, urbanization and social processes [3]. Floods and landslides are two types of disasters often triggered when these areas are subjected to heavy and prolonged rainfall. Total runoff, a key hydrological variable in assessing urban flood risks, is primarily related to the rainfall intensity and land use. Over urban-based impervious land surfaces, most rainfall is converted into runoff, and without proper flood management measures, such as river restoration measures [4], floods may occur. Landslides can occur within steep terrain where unconsolidated sediments or fractured rock leads to surface instability. In areas of rapid growth, such as in the impoverished areas on the outskirts or within megacities, landslide disasters can be further exacerbated by improper building of roads, houses, poor drainage, and illegal or improper digging [5–7]. Intense and/or prolonged rainfall is the predominate triggering factor in landslide-prone areas where increased soil saturation and runoff can serve to mobilize the downstream transport of debris, rock, and soil. Disaster risk mitigation in urban areas can be addressed by reducing the hazard and/or the system vulnerability through both structural and non-structural measures [2,4]. Disaster monitoring and forecasting are examples of non-structural measures that can be used to provide stakeholders with information on how to implement or improve structural measures such as river canalization, dykes and dams, or slope retrofitting as well as other non-structural measures such as flood and landslide mapping for improved characterization of past hazards, land use regulation and early warning systems.

Recent studies have proposed near real time and even real time urban flood monitoring systems through ground-based observations of water levels [8] and meteorological variables [9], weather radar combined with ground-based sensors [10], and using crowdsourcing techniques [11]. Specialized hydrological and hydrodynamic models have also been widely used to monitor urban floods [12–18]. High-resolution remote sensing data has been adopted as another way to monitor urban flood extent and depth [19,20] and used as inputs in computational models or as reference for model evaluation [21]. The majority of these modeling studies indicate that runoff accuracy is key for precisely modeling urban floods.

Near real time or early warning landslide systems primarily rely on in situ rain gauge networks to rapidly identify the rainfall necessary to potentially trigger a landslide. In the city of Rio de Janeiro, an early warning system creates alerts when peak rainfall intensities reach a threshold in order to highlight across the city where landslides are more likely to occur (http://alertario.rio.rj.gov.br/). Similar systems in Italy have employed a network of automated rain gauges to highlight areas at higher hazard due to landslides at the municipal and regional scale [22]. In Norway, a combination of landslide susceptibility and rainfall thresholds is used to dynamically identify potential levels of landslide hazard across different municipalities [23]. A landslide forecasting approach based on the identification of rainfall intensity–duration thresholds is proposed for the mountainous region of Rio de Janeiro State [24]. These systems typically provide information to emergency managers, civil protection and weather forecasters to alert communities of elevated landslide risk. Nevertheless, nearly all of the current landslide early warning systems in place rely on in situ gauge or radar networks to inform landslide hazard [25,26]. Kirschbaum and Stanley [27] provide a global framework for characterizing rainfall-triggered landslide hazards using satellite rainfall. However, the topic of how satellite precipitation estimates may be used to resolve extreme rainfall relevant for triggering hydrometeorological hazards at an urban scale is an area of open research.

Over twenty years of satellite-based global rainfall estimates though the Tropical Rainfall Measuring Mission (TRMM) [28] and the Global Precipitation Measurement (GPM) [29,30] mission have improved our knowledge on the global water cycle and its impacts on society [31]. Several studies have focused on the evaluation of different satellite products [32–35] and their usefulness as a model input to simulate the water and energy balance [36–38]. Although satellite-based rainfall estimates have been widely evaluated at large scales [39–41], few studies have focused at local scale hydrometeorological extremes. An evaluation of TRMM Multi-satellite Precipitation Analysis (TMPA) [28] and Integrated Multi-satellite Retrievals for GPM (IMERG) [42] rainfall estimates over Singapore, using observations at 48 pluviometric stations as reference, concluded that daily satellite products show moderate correlation with gauge data and underestimate heavy rainfall events [43]. Another evaluation of one storm event over Attica, Greece, showed that IMERG underestimated both rainfall and its subsequent flood event [18]. These studies suggest that such products are generally not adapted to monitor local scale rainfall heterogeneities at shorter timescales (i.e., sub-daily and daily), which are essential for urban disaster monitoring. However, such a suggestion is inconclusive, due to the restricted number of studies and limited coverage. Furthermore, there remains an open question in the field as to how satellite precipitation estimates may be leveraged to identify and characterize high-impact hydrometeorological events at a city scale.

The focus for this study involves the municipality of Rio de Janeiro (RJ), Brazil, and has been developed as part of a pioneering partnership between NASA and the city to advance natural disaster monitoring capabilities at the city scale. As part of this partnership, RJ has been providing critical in situ data used in combination with NASA datasets and tools to further parameterize and test modeling frameworks designed to monitor and forecast floods and landslides. Spanning 1204 km^2 in area, RJ is located in Southeastern Brazil (Figure 1a) and stands as the second largest city in the country and ranked fifth in Latin America. The city has witnessed a significant urbanization rate in the past few decades, where 6.72 million people live. Part of this urban expansion is composed of informal settlements, mostly in steep hills within the city. More recently, informal settlements have also been spreading in flat flood-prone areas. RJ has historically suffered from massive rainstorms and subsequent floods and landslides, all of which have caused casualties and adverse socioeconomic impacts. For example, heavy storms with rain accumulations of 90 mm within one hour and wind gusts of around 110 km/h impacted the city between 6–7 February and 7–8 April 2019, causing floods and landslides, and killing numerous people [44–47].

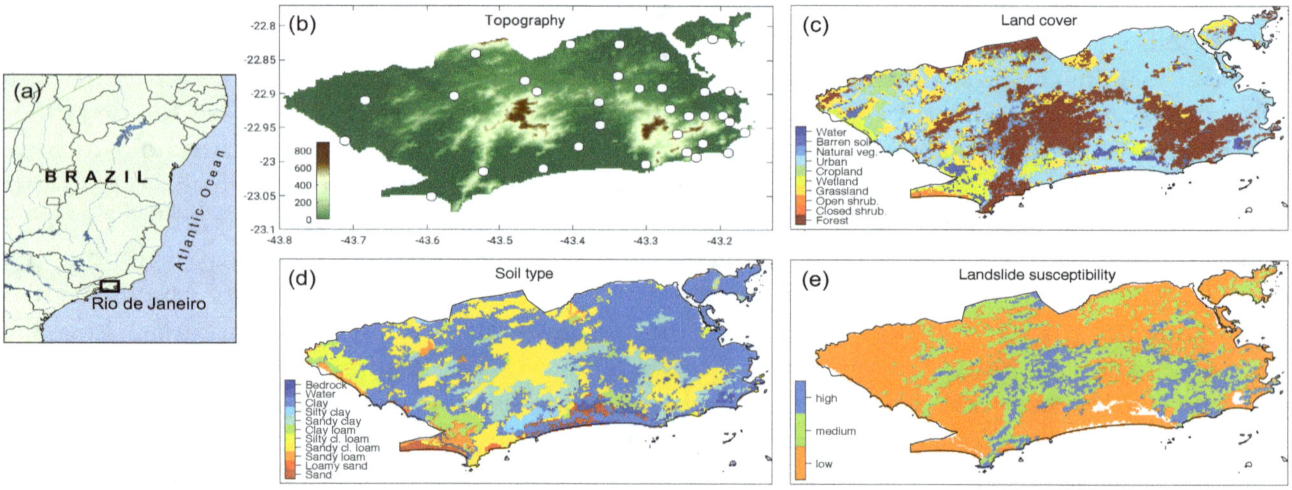

Figure 1. Study area/geographic location, and land surface datasets available for Rio de Janeiro city: (**b**) topography, (**c**) land cover, (**d**) soil type and (**e**) landslide susceptibility maps. The spatial distribution of pluviometric stations is also provided in panel (**a**).

This study is motivated by this NASA-RJ partnership's efforts to develop a near real-time disaster monitoring system for Rio de Janeiro, and our current limited understanding on satellite capabilities to monitor local scale rainfall heterogeneities. Here, we aim to test the concept of how IMERG precipitation estimates can inform hydrometeorological modeling capabilities for natural disasters at a city scale, and their impact on simulating processes that influence floods and landslides within urban areas. We define urban rainfall-related disasters as floods and landslides, and their hydrometeorological related triggers as total runoff (TR) and rootzone soil moisture (RZSM). Here, RZSM is defined as the soil layer one meter below the land surface. A state-of-the-art land surface model (LSM) is used to simulate hydrological processes over the city, and rainfall fields, derived from the interpolation of ground-based observations, are used as the reference for the evaluation. We first investigate how well IMERG products and gauges match across the city, and whether they are capable of detecting major events that occurred in the region during the wet seasons from 2001 to 2019. Wet seasons are defined as the period between November and April, when most storms take place. Then, we evaluate the potential of IMERG products as a predictor of TR and RZSM variables, derived from modeling experiments using LSM outputs. This evaluation is performed across the city and over urban and landslide-prone areas. Finally, we evaluate IMERG precipitation during two recent extreme events, which occurred in February and April 2019. Details on the datasets and methods adopted in this study, results, and conclusions are provided in the following sections.

2. Datasets and Methods

2.1. Modeling Framework

The modeling framework designed for RJ used the NASA Land Information System (LIS) [48] where multiple LSMs are implemented. LIS allows the integration and assimilation of multiple satellite datasets into LSMs, including IMERG rainfall estimates. Land model parameters were processed using the Land surface Data Toolkit (LDT) [49]. The LSM, as well as datasets used in this study, experimental design and evaluation procedure are described next.

Noah-MP

The Noah Multi-Parameterization (Noah-MP) [50,51] land surface model, version 4.0.1, is used to simulate the vertical water and energy balances over the city. The Noah-MP LSM builds upon the well-known Noah LSM [52], which has been used in a variety of operational models, applications and research studies. Noah-MP contains four soil layers and different parameterization and physics options, which include different static vegetation and dynamic vegetation schemes, canopy resistance effects, radiation transfer (e.g., two-stream approximation), runoff and groundwater schemes, snow model options, and even crop and urban canopy schemes. We apply the dynamic vegetation scheme, which employs the Ball-Berry canopy and stomatal-resistance option [53] and Noah-based soil moisture factor for stomatal resistance. The TOPMODEL simulated groundwater scheme [54] is selected, and the Noah-based lower boundary of soil temperature option is applied. For the crop and urban physics schemes, these were not applied at this time, since their physics require more review prior to application in LIS. However, to represent more realistic impervious surface responses, land use and soil types related to urban grid-cells were treated as "bedrock", based on suggested parameters in [55] and assigning a slightly higher porosity for these points. According to Schueler [56], the porosity of urban soils varies between 17% to almost 40%. In this sense, an average porosity of 28% to represent urban areas is a valid assumption. The soil configuration simply used the dominant soil texture type map. Other climatology-based vegetation and albedo parameter maps include monthly greenness fraction [57] and global (snow-free) albedo [57]. Table 1 summarizes the main schemes used in Noah-MP.

Table 1. Noah-MP options adopted in this study to represent physical processes.

Physical Process	Noah-MP 4.0.1 Options	References
Vegetation	Dynamic simulation of carbon uptake and calculated by partitioning LAI and shade fraction (Option 2)	[51]
Stomatal resistance	Ball-Berry (Option 1)	[53]
Soil moisture factor for stomatal resistance	Noah-type based on soil moisture (Option 1)	[58]
Runoff & groundwater	SIMGM: based on TOPMODEL (Option 1)	[54]
Surface layer drag coefficient	Monin-Obukhov (Option 1)	[59]
Super-cooled liquid water	Standard freezing point depression (Option 1)	[60]
Radiation transfer	Modified two-stream scheme (Option 1)	[61]

Note: Cold-season related processes and options (e.g., snow fall, accumulation and depth) are not included here.

2.2. Meteorological Datasets

Noah-MP was forced with the operational Global Data Assimilation System (GDAS) meteorological analysis dataset, IMERG and rainfall fields derived from pluviometric stations. These datasets are described below.

IMERG

IMERG is an algorithm that fuses information from multiple sources, including satellite microwave and infrared precipitation estimates and ground gauge information. The IMERG algorithm uses several passive microwave sensors to assemble and intercalibrate precipitation estimates. However, due to the limited sampling of passive microwave sensors on low-earth-orbit platforms, the gaps are filled by microwave-adjusted infrared estimates [62]. The IMERG algorithm is run three different times, generating the following multi-satellite products: (i) IMERG Early (IMERG-E), available ~4 h after the satellite observation time, allowing quick assessment for flood and landslide prediction; (ii) Late, available ~12 h after the observation time, mainly for agricultural applications; and (iii) Final (IMERG-F), available ~2 months after observation, once monthly gauge data become available, for research applications. IMERG Early and Late are calibrated with climatological sets of coefficients that vary by month and location, and IMERG Final is adjusted with monthly gauge observations. Global data fields are available for 2001-present and provided by the NASA Precipitation Processing System at 0.1° spatial resolution and at a 30-min time step. About 20 IMERG pixels covering the city were spatially interpolated to 0.002 degrees (approximately 200 m) using the budget-bilinear approach.

Pluviometric Stations

Rainfall has been measured operationally every 15 min since 1997 at 33 pluviometric stations within the city. Their spatial distribution is heterogeneous, mainly concentrated in the more developed eastern portion of the city where there is a higher concentration of informal settlements on hills, hence more prone to landslides. Fewer stations are located in the less developed western side (locations are shown in Figure 1b), where informal settlements occupy flat and often flood-prone areas. The inverse distance squared weighting approach [63] was used to spatially distribute point-based rainfall observations throughout the city at a 0.002-degree spatial resolution and 30-min time step. Morphology was taken into account following a similar approach suggested by Salvatici et al. [64], where the domain is divided into sub-regions composed of catchments limited by geomorphological features and the interpolation is performed using only rainfall observations within sub-regions. Here, the city was ***divided*** into four catchment-based sub-regions, which are representative of the city's morphology: West, South, Southeast and Northeast.

2.3. *Land Surface Parameterization*

Key land surface parameters used in Noah-MP, in particular, topography, land cover and soil type, were acquired from local measurements and high-resolution airborne and satellite imagery, and they are described below and shown in Figure 1. A landslide susceptibility map, used in the analysis of results, is also described.

Topography

Lidar data used for the model was acquired through an aerial survey in the second half of 2019, with a raw density of eight points per square meter. Point-based cloud data was filtered for ground-only returns and interpolated as a 15-cm spatial resolution digital terrain model map. For the purpose of this study, the hyper-resolution data were resampled to 0.002 degrees.

Land Cover

Local land cover data were acquired from the City's Government Environment Secretariat, reference year 2016, as a vectorial database at a 1:10,000 cartographic scale. The vectorial database was reclassified following the International Geosphere–Biosphere Programme (IGBP) [65] land cover classification and converted to a 0.002-degree raster map. According to the resulting map, urban areas represent 50% of the city's land use.

Soil Type

The Rio de Janeiro soil map was derived from 41 soil mapping units at the scale of 1:50,000, updated by Gomes et al. [66]. According to this study, Ultisol (also called Alfisol) soil class predominates in the domain, followed by Oxisols [67] in, respectively, 40% and 17% of the region. A representative soil textural class, according to the American textural classification system [68], was assigned to each of the mapping units, taking into account granulometric fractions information (sand, silt and clay content) of representative soil profiles obtained according to Embrapa [69]. The sandy loam class from indiscriminate mangrove soils was assigned based on the literature on these soils. This study allowed the estimation of the different hydraulic parameters for RJ soils required by the model, taking its standard tables that associate textural classes with the referred hydraulic parameters as reference.

Landslide Susceptibility Map

Rio de Janeiro's landslide susceptibility map is available at a 5-meter spatial resolution as a result of a multicriteria spatial analysis based on the overlapping of the following layers: lithology, soil type, slope, land cover, geotechnics and geomorphology. The susceptibility map was developed by GEORIO, the municipal geological service, and is based on site investigations and expert analysis [70]. Three classes are defined in the map: low, medium and high susceptibilities. Here, we adopted classes medium and high as landslide-prone areas, corresponding to 42% of the city's surface area.

2.4. *Experimental Design*

To determine the potential of IMERG products to monitor TR and RZSM throughout the city, three experiments were designed: one using ground-based precipitation (called the reference experiment hereafter), and two using IMERG datasets, called IMERG-E and IMERG-F experiments hereafter. Since our goal is to explore the potential for IMERG to monitor natural disasters at the near real-time and possibly to be used to generate initial hydrological conditions for forecasts, IMERG-F is considered here as a benchmark for IMERG-E, due to its relatively long latency and inclusion of ground-based observations. In order to isolate the impact of IMERG products on the hydrological modeling, all experiments were composed of the same land surface parameters and GDAS meteorological forcings, except for the rainfall data. Experiments were run for 19 years, from 2001 to 2019, where all meteorological datasets overlap, on a 30-min time step and 0.002-degree spatial resolution. GDAS data was time-interpolated using linear weights to bring it down to the 30-min time step. Land surface parameters described above were also rescaled to 0.002 degrees, matching the spatial resolution of the modeling system. The spatial and temporal resolutions adopted here are sufficiently fine to represent fine-scale physical processes of interest. Experiments were spun up for 50 years, allowing modeled water storage states to stabilize.

2.5. Evaluation Procedure

The lack of in situ data and insufficiently high-resolution satellite-based soil moisture and runoff observations prevents us from evaluating model outputs against an observable truth. In the absence of such observations, we compared IMERG experiments against the reference experiment, as it could be considered as the most accurate representation of precipitation over the domain. Floods and landslides are mostly triggered by extreme rainfall events, mostly common during the wet season. Here, we defined the months from November to April as the wet season and focused all analyses on that period, neglecting events that may have occurred in other months. The IMERG evaluation was performed daily at both point-based and city-wide scales, using metrics such as bias, root mean square error (RMSE), and correlation. The evaluation of extreme event detection skill was performed by using two categorical metrics: the probability of detection (POD) and false alarm rate (FAR). POD defines the rate of correct satellite-based rainfall estimates, while FAR is the rate of false ones, as defined below:

$$\mathrm{POD} = \frac{\mathrm{a}}{\mathrm{a} + \mathrm{c}} \tag{1}$$

$$\mathrm{FAR} = \frac{\mathrm{b}}{\mathrm{b} + \mathrm{d}} \tag{2}$$

where a is the number of simulated events above the threshold correctly observed, b stands for the number of false alarms (simulated events above the threshold detected but not observed), c is the number of simulated events above the threshold not observed, and d is the sum of events above the threshold neither observed nor simulated. POD values range from zero to 1, where zero corresponds to no skill in correctly detecting events and 1 stands for a skillful system correctly detecting all events. FAR values also range from zero to one, where zero is the ideal case, meaning that no events were wrongly detected, and 1 indicates that events were wrongly detected throughout the study period.

For the point-based evaluation, we categorized the extreme events into three classes according to different rainfall thresholds [71,72]: moderate rain (≥10 mm/day), heavy rain (≥25 mm/day) and torrential rain (≥50 mm/day). Next, we evaluated the accuracy of spatially-distributed precipitation estimates and their impact on simulated rootzone soil moisture and total runoff by means of correlation, root mean square error (RMSE), POD and FAR. For the spatial analysis, thresholds were defined as percentiles rather than absolute values. Percentile thresholds are commonly used for extreme event monitoring and forecast [73–76]. Percentiles were computed for each pixel from each experiment, individually, using the full 2001–2019 period (i.e., dry seasons were also considered). By considering the full period, which includes a large quantity of low or no rain events during the dry seasons, it can impact percentiles in a way that extreme events become rarer, thus higher percentile values. Daily rainfall, RZSM and TR over the full period (i.e., a total of 6939 daily events per grid cell) were ranked and converted to their corresponding percentiles, from 1% (least severe wet events) to 100% (most severe wet events). Finally, five percentile thresholds of varying likelihood of occurrence were defined: abnormally wet (≥70%), moderately wet (≥80%), severely wet (≥90%), extremely wet (≥95%), and exceptionally wet (≥98%).

3. Results

3.1. Point-Based IMERG Evaluation

Point-based comparisons at 33 pluviometric stations indicate that both IMERG products overestimate rainfall estimates during wet seasons throughout the city, with median biases of 0.66 mm/day for IMERG-E (Figure 2a) and 1.6 mm/day for IMERG-F (Figure 2d). The higher bias obtained with IMERG-F could be explained by corrections using ground-based observations. Brazil's nationwide rain gauge network [77], normally available for the international scientific community, is not available in Rio de Janeiro and remains scarce in the surrounding region. This network also excludes the municipal rain gauge network used in this study. Such a geographic limitation may result in biases over coastal areas where orographic and other meteorological processes dominate precipitation patterns and where the satellite retrieval algorithms may be less sensitive to land-ocean background states. Biases with respect to rain gauges could result from the positioning of the gauges lower down on the slopes where peak intensities may not be resolved, or from gauge under catch resulting from wind or debris on the gauge. One station stands out with a high negative bias (i.e., satellite estimates are lower than observations) of −2.66 mm/day and −1.78 mm/day, respectively, in Southeast RJ. That station is situated at 355-m high on steep hills near the coast where landslides are frequent, and neighbourhoods downstream often flood during storms. RMSE values are high relative to the average rainfall, varying from 10.4 to 19.5 mm/day, depending on the station and satellite product (Figure 2b,e). Daily IMERG products show moderate correlation with gauge data, varying from 0.38 and 0.58 (Figure 2c,f) and median correlation values of 0.49 (IMERG-E) and 0.54 (IMERG-F). These results are consistent with the point-based skill evaluation, revealing a moderate POD for events greater than or equal to 10 mm/day, averaging 0.53 and 0.6, respectively (Figure 3a,d). Overall, IMERG-F indicates more skill in detecting extreme events than IMERG-E, although it has higher bias and RMSE values. POD values degrade for higher rainfall thresholds greater than or equal to 25 and 50 mm/day, averaging 0.33 and 0.16 for IMERG-E and 0.44 and 0.24 for IMERG-F. As shown in Figure 3g–l, false alarm rates of extreme events are high for both satellite products and all selected thresholds, with FAR averages varying from 0.57 (rainfall ≥ 10 mm/day) to 0.8 (rainfall ≥ 50 mm/day). High FAR values indicate that high rainfall rates are detected by IMERG, but not observed, which could explain the positive bias and high RMSE discussed above.

3.2. IMERG Impacts on Total Runoff and Rootzone Soil Moisture Simulations

Figure 4 shows the spatially-distributed averages of rainfall, RZSM and TR over the city during wet seasons derived from the spatial distribution of reference pluviometric data and IMERG products, and corresponding monthly climatologies. Based on the reference rainfall, the average precipitation over RJ is 4.43 mm/day, with a heterogeneous spatial distribution, varying from 3.65 to 7.47 mm/day. The spatial variability of RZSM derived from the reference rainfall reflects the land cover classification and soil composition, with low average values over urban areas (0.13 m^3/m^3), explained by the low infiltration rates as a result of the impervious urban land cover, and high over landslide-prone areas (0.26 m^3/m^3). RZSM averages 0.21 m^3/m^3 over the city. Generally, high soil moisture variability means more infiltration during rain events, resulting in more water available for evapotranspiration, hence less available surface water to run off. This is evident in the model simulation, where average TR is 4.33 mm/day (or 97% of rainfall) over urban areas, and 0.65 mm/day (15% of rainfall) over landslide-prone areas.

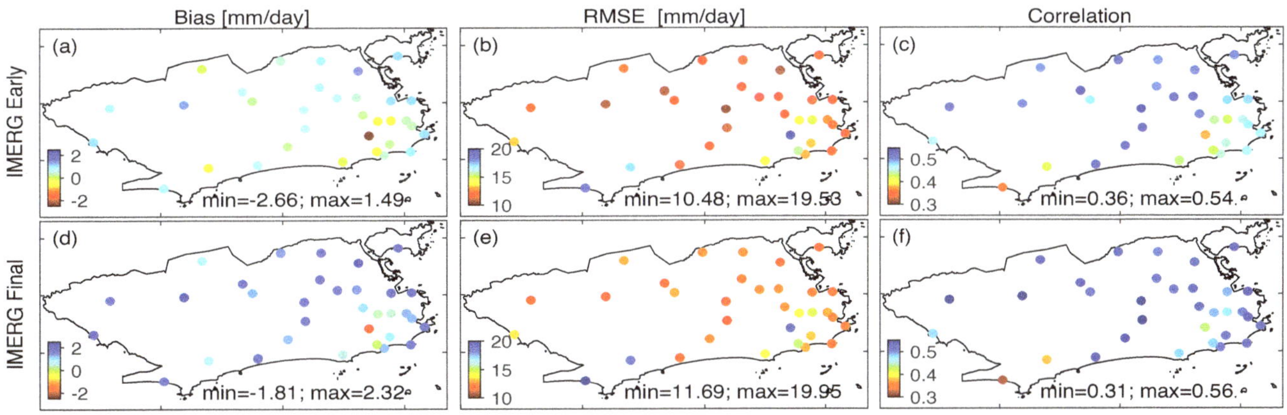

Figure 2. Daily bias, root mean square error (RMSE) and correlation for IMERG Early (**a**–**c**) and IMERG Final (**d**–**f**) at 33 pluviometric stations for 2001–2019 wet seasons (November-April). Limits of Rio de Janeiro city are defined by the black line.

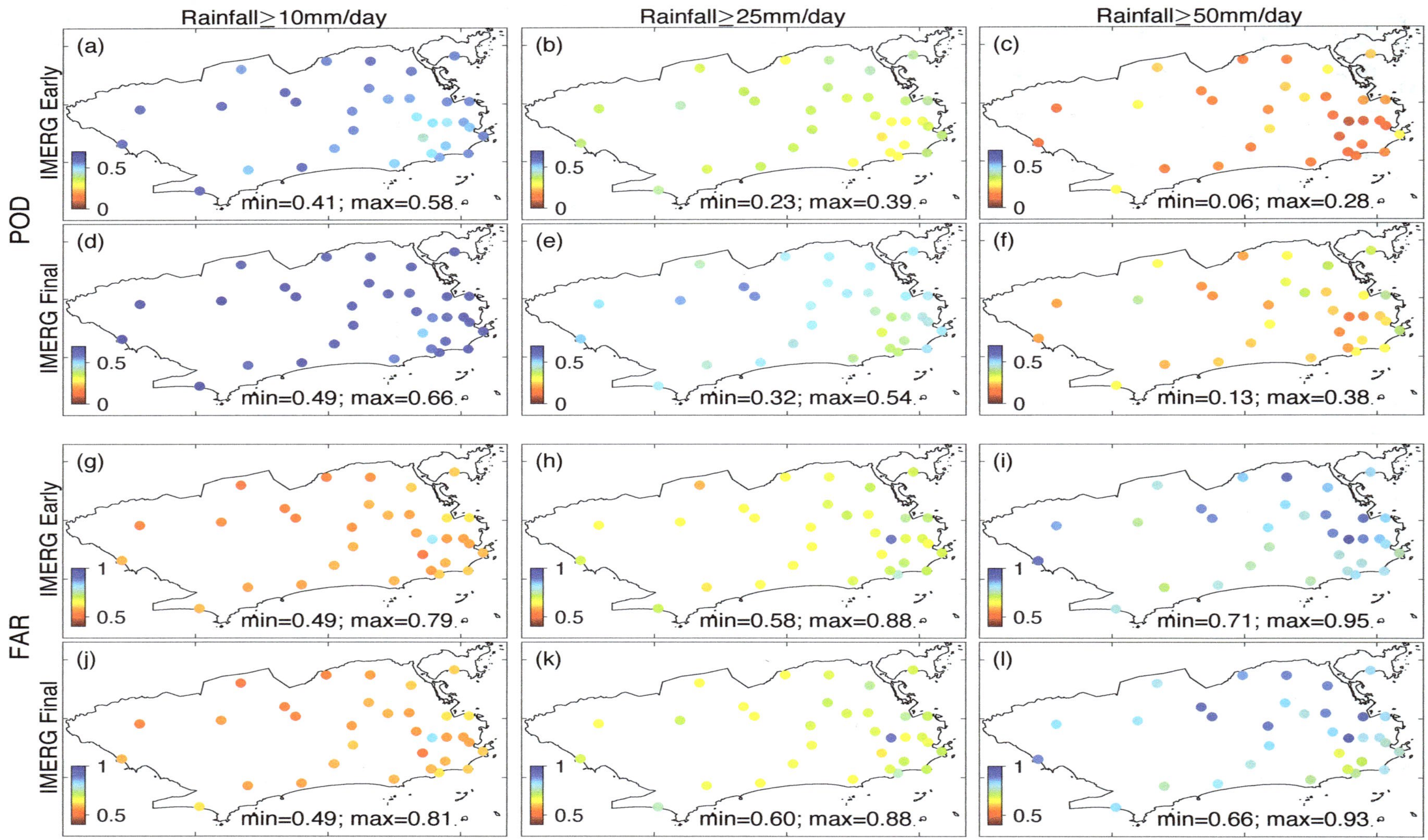

Figure 3. Probability of detection (POD) and false alarm rate (FAR) maps at three rainfall thresholds (≥10 mm/day, ≥25 mm/day and ≥50 mm/day) for IMERG Early (**a–c**,**g–i**) and IMERG Final (**d–f**, **j–l**) at 33 pluviometric stations for 2001–2019 wet seasons.

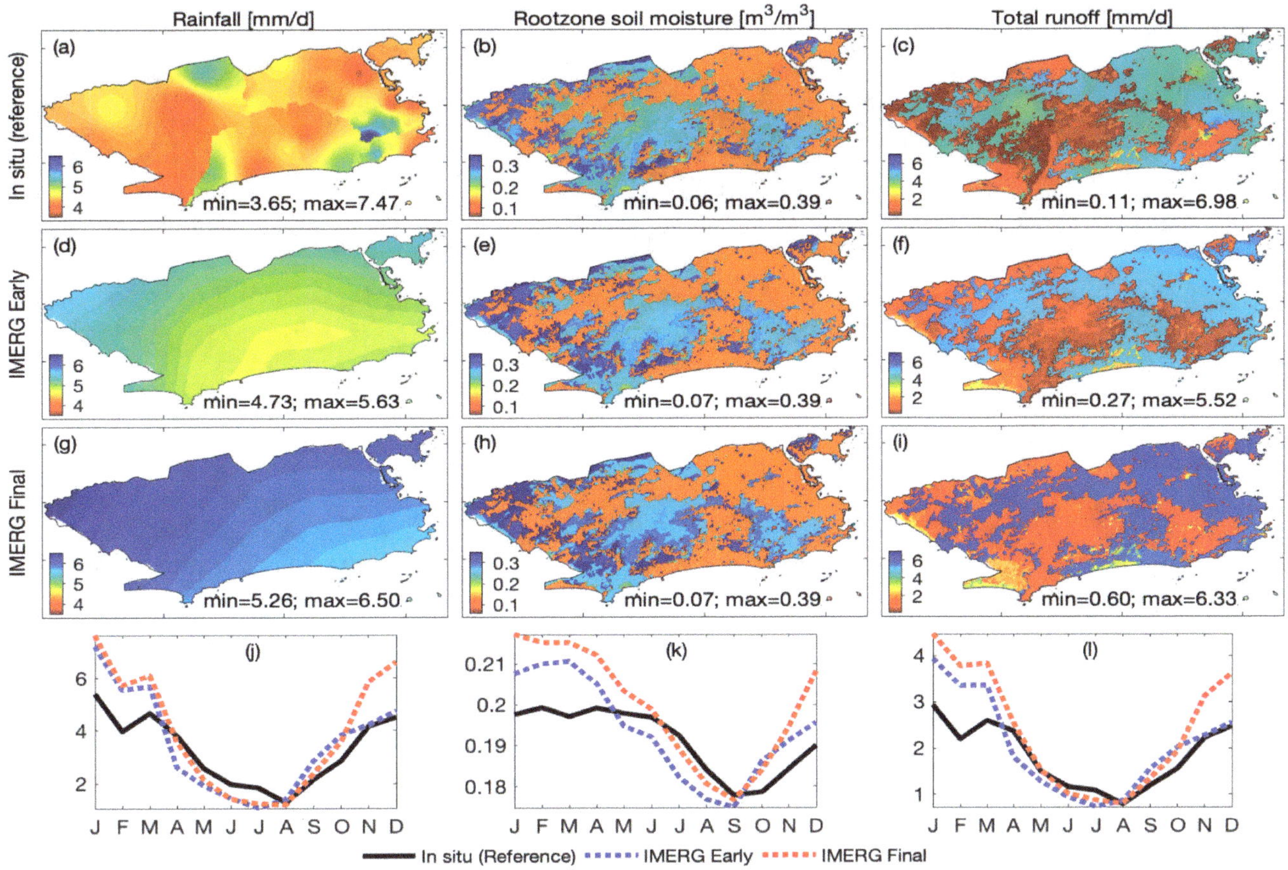

Figure 4. Spatially distributed rainfall, rootzone soil moisture (RZSM) and total runoff (TR) averaged over 2001–2019 wet seasons derived from the experiments using ground-based (**a**–**c**), IMERG Early (**d**–**f**) and IMERG Final (**g**–**i**) rainfall estimates. Spatially-averaged monthly climatology time series are shown in the bottom (**j**–**l**).

Although IMERG products overestimate average rainfall over the city during wet seasons by about 12% or 5.01 mm/day (IMERG-E), and 35% or 5.93 mm/day (IMERG-F), they both show low spatial variability. This could be a result of IMERG's insufficiently fine spatial resolution to represent local heterogeneities. Monthly climatologies show that IMERG overestimates rainfall during most of the wet season, except April. Such differences in rainfall have little impact on average RZSM, but significantly impacts TR, especially over urban areas, with average increases ranging from 14% to 34%. As a result, RMSE for the spatially distributed rainfall is high, varying from 9 to 19.4 mm/day (Figure 5a,d), similarly to those found for the point-based evaluation. RMSE is particularly high in the city's coastal south-eastern hilly extents, a rainfall hotspot where observations are much higher than the city average. IMERG products also show the lowest correlation with the reference rainfall over that location, varying between 0.38–0.45.

Except for the rainfall hotspot, correlation over most of the city ranges from 0.5 to 0.64, with IMERG-F performing generally better than IMERG-E. Despite the moderate IMERG performance in providing spatially distributed rainfall estimates over the city, RZSM is not as degraded, with relatively low RMSE values, ranging from zero to 0.09 m^3/m^3, and higher correlation, varying from 0.41 and 0.86, depending on the satellite product and location. On the other hand, TR is highly degraded, in particular in the south-eastern extent of the city, with RMSE values as high as 16.7 mm/day and correlation as low as 0.23.

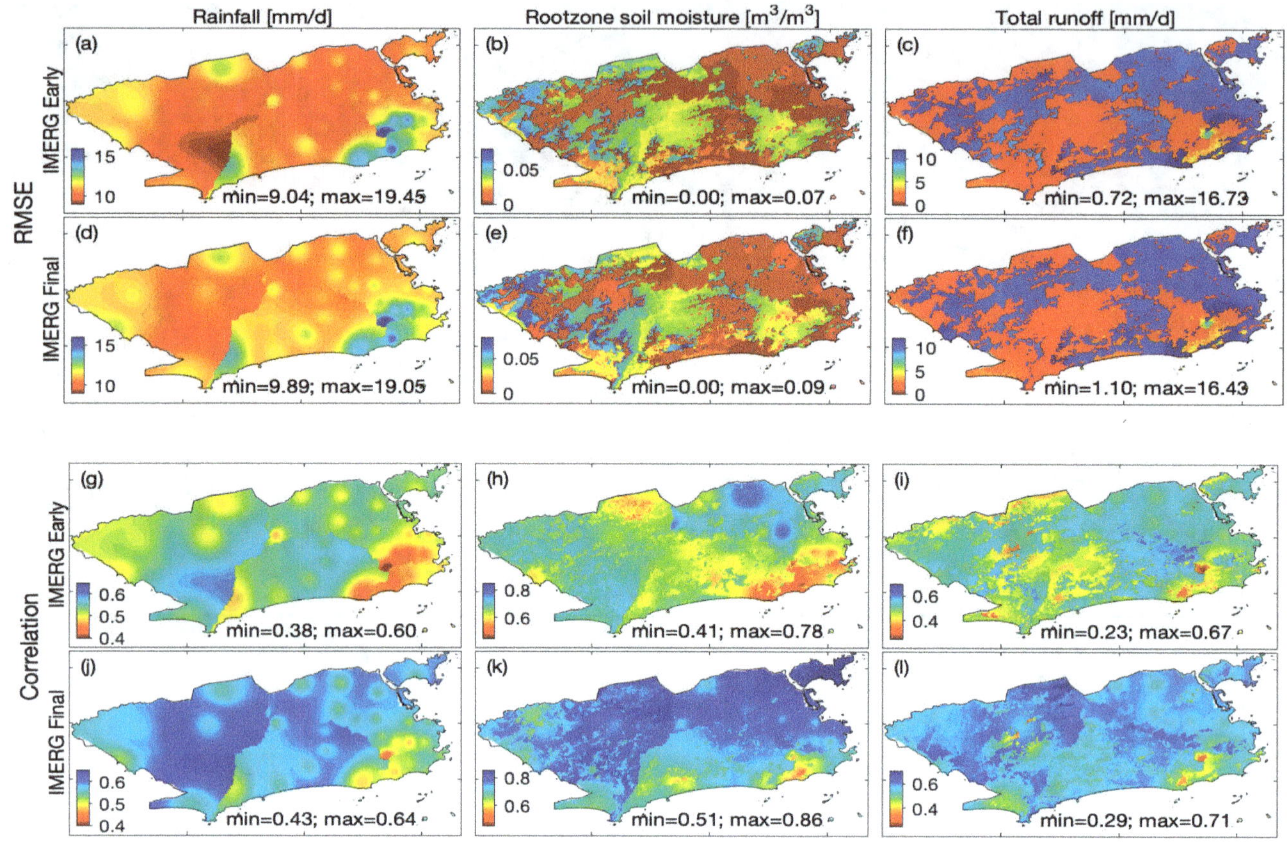

Figure 5. Spatially distributed RMSE and correlation of absolute rainfall, rootzone soil moisture and total runoff estimates over 2001–2019 wet seasons. Results are shown for experiments with IMERG Early (**a**–**c**,**g**–**i**) and IMERG Final (**d**–**f**,**j**–**l**).

Figure 6 summarizes the results for the full domain, landslide-prone and urban areas. A comparison of performances between satellite products reveals that IMERG-E has lower RMSE values over the three regions but low correlation, indicating that it has better skill in estimating rainfall magnitudes and not in rainfall timing. RMSE is basically the same for both IMERG-based RZSM estimates and very similar for TR estimates, while IMERG-F shows an overall better performance in terms of correlation. Over landslide-prone areas, where RZSM is a disaster trigger, RMSE is low relative to the average soil moisture and correlation is higher than those obtained for rainfall estimates. On the other hand, TR over urban areas does not perform as well. Although the western extent of the city is poorly monitored, it shows an overall better performance in term of RMSE and correlation, possibly due to the lower irregular geomorphology.

3.3. Evaluation of Selected Events

We selected two major storms to further evaluate the potential of IMERG in detecting natural disaster triggers in urban areas. They occurred on 5–7 February 2019 (Event 1) and 9–10 April 2019 (Event 2) and resulted in floods and landslides, causing major socio-economic damage in the city. Figure 7 shows hourly rainfall from the reference and IMERG products and their impact on TR over urban and RZSM over landslide-prone areas for both events. In Event 1, two rainfall peaks are observed two days apart, reaching 15 mm/h. Although the first one was detected by IMERG, both satellite products underestimate the total rainfall by about half. The second peak is missed by IMERG, showing smaller rainfall events distributed over 24 h instead. Over urban areas, the total observed rainfall during Event 1 was 128 mm, as opposed to a much lower amount of 71 and 62 mm detected by IMERG-E and IMERG-F, respectively. Such low IMERG-based rainfall rates resulted in 45–51% TR underestimation. Over landslide-prone areas, the reference experiment shows two consecutive RZSM raises within 2 days, occurring during each rainfall peak and totalling a soil moisture increase of 65%. The soil saturation caused by the second raise occurred in 6–7 February concomitant with the surging of numerous landslides over the city [45]. High RZSM levels derived from IMERG-based experiments at the beginning of Event 1 are explained by an early storm detected by IMERG, peaking ~10 mm/h, but not observed (not shown). As a result, the low rainfall peak was enough for RZSM to reach the reference level at the end of the first peak, but not to rise to the maximum reference moisture at the end of the second peak. Event 2 was characterized by a 2-day-long storm, totalling 201 mm of rainfall across the city (193 mm over urban areas and 213 mm over landslide-prone areas), with 183 mm during the first 24 h. This storm had three peaks with decreasing magnitudes over time. Both satellite products successfully detected the two major ones, but IMERG-F performed better, with a total rainfall of 122 mm, as opposed to 69mm with IMERG-E. The underestimated rainfall resulted in 36–64% lower TR over urban areas. Although slightly underestimated, soil moisture signals were properly represented, in particular by IMERG-F rainfall. The results obtained for these two cases suggest that, although IMERG products overestimate the average rainfall across the city, some severe storms might still be underestimated.

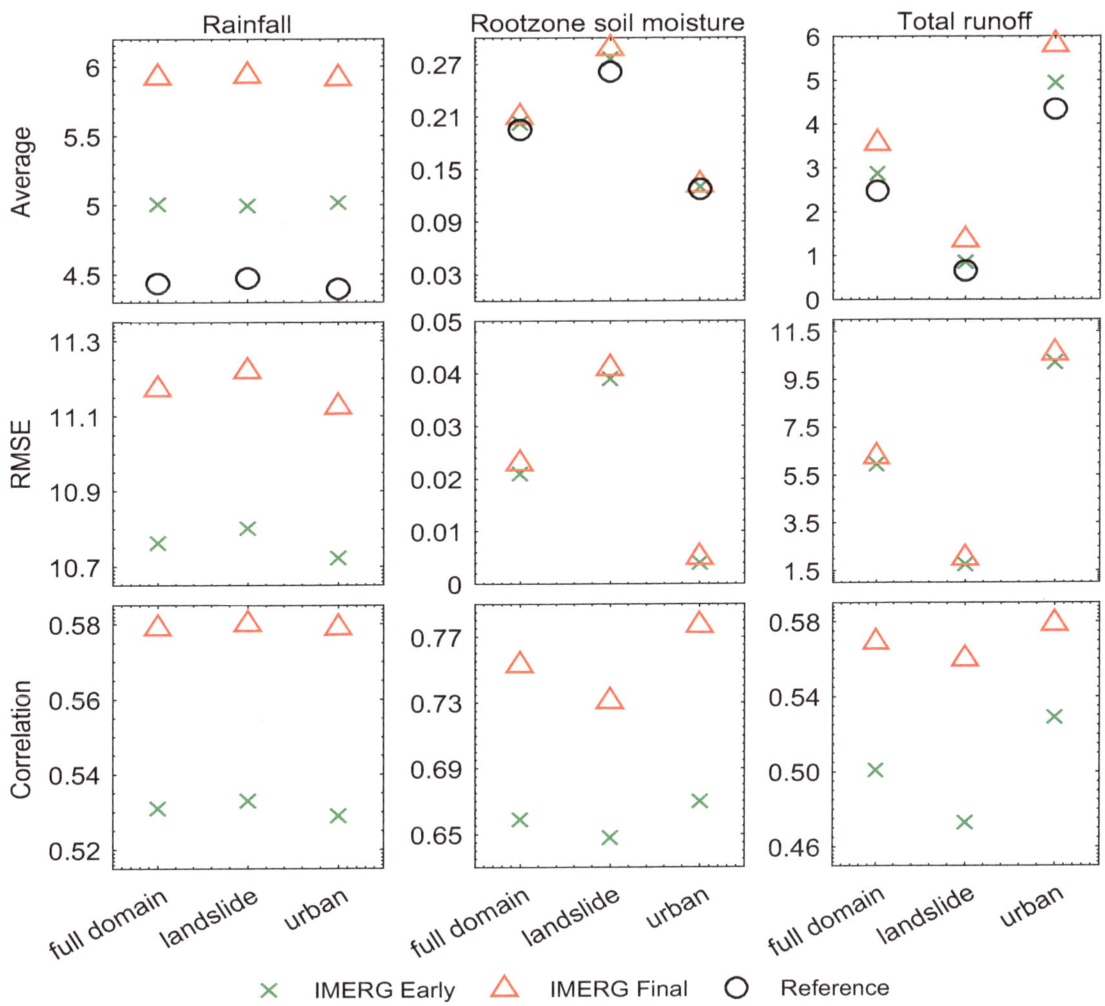

Figure 6. Average, RMSE and correlation of precipitation, rootzone soil moisture and total runoff derived from experiments using ground-based, IMERG Early and IMERG Final precipitation estimates. Results are shown as medians for 2001–2019 wet seasons over the full domain, and landslide-prone and urban areas. Average and RMSE units are mm/day for rainfall and total runoff, and m^3/m^3 for rootzone soil moisture.

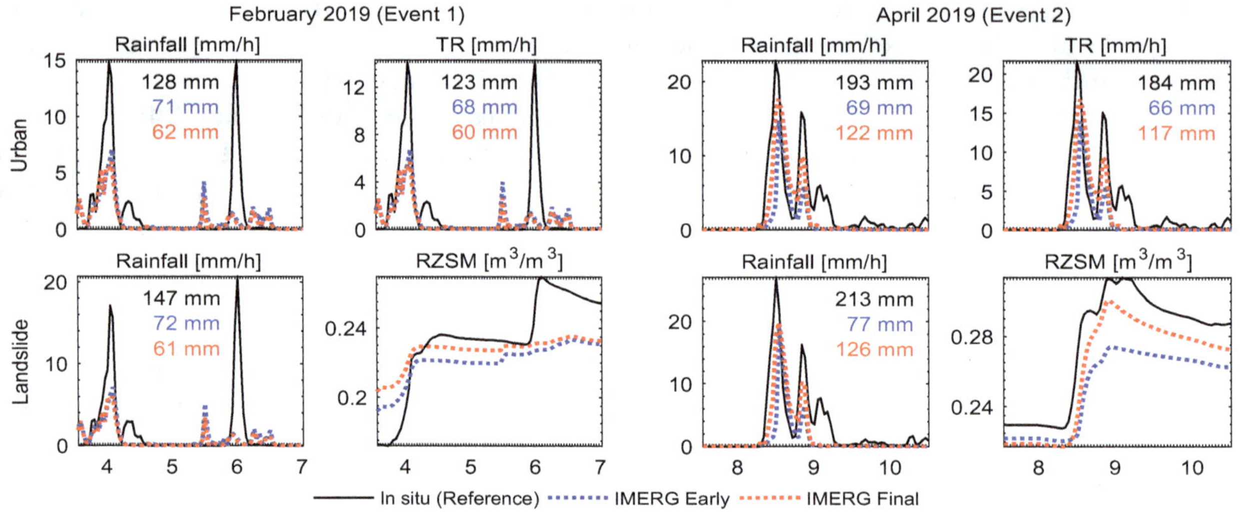

Figure 7. Selected storm events in Rio de Janeiro city and impacts on total runoff (TR) and rootzone soil moisture (RZSM) over urban (top) and landslide-prone (bottom) areas. Left panels show the evolution of the storm occurred from 5–7 February 2019 (Event 1), and right panels show the storm that occurred in 9–10 April 2019 (Event 2). Results are shown at the hourly time step and x-axes indicate days of the month.

3.4. Detection Skill of Extreme Events in Terms of Percentiles

Based on the evidence of an overall mismatch between reference-based and IMERG-based extreme events in terms of the total mm/day, we considered event detection evaluation in terms of percentiles. Using percentiles is a way to normalize time series independently, based on the magnitude of the events, and remove any biases found between two independent sources. In this way, it helps us understand IMERG's skill to detect daily extreme categorical events, even if the magnitudes differ. As an example, Figure 8 shows the spatial distribution of POD and FAR for normalized rainfall, RZSM and TR at a percentile threshold ≥70%, i.e., panels show the probability of IMERG-based hydrometeorological variables to detect or provide a false alarm of an event that is 70% or higher than all events in the time series at each pixel (POD and FAR maps for percentile thresholds ≥80%, ≥90%, ≥95% and ≥98% are shown in Figures S1–S4). Based on similar maps for all five thresholds, as defined in Section 2.5, we computed medians of percentile-based POD and FAR values over urban and landslide-prone areas for each IMERG-based experiment. As summarized in Figure 9, we can observe that percentile-based POD and FAR are significantly higher than those obtained using absolute values of rainfall, TR and RZSM. This demonstrates that, although IMERG-based experiments moderately detect event magnitudes, it shows higher skill in detecting occurrences. However, both POD and FAR degrade as a function of rainfall intensity. Median POD values range from 0.75 to 0.84 for abnormal (≥70%) hydrometeorological events (i.e., rainfall, TR and RZSM) occurring over urban and landslide-prone areas and degrade to 0.28–0.42 for exceptional events (≥98%). This means that the more exceptional the event is, less likely IMERG will detect it. IMERG-F skill to detect extreme events is slightly higher. Median FAR values are generally low, ranging from 0.2 to 0.29 for abnormal events, improving to 0.02 for exceptional events. Such an improvement is explained by the fact that the more exceptional the event is, larger the ensemble of non-observed events will be. All thresholds below 95% show that IMERG experiments provide fewer false alarms for rainfall than RZSM over landslide-prone areas, indicating that IMERG generally overestimates low rain events during the wet seasons, resulting in more RZSM extreme events than observed. Rainfall and TR show similar FAR for all thresholds, confirming the high correlation between these two variables over urban areas.

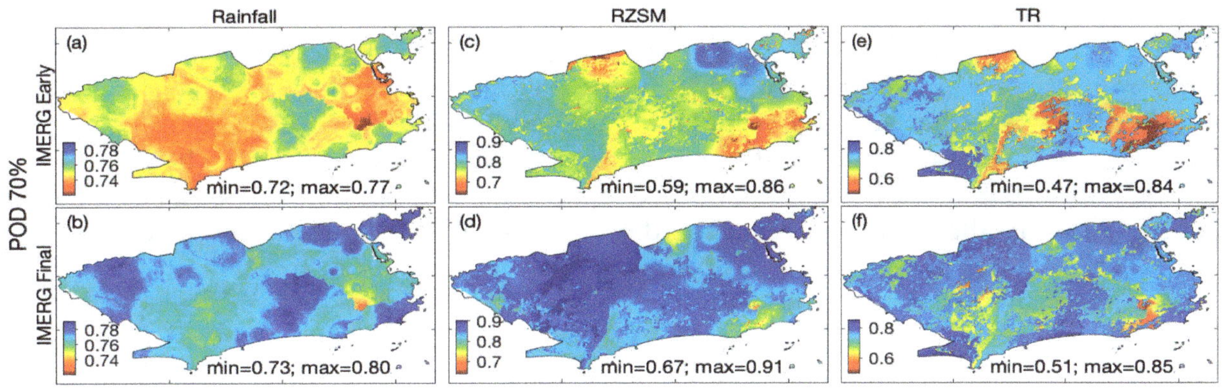

Figure 8. Probability of detection (POD) and false alarm rate (FAR) maps for rainfall (**a**,**b**), rootzone soil moisture (RZSM) (**c**,**d**) and total runoff (TR) (**e**,**f**) events derived from experiments with IMERG Early and IMERG Final at the percentile threshold 70% chance of occurrence.

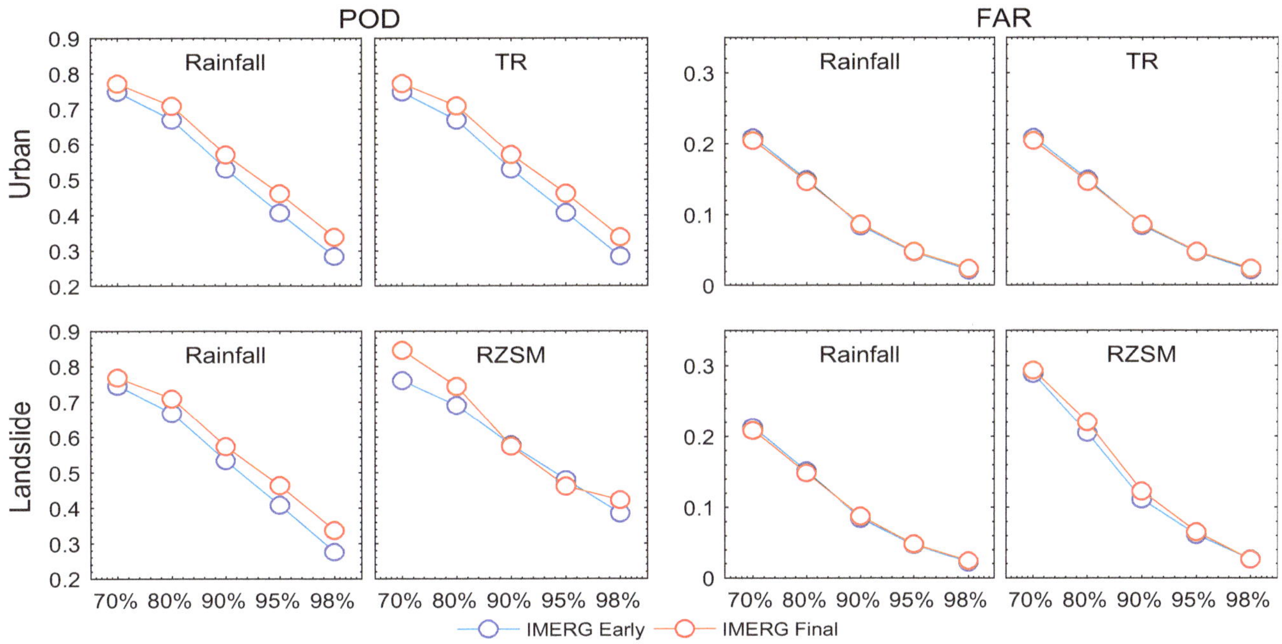

Figure 9. Median POD and FAR for rainfall, rootzone soil moisture (RZSM) and total runoff (TR) rates derived from IMERG-E and IMERG-F experiments at percentile thresholds of ≥70%, ≥80%, ≥90%, ≥95% and ≥98% over the urban and landslide-prone areas.

4. Discussion

It is important to highlight that although we used ground-based rainfall observations as the reference for this analysis, the gauging network spatial distribution does not represent all climatologic and geographic heterogeneities found in the city. The complex topography within RJ creates microclimates that cause rainfall accumulations to significantly vary across the city. Rainfall totals can differ based on the prevailing wind patterns of the storms and where the storms originate and move, further complicating the ability to systematically define local maxima in rainfall. The orographic enhancement of these systems resulting from the complex topography and microclimates within the city further complicate the ability to get a complete and accurate representation of total rainfall across the city. The geographic distribution of the 33 pluviometric zones are primarily located in low elevations, which could result in an underestimation of the actual precipitation over the city. There is also a sparser gauge network in the western portion of the municipality, which may impact the total accumulations of gauge-based estimates when comparing averages across the city. However, there are also known challenges in comparing satellite estimates with individual observations from rainfall gauges for a number of reasons. First, the IMERG data is composed of integrating passive microwave estimates from a constellation of approximately 10 satellites that are in non-synchronous or polar orbits. These observations are then merged, morphed and filled in with infrared date from geostationary satellites. If there is a short convective storm that has high rainfall intensities, but which develops and dissipates within 30 min, unless there is a passive microwave sensor observing that storm directly, there is a chance that the satellite product could partly or entirely miss the event. This likely is why we observe IMERG resolving a portion of the extreme rainfall but not resolving the entire event, hence underestimating the total storm volumes.

The sparse number of gauges in the western part of the city and the interpolation technique used to generate the reference rainfall fields are other factors impacting the accuracy of the reference rainfall fields. Here, we assumed that the inverse distance squared weighting approach constrained by catchment-based sub-regions was sufficient to take morphology into account and accurately spatialize point-based rainfall observations. However, we acknowledge that this technique has restrictions by not considering all topographic and coastal effects. Such restrictions, in addition to limitations in the land parameterization and numerical representation of physical processes in Noah-MP, may result in inaccurate estimates of water storages and fluxes. For example, due to the unavailability of an urban hydrology module in the Noah-MP available in LIS, we used the "bedrock" soil class to represent urban imperviousness. Although this could emulate the impacts of urbanization on water fluxes and storage, it is not ideal to represent impacts buildings on the energy balance [78]. It also is important to note that, since Noah-MP does not store TR from a time step to another, surface water is not available for further evaporation after rain events. Hence, the actual TR could be lower during prolonged floods.

5. Summary and Final Considerations

Previous studies have successfully demonstrated that satellite precipitation can be used at large and mesoscale hydrometeorological monitoring and modeling with sufficient accuracy [79–81]. Motivated by the current limited understanding of how satellite-based rainfall estimates perform at local scales, in particular over urban areas, and taking advantage of a partnership between NASA and Rio de Janeiro city, we explored the potential of IMERG products to monitor natural disaster triggers in the city. To derive urban flood and landslide triggering indicators from IMERG rainfall estimates, we used a state-of-the-art modeling system composed of the Noah-MP LSM, version 4.0.1, and high-resolution land surface parameters. We analyzed IMERG Early and IMERG Final rainfall estimates and rainfall-driven urban flood and landslide triggers (i.e., rootzone soil moisture and total runoff) and events that occurred during the wet season from 2001 to 2019 over urban and landslide-prone areas, using ground-based rainfall observations as reference. We found that long-term averaged IMERG products overestimate rainfall throughout the city and that impacts on TR and RZSM vary spatially as a function of land cover and soil types. We also noticed that IMERG-F generally outperforms IMERG-E, which could be explained by the additional adjustments with monthly gauge observations. However, due to its 2-month latency, IMERG-F was considered here as a benchmark for IMERG-E rather than a product that could be used for near real-time hazard warning systems. IMERG-based experiments resulted in a moderate performance to reproduce the spatial and temporal distribution of rainfall, TR and RZSM, and to quantitatively reproduce extreme events.

Based on our results and, more broadly, on previous studies using satellite precipitation data at a city scale, one can conclude that, IMERG's 10-km spatial resolution is insufficient to capture local heterogeneities in cities with complex reliefs such as Rio de Janeiro. The inability to represent these complexities can have an enormous impact on the accuracy of landslide and urban flood early warning systems, since these occur at a very local scale. On the other hand, the temporal resolution and low latency (~4 h for IMERG Early) appear to be sufficient for rainfall-triggered disaster monitoring, particularly if the relative (e.g., percentile) distributions, rather than actual values, are considered. However, our results showed potential in qualitatively detecting city-wide extreme events after normalizing signals through percentiles. POD values were above 0.5 for percentile thresholds as high as ≥90%, indicating that normalized IMERG-based experiments could be used to detect certain extreme events, supporting early warning systems and decision making related to natural disasters in the city.

This study evaluated the potential of IMERG to monitor natural disaster triggers in Rio de Janeiro. The next steps of this ongoing research will focus on evaluating its potential to support the operational monitoring and forecast actual natural disasters, in particular urban flood events across the city. It is also suggested that future work should focus on the improvement of algorithms and sensors to refine the detection of local heterogeneities, as well as reducing the latency of merged products.

Supplementary Materials:
Figure S1: POD and FAR maps for rainfall, RZSM and TR events derived from experiments with IMERG Early and IMERG Final at the percentile threshold ≥80% chance of occurrence, Figure S2: POD and FAR maps for rainfall, RZSM and TR events derived from experiments with IMERG Early and IMERG Final at the percentile threshold ≥90% chance of occurrence, Figure S3: POD and FAR maps for rainfall, RZSM and TR events derived from experiments with IMERG Early and IMERG Final at the percentile threshold ≥95% chance of occurrence, Figure S4: POD and FAR maps for rainfall, RZSM and TR events derived from experiments with IMERG Early and IMERG Final at the percentile threshold ≥98% chance of occurrence.

Author Contributions: Conceptualization, A.G.; methodology, A.G., D.K.; software, A.G., K.A.; validation, A.G., D.K., S.K.; formal analysis, A.G.; investigation, A.G.; resources, A.G., D.K.; data curation, A.G., F.M., M.O.; writing—original draft preparation, A.G.; writing—review and editing, A.G., D.K., F.M., M.O., S.K., K.A.; visualization, A.G.; supervision, A.G.; project administration, A.G., D.K., F.M.; funding acquisition, D.K. All authors have read and agreed to the published version of the manuscript.

Acknowledgments: We would like to thank J.F. Lumbreras (Embrapa Soils, Brazil) for the technical support in the Rio de Janeiro soil texture classification, and to Rio de Janeiro's Municipal Environment Secretariat and Pereira Passos Institute staffs for the technical support on land cover data processing. Computing resources supporting this work were provided by the NASA High-End Computing (HEC) Program through the NASA Center for Climate Simulation (NCCS) at NASA's Goddard Space Flight Center.

References

1. Mitchell, J.K. Megacities and natural disasters: A comparative analysis. *GeoJournal* **1999**, *49*, 137–142. [CrossRef]
2. Ogie, R.I.; Adam, C.; Perez, P. A review of structural approach to flood management in coastal megacities of developing nations: Current research and future directions. *J. Environ. Plan. Manag.* **2020**, *63*, 127–147. [CrossRef]
3. ELLA. *Urban Disaster Risk Management in Latin American Cities*; Evidence and Lessons from Latin America: Lima, Peru, 2017; Volume 13.
4. Miguez, M.G.; Veról, A.P.; Bianchini, L. Flood risk assessment and management: A case study in Rio de Janeiro. *WIT Trans. Built Environ.* **2013**, *134*, 29–41. [CrossRef]
5. Diaz, V.J. Landslides in the squatter settlements of Caracas; towards a better understanding of causative factors. *Environ. Urban.* **1992**, *4*, 80 89. [CrossRef]
6. Haigh, M.; Rawat, J.S. Landslide causes: Human impacts on a Himalayan landslide swarm. *Belgeo* **2011**, 201–220. [CrossRef]
7. Cui, Y.; Cheng, D.; Choi, C.E.; Jin, W.; Lei, Y.; Kargel, J.S. The cost of rapid and haphazard urbanization: Lessons learned from the Freetown landslide disaster. *Landslides* **2019**, *16*, 1167–1176. [CrossRef]
8. Lo, S.-W.; Wu, J.-H.; Lin, F.-P.; Hsu, C.-H. Visual Sensing for Urban Flood Monitoring. *Sensors* **2015**, *15*, 20006–20029. [CrossRef]
9. Garcia, F.C.C.; Retamar, A.E.; Javier, J.C. A real time urban flood monitoring system for metro Manila. In Proceedings of the TENCON 2015—2015 IEEE Region 10 Conference, Macao, China, 1–4 November 2015; IEEE: Piscataway, NJ, USA, 2015; pp. 1–5.
10. Hsu, S.-Y.; Chen, T.-B.; Du, W.-C.; Wu, J.-H.; Chen, S.-C. Integrate Weather Radar and Monitoring Devices for Urban Flooding Surveillance. *Sensors* **2019**, *19*, 825. [CrossRef]
11. Wang, R.-Q.; Mao, H.; Wang, Y.; Rae, C.; Shaw, W. Hyper-resolution monitoring of urban flooding with social media and crowdsourcing data. *Comput. Geosci.* **2018**, *111*, 139–147. [CrossRef]
12. Chen, Y.; Zhou, H.; Zhang, H.; Du, G.; Zhou, J. Urban flood risk warning under rapid urbanization. *Environ. Res.* **2015**, *139*, 3–10. [CrossRef]

13. Leitão, J.P.; Simões, N.E.; Maksimović, Č.; Ferreira, F.; Prodanović, D.; Matos, J.S.; Sá Marques, A. Real-time forecasting urban drainage models: Full or simplified networks? *Water Sci. Technol.* **2010**, *62*, 2106–2114. [CrossRef] [PubMed]
14. Ozdemir, H.; Sampson, C.C.; De Almeida, G.A.M.; Bates, P.D. Evaluating scale and roughness effects in urban flood modelling using terrestrial LIDAR data. *Hydrol. Earth Syst. Sci.* **2013**, *17*, 4015–4030. [CrossRef]
15. Sampson, C.C.; Bates, P.D.; Neal, J.C.; Horritt, M.S. An automated routing methodology to enable direct rainfall in high resolution shallow water models. *Hydrol. Process.* **2013**, *27*, 467–476. [CrossRef]
16. Newman, A.J.; Clark, M.P.; Sampson, K.; Wood, A.; Hay, L.E.; Bock, A.; Viger, R.J.; Blodgett, D.; Brekke, L.; Arnold, J.R.; et al. Development of a large-sample watershed-scale hydrometeorological data set for the contiguous USA: Data set characteristics and assessment of regional variability in hydrologic model performance. *Hydrol. Earth Syst. Sci.* **2015**, *19*, 209–223. [CrossRef]
17. Miguez, M.G.; Battemarco, B.P.; De Sousa, M.M.; Rezende, O.M.; Veról, A.P.; Gusmaroli, G. Urban Flood Simulation Using MODCEL—An Alternative Quasi-2D Conceptual Model. *Water* **2017**, *9*, 445. [CrossRef]
18. Varlas, G.; Anagnostou, M.N.; Spyrou, C.; Papadopoulos, A.; Kalogiros, J.; Mentzafou, A.; Michaelides, S.; Baltas, E.; Karymbalis, E.; Katsafados, P. A Multi-Platform Hydrometeorological Analysis of the Flash Flood Event of 15 November 2017 in Attica, Greece. *Remote Sens.* **2018**, *11*, 45. [CrossRef]
19. Schumann, G.J.-P.; Neal, J.C.; Mason, D.C.; Bates, P.D. The accuracy of sequential aerial photography and SAR data for observing urban flood dynamics, a case study of the UK summer 2007 floods. *Remote Sens. Environ.* **2011**, *115*, 2536–2546. [CrossRef]
20. Feng, Q.; Liu, J.; Gong, J. Urban Flood Mapping Based on Unmanned Aerial Vehicle Remote Sensing and Random Forest Classifier—A Case of Yuyao, China. *Water* **2015**, *7*, 1437–1455. [CrossRef]
21. Neal, J.; Schumann, G.; Fewtrell, T.; Budimir, M.; Bates, P.; Mason, D. Evaluating a new LISFLOOD-FP formulation with data from the summer 2007 floods in Tewkesbury, UK. *J. Flood Risk Manag.* **2011**, *4*, 88–95. [CrossRef]
22. Segoni, S.; Battistini, A.; Rossi, G.; Rosi, A.; Lagomarsino, D.; Catani, F.; Moretti, S.; Casagli, N. Technical Note: An operational landslide early warning system at regional scale based on space–time-variable rainfall thresholds. *Nat. Hazards Earth Syst. Sci.* **2015**, *15*, 853–861. [CrossRef]
23. Piciullo, L.; Dahl, M.-P.; Devoli, G.; Colleuille, H.; Calvello, M. Adapting the EDuMaP method to test the performance of the Norwegian early warning system for weather-induced landslides. *Nat. Hazards Earth Syst. Sci.* **2017**, *17*, 817–831. [CrossRef]
24. Rosi, A.; Canavesi, V.; Segoni, S.; Dias Nery, T.; Catani, F.; Casagli, N. Landslides in the Mountain Region of Rio de Janeiro: A Proposal for the Semi-Automated Definition of Multiple Rainfall Thresholds. *Geosciences* **2019**, *9*, 203. [CrossRef]
25. Segoni, S.; Piciullo, L.; Gariano, S.L. A review of the recent literature on rainfall thresholds for landslide occurrence. *Landslides* **2018**, *15*, 1483–1501. [CrossRef]
26. Guzzetti, F.; Gariano, S.L.; Peruccacci, S.; Brunetti, M.T.; Marchesini, I.; Rossi, M.; Melillo, M. Geographical landslide early warning systems. *Earth-Science Rev.* **2020**, *200*, 102973. [CrossRef]
27. Kirschbaum, D.; Stanley, T. Satellite-Based Assessment of Rainfall-Triggered Landslide Hazard for Situational Awareness. *Earth's Futur.* **2018**, *6*, 505–523. [CrossRef]
28. Huffman, G.J.; Bolvin, D.T.; Nelkin, E.J.; Wolff, D.B.; Adler, R.F.; Gu, G.; Hong, Y.; Bowman, K.P.; Stocker, E.F. The TRMM Multisatellite Precipitation Analysis (TMPA): Quasi-Global, Multiyear, Combined-Sensor Precipitation Estimates at Fine Scales. *J. Hydrometeorol.* **2007**, *8*, 38–55. [CrossRef]
29. Hou, A.Y.; Kakar, R.K.; Neeck, S.; Azarbarzin, A.A.; Kummerow, C.D.; Kojima, M.; Oki, R.; Nakamura, K.; Iguchi, T. The Global Precipitation Measurement Mission. *Bull. Am. Meteorol. Soc.* **2014**, *95*, 701–722. [CrossRef]
30. Skofronick-Jackson, G.; Petersen, W.A.; Berg, W.; Kidd, C.; Stocker, E.F.; Kirschbaum, D.B.; Kakar, R.; Braun, S.A.; Huffman, G.J.; Iguchi, T.; et al. The Global Precipitation Measurement (GPM) Mission for Science and Society. *Bull. Am. Meteorol. Soc.* **2017**, *98*, 1679–1695. [CrossRef]
31. Kirschbaum, D.B.; Huffman, G.J.; Adler, R.F.; Braun, S.; Garrett, K.; Jones, E.; McNally, A.; Skofronick-Jackson, G.; Stocker, E.; Wu, H.; et al. NASA's Remotely Sensed Precipitation: A Reservoir for Applications Users. *Bull. Am. Meteorol. Soc.* **2017**, *98*, 1169–1184. [CrossRef]
32. Khan, S.; Maggioni, V.; Kirstetter, P.-E. Investigating the Potential of Using Satellite-Based Precipitation Radars as Reference for Evaluating Multisatellite Merged Products. *J. Geophys. Res. Atmos.* **2018**, *123*, 8646–8660. [CrossRef]
33. Skofronick-Jackson, G.; Kirschbaum, D.; Petersen, W.; Huffman, G.; Kidd, C.; Stocker, E.; Kakar, R. The Global

Precipitation Measurement (GPM) mission's scientific achievements and societal contributions: Reviewing four years of advanced rain and snow observations. *Q. J. R. Meteorol. Soc.* **2018**, *144*, 27–48. [CrossRef]

34. Kim, K.; Park, J.; Baik, J.; Choi, M. Evaluation of topographical and seasonal feature using GPM IMERG and TRMM 3B42 over Far-East Asia. *Atmos. Res.* **2017**, *187*, 95–105. [CrossRef]
35. Khodadoust Siuki, S.; Saghafian, B.; Moazami, S. Comprehensive evaluation of 3-hourly TRMM and half-hourly GPM-IMERG satellite precipitation products. *Int. J. Remote Sens.* **2017**, *38*, 558–571. [CrossRef]
36. Zubieta, R.; Getirana, A.; Espinoza, J.C.; Lavado-Casimiro, W.; Aragon, L. Hydrological modeling of the Peruvian-Ecuadorian Amazon Basin using GPM-IMERG satellite-based precipitation dataset. *Hydrol. Earth Syst. Sci.* **2017**, *21*, 3543–3555. [CrossRef] [PubMed]
37. Boone, A.; De Rosnay, P.; Balsamo, G.; Beljaars, A.; Chopin, F.; Decharme, B.; Delire, C.; Ducharne, A.; Gascoin, S.; Grippa, M.; et al. The AMMA land surface model intercomparison project (ALMIP). *Bull. Am. Meteorol. Soc.* **2009**, *90*, 1865–1880. [CrossRef]
38. Getirana, A.; Espinoza, J.; Ronchail, J.; Rotunno Filho, O.C. Assessment of different precipitation datasets and their impacts on the water balance of the Negro River basin. *J. Hydrol.* **2011**, *404*, 304–322. [CrossRef]
39. Gebregiorgis, A.S.; Kirstetter, P.; Hong, Y.E.; Gourley, J.J.; Huffman, G.J.; Petersen, W.A.; Xue, X.; Schwaller, M.R. To What Extent is the Day 1 GPM IMERG Satellite Precipitation Estimate Improved as Compared to TRMM TMPA-RT? *J. Geophys. Res. Atmos.* **2018**, *123*, 1694–1707. [CrossRef]
40. Prakash, S.; Mitra, A.K.; AghaKouchak, A.; Liu, Z.; Norouzi, H.; Pai, D.S. A preliminary assessment of GPM-based multi-satellite precipitation estimates over a monsoon dominated region. *J. Hydrol.* **2018**, *556*, 865–876. [CrossRef]
41. Rozante, J.; Vila, D.; Barboza Chiquetto, J.; Fernandes, A.; Souza Alvim, D. Evaluation of TRMM/GPM Blended Daily Products over Brazil. *Remote Sens.* **2018**, *10*, 882. [CrossRef]
42. Huffman, G.; Bolvin, D.T.; Braithwaite, D.; Hsu, K.; Joyce, R.; Kidd, C.; Nelkin, E.J.; Xie, P. *NASA Global Precipitation Measurement (GPM) Integrated Multi-satellitE Retrievals for GPM (IMERG) Prepared for: Global Precipitation Measurement (GPM) National Aeronautics and Space Administration (NASA)*; Algorithm Theoretical Basis Document Version 4.5; National Aeronautics and Space Administration: Washington, DC, USA, 2015; 26p.
43. Tan, M.; Duan, Z. Assessment of GPM and TRMM Precipitation Products over Singapore. *Remote Sens.* **2017**, *9*, 720. [CrossRef]
44. Floodlist Brazil—10 Dead After Torrential Rain in Rio de Janeiro. Available online: http://floodlist.com/america/brazil-rio-floods-april-2019 (accessed on 29 June 2020).
45. Floodlist Brazil—Deadly Floods and Landslides in Rio de Janeiro After 90 mm of Rain in 1 Hour. Available online: http://floodlist.com/america/brazil-rio-de-janeiro-flood-storm-february-2019 (accessed on 29 June 2020).
46. Fonseca, P.; Gaier, R.V. Powerful, "Abnormal" Rains Lash Rio de Janeiro, at Least Six Dead. Available online: https://www.reuters.com/article/us-brazil-weather/powerful-abnormal-rains-lash-rio-de-janeiro-at-least-six-dead-idUSKCN1RL20U (accessed on 29 June 2020).
47. Jeantet, D. Heavy Rains Cause Deadly Floods in Rio de Janeiro. Available online: https://www.pbs.org/newshour/world/heavy-rains-cause-deadly-floods-in-rio-de-janeiro (accessed on 29 June 2020).
48. Kumar, S.V.; Peters-Lidard, C.D.; Tian, Y.; Houser, P.R.; Geiger, J.; Olden, S.; Lighty, L.; Eastman, J.L.; Doty, B.; Dirmeyer, P.; et al. Land information system: An interoperable framework for high resolution land surface modeling. *Environ. Model. Softw.* **2006**, *21*, 1402–1415. [CrossRef]
49. Arsenault, K.R.; Kumar, S.V.; Geiger, J.V.; Wang, S.; Kemp, E.; Mocko, D.M.; Beaudoing, H.K.; Getirana, A.; Navari, M.; Li, B.; et al. The Land surface Data Toolkit (LDT v7.2)—A data fusion environment for land data assimilation systems. *Geosci. Model Dev.* **2018**, *11*, 3605–3621. [CrossRef]
50. Yang, Z.-L.; Niu, G.-Y.; Mitchell, K.E.; Chen, F.; Ek, M.B.; Barlage, M.; Longuevergne, L.; Manning, K.; Niyogi, D.; Tewari, M.; et al. The community Noah land surface model with multiparameterization options (Noah-MP): 2. Evaluation over global river basins. *J. Geophys. Res. Atmos.* **2011**, *116*. [CrossRef]
51. Niu, G.Y.; Yang, Z.L.; Mitchell, K.E.; Chen, F.; Ek, M.B.; Barlage, M.; Kumar, A.; Manning, K.; Niyogi, D.; Rosero, E.; et al. The community Noah land surface model with multiparameterization options (Noah-MP): 1. Model description and evaluation with local-scale measurements. *J. Geophys. Res. Atmos.* **2011**, *116*, 1–19. [CrossRef]
52. Ek, M.B.; Mitchell, K.E.; Lin, Y.; Rogers, E.; Grunmann, P.; Koren, V.; Gayno, G.; Tarpley, J.D. Implementation of Noah land surface model advances in the National Centers for Environmental Prediction operational mesoscale Eta model. *J. Geophys. Res. Atmos.* **2003**, *108*. [CrossRef]

53. Ball, J.T.; Woodrow, I.E.; Berry, J.A. A Model Predicting Stomatal Conductance and its Contribution to the Control of Photosynthesis under Different Environmental Conditions. In *Progress in Photosynthesis Research*; Springer: Dordrecht, The Netherlands, 1987; pp. 221–224.
54. Niu, G.-Y.; Yang, Z.-L.; Dickinson, R.E.; Gulden, L.E.; Su, H. Development of a simple groundwater model for use in climate models and evaluation with Gravity Recovery and Climate Experiment data. *J. Geophys. Res.* **2007**, *112*, D07103. [CrossRef]
55. Campbell, P.C.; Bash, J.O.; Spero, T.L. Updates to the Noah Land Surface Model in WRF-CMAQ to Improve Simulated Meteorology, Air Quality, and Deposition. *J. Adv. Model. Earth Syst.* **2019**, *11*, 231–256. [CrossRef]
56. Schueler, T.R. The Compaction of Urban Soils. In *The Practice of Watershed Protection*; Schueler, T.R., Holland, H.K., Eds.; Center for Watershed Protection: Ellicott City, MD, USA, 2000; pp. 661–665.
57. Csiszar, I.; Gutman, G. Mapping global land surface albedo from NOAA AVHRR. *J. Geophys. Res. Atmos.* **1999**, *104*, 6215–6228. [CrossRef]
58. Chen, F.; Mitchell, K.; Schaake, J.; Xue, Y.; Pan, H.-L.; Koren, V.; Duan, Q.Y.; Ek, M.; Betts, A. Modeling of land surface evaporation by four schemes and comparison with FIFE observations. *J. Geophys. Res. Atmos.* **1996**, *101*, 7251–7268. [CrossRef]
59. Monin, A.S.; Obukhov, A.M. Basic laws of turbulent mixing in the surface layer of the atmosphere. *Tr. Akad. Nauk. SSSR Geophiz. Inst.* **1954**, *24*, 163–187.
60. Niu, G.-Y.; Yang, Z.-L. Effects of Frozen Soil on Snowmelt Runoff and Soil Water Storage at a Continental Scale. *J. Hydrometeorol.* **2006**, *7*, 937–952. [CrossRef]
61. Niu, G.-Y.; Yang, Z.-L. Effects of vegetation canopy processes on snow surface energy and mass balances. *J. Geophys. Res. Atmos.* **2004**, *109*. [CrossRef]
62. Cosgrove, B.A.; Lohmann, D.; Mitchell, K.E.; Houser, P.R.; Wood, E.F.; Schaake, J.C.; Robock, A.; Marshall, C.; Sheffield, J.; Duan, Q.; et al. Real-time and retrospective forcing in the North American Land Data Assimilation System (NLDAS) project. *J. Geophys. Res. Atmos.* **2003**, *108*, 2002JD003118. [CrossRef]
63. Shepard, D. A two-dimensional interpolation function for irregularly-spaced data. In Proceedings of the 1968 23rd ACM national conference, Las Vegas, NV, USA, 27–29 August 1968; ACM Press: New York, NY, USA, 1968; pp. 517–524.
64. Salvatici, T.; Tofani, V.; Rossi, G.; D'Ambrosio, M.; Tacconi Stefanelli, C.; Masi, E.B.; Rosi, A.; Pazzi, V.; Vannocci, P.; Petrolo, M.; et al. Application of a physically based model to forecast shallow landslides at a regional scale. *Nat. Hazards Earth Syst. Sci.* **2018**, *18*, 1919–1935. [CrossRef]
65. Wickland, D.E. Mission to Planet Earth: The Ecological Perspective. *Ecology* **1991**, *72*, 1923–1933. [CrossRef]
66. Gomes, J.B.V.; Lumbreras, J.F.; Palmieri, F.; Zaroni, M.J.; Oliveira, R.P.; Bhering, S.B.; Calderano, S.B.; Santos, H.G.; Cunha, T.J.F.; Aglio, M.L.D. Atualização do Levantamento Semidetalhado de Solos do Município do Rio de Janeiro, RJ. In *Mapeamento Pedológico e Interpretações Úteis ao Planejamento Ambiental do Município do Rio de Janeiro*; Lumbreras, J.F., Gomes, J.B.V., Eds.; Embrapa Solos: Rio de Janeiro, Brazil, 2004; pp. 7–270, ISBN 85-85864-16-8.
67. Soil Survey Staff. *Keys to Soil Taxonomy*; Department of Agriculture, Natural Resources Conservation Service: Washington, DC, USA, 2014; Volume 12.
68. *Soil Science Division Staff Soil Survey Manual*; USDA Handbook 18; USDA: Washington, DC, USA, 2017; pp. 145–148. [CrossRef]
69. Embrapa. *Mapeamento Pedológico e Interpretações Úteis ao Planejamento Ambiental do Município do Rio de Janeiro*; Embrapa: Brasilia, Brazil, 2004; ISBN 8585864168.
70. GEORIO Rio de Janeiro City Landslide Susceptibility Map. Available online: https://www.data.rio/app/suscetibilidade-a-deslizamentos (accessed on 17 November 2020).
71. Xu, R.; Tian, F.; Yang, L.; Hu, H.; Lu, H.; Hou, A. Ground validation of GPM IMERG and TRMM 3B42V7 rainfall products over southern Tibetan Plateau based on a high-density rain gauge network. *J. Geophys. Res. Atmos.* **2017**, *122*, 910–924. [CrossRef]
72. Jiang, Q.; Li, W.; Wen, J.; Qiu, C.; Sun, W.; Fang, Q.; Xu, M.; Tan, J. Accuracy Evaluation of Two High-Resolution Satellite-Based Rainfall Products: TRMM 3B42V7 and CMORPH in Shanghai. *Water* **2018**, *10*, 40. [CrossRef]
73. Arsenault, K.R.; Shukla, S.; Hazra, A.; Getirana, A.; McNally, A.; Kumar, S.V.; Koster, R.D.; Peters-Lidard, C.D.; Zaitchik, B.F.; Badr, H.; et al. The NASA hydrological forecast system for food and water security applications. *Bull. Am. Meteorol. Soc.* **2020**. [CrossRef]
74. Getirana, A.; Rodell, M.; Kumar, S.; Beaudoing, H.K.; Arsenault, K.; Zaitchik, B.; Save, H.; Bettadpur, S. GRACE Improves Seasonal Groundwater Forecast Initialization over the United States. *J. Hydrometeorol.* **2020**, *21*, 59–71. [CrossRef]

75. Li, B.; Rodell, M.; Kumar, S.; Beaudoing, H.K.; Getirana, A.; Zaitchik, B.F.; Goncalves, L.G.; Cossetin, C.; Bhanja, S.; Mukherjee, A.; et al. Global GRACE Data Assimilation for Groundwater and Drought Monitoring: Advances and Challenges. *Water Resour. Res.* **2019**, 1–23. [CrossRef]
76. Svoboda, M.; LeComte, D.; Hayes, M.; Heim, R.; Gleason, K.; Angel, J.; Rippey, B.; Tinker, R.; Palecki, M.; Stooksbury, D.; et al. THE DROUGHT MONITOR. *Bull. Am. Meteorol. Soc.* **2002**, *83*, 1181–1190. [CrossRef]
77. Rozante, J.R.; Moreira, D.S.; de Goncalves, L.G.G.; Vila, D.A. Combining TRMM and Surface Observations of Precipitation: Technique and Validation over South America. *Weather Forecast.* **2010**, *25*, 885–894. [CrossRef]
78. Salamanca, F.; Zhang, Y.; Barlage, M.; Chen, F.; Mahalov, A.; Miao, S. Evaluation of the WRF-Urban Modeling System Coupled to Noah and Noah-MP Land Surface Models Over a Semiarid Urban Environment. *J. Geophys. Res. Atmos.* **2018**, *123*, 2387–2408. [CrossRef]
79. Paiva, R.C.D.; Buarque, D.C.; Collischonn, W.; Bonnet, M.P.; Frappart, F.; Calmant, S.; Bulhões Mendes, C.A. Large-scale hydrologic and hydrodynamic modeling of the Amazon River basin. *Water Resour. Res.* **2013**, *49*, 1226–1243. [CrossRef]
80. Harris, A.; Rahman, S.; Hossain, F.; Yarborough, L.; Bagtzoglou, A.; Easson, G. Satellite-based Flood Modeling Using TRMM-based Rainfall Products. *Sensors* **2007**, *7*, 3416–3427. [CrossRef]
81. Wright, D.B. Rainfall Information for Global Flood Modeling. In *Global Flood Hazard: Applications in Modeling, Mapping, and Forecasting*; Schumann, G.J., Bates, P.D., Apel, H., Aronica, G.T., Eds.; American Geophysical

3

Multi-Scale Analysis of the Relationship between Land Subsidence and Buildings: A Case Study in an Eastern Beijing Urban Area using the PS-InSAR Technique

Qin Yang [1,2], Yinghai Ke [1,2,3,*], Dongyi Zhang [1,2,3], Beibei Chen [1,2,3], Huili Gong [1,2,3], Mingyuan Lv [1,2,3], Lin Zhu [1,2,3] and Xiaojuan Li [1,2,3]

1 College of Resources Environment and Tourism, Capital Normal University, 105 North Road of the 3rd Ringroad, Haidian District, Beijing 100048, China; yqinss@163.com (Q.Y.); dongyi1212@foxmail.com (D.Z.); cnucbb@yeah.net (B.C.); gonghl_1956@126.com (H.G.); mingyuanlv@126.com (M.L.); hi-zhulin@163.com (L.Z.); lixiaojuan@cnu.edu.cn (X.L.)
2 Laboratory Cultivation Base of Environment Process and Digital Simulation, Beijing 100048, China
3 Beijing Laboratory of Water Resources Security, Beijing 100048, China
* Correspondence: yke@cnu.edu.cn.

Abstract: Beijing is severely affected by land subsidence, and rapid urbanisation and building construction might accelerate the land subsidence process. Based on 39 Envisat Advanced Synthetic Aperture Radar (ASAR) images acquired between 2003–2010, 55 TerraSAR-X images acquired between 2010–2016, and urban building information, we analysed the relationship between land subsidence and buildings at the regional, block, and building scales. The results show that the surface displacement rate in the Beijing urban area ranged from −109 mm/year to +13 mm/year between 2003–2010, and from −151 mm/year to +19 mm/year between 2010–2016; two subsidence bowls were mainly distributed in the eastern part of the Chaoyang District. The displacement rate agreed well with the levelling measurements, with an average bias of less than six mm/year. At the regional scale, the spatial pattern of land subsidence was mainly controlled by groundwater extraction, compressible layer thickness, and geological faults. Subsidence centres were located in the area around ground water funnels with a compressible layer depth of 50–70 m. The block-scale analysis demonstrated a clear correlation between the block construction age and the spatial unevenness of subsidence. The blocks constructed between 1998–2005 and after 2005 showed considerably more subsidence unevenness and temporal instability than the blocks constructed before 1998 during both time periods. The examination of the new blocks showed that the spatial unevenness increased with building volume variability. For the 16 blocks with a high building volume, variability, and subsidence unevenness, the building-scale analysis showed a positive relationship between building volume and settlement in most blocks, although the R^2 was lower than 0.5. The results indicate that intense building construction in urban areas could cause differential settlement at the block scale in Beijing, while the settlement of single buildings could be influenced by the integrated effects of building volume, foundation structures, and the hydrogeological background.

Keywords: land subsidence; PS-InSAR; uneven settlement; building construction; Beijing urban area

---

1. Introduction

Land subsidence, a phenomenon of gradual land surface settling, is a geological hazard caused mainly by anthropogenic activities such as subsurface fluid extraction, underground mining, and engineering construction. Currently, more than 50 cities in China have faced land subsidence issues, and of these, Beijing has been among the most seriously affected cities since the 1950s [1]. Uneven land surface settlement has been reported to cause damages to urban infrastructures, such as wall cracks and pipeline ruptures, leading to losses in the national economy [1]. It is known that land subsidence on the Beijing plains is mainly caused by an excessive withdrawal of groundwater and is controlled by lithological and geological structures such as clay layer thickness, active faults, and aquifer types. Chen et al. [2] investigated land subsidence processes between 2003–2010 in the Beijing plains area and reported that land subsidence in this area is greatest where the compressible layer thickness is approximately 50–70 m, and that the distribution of subsidence bowls is controlled by Quaternary faults. Chen et al. [3] and Lei et al. [4] reported that groundwater-level variations in the second confined aquifer had the greatest impact on the development of land subsidence, and significant differences in the deformation gradient were found on both sides of the faults. Zhou et al. [5,6] investigated the relationship between land-use types and subsidence rates and found that serious subsidence occurred mainly in wetland, paddy fields, upland soils, vegetable land, and peasant-inhabited land. Gao et al. [7] revealed that the inelastic and permanent compaction of the Beijing aquifer system was due to the continuous decline in the water level of the northern subsidence area in Beijing, and elastic deformation outside the subsidence area.

Although subsurface fluid extraction, e.g., groundwater over-exploitation, is a major factor contributing to regional land subsidence in many cities in China and other countries [8,9], research has shown that rapid urbanisation can also be a contributing factor, as building construction and space utilisation usually develop consolidation processes [10]. A case study by Solari et al. [11] demonstrated the correlation between the age of the construction of buildings and the subsidence rates in two small urban areas in Pisa, Italy. Pretesi et al. [12] detected ground subsidence that was caused by the soil consolidation process of a newly constructed building in Florence, Italy. Chen et al. [13] found that land subsidence in the Loess Plateau region of China exhibited a high correlation with the distribution of building land, and that subsidence rates increased with building density.

The investigation of the relationship between buildings and land subsidence requires the continuous monitoring of land surface displacement over a large area. Compared to traditional geodetic methods such as Global Positioning System (GPS) and levelling and deep soil settlement surveys, the permanent scatterer interferometric synthetic aperture radar (PS-InSAR) technique has become widely recognised for monitoring ground displacement over large areas in a timely and cost-efficient way [14–16]. Studies have also found that the PS-InSAR technique has the capability to obtain high-density and precise measurements of the construction area and linear objects in urban areas based on high-resolution synthetic aperture radar (SAR) data. Liao et al. [17] determined the long-term subtle deformation of large man-made structures in Shanghai with millimetre-scale accuracy using the Persistent Scatterer Interferometry (PSI) method based on TerraSAR-X images. Qin et al. [18] applied the PS-InSAR technique and 26 TerraSAR-X images to monitor surface deformation along a rail transit in Shanghai. Chen et al. [19] monitored land surface deformation around the Beijing Subway Line 6 area based on Radarsat-2 data between 2009–2012. Solari et al. [20] derived a time series of ground subsidence in an urban area in Pisa, Italy during 1992 and 2010 based on European Remote Sensing satellites (ERS) 1/2 and Envisat Advanced Synthetic Aperture Radar (ASAR) datasets and identified the deformation of city buildings. Tapete et al. [21] monitored the deformation of linear structures using measurement points from the InSAR technique based on Radarsat-1 data and GPS measurements, and analysed the uncertainty of the measurements. The utilisation of multi-source datasets has demonstrated the capability of facilitating an analysis of the long-term temporal evolution of ground deformation over urban areas.

During the last few decades in Beijing, urban areas have been rapidly expanding, and the land surface has been intensely developed and utilised. Buildings impose extra loads on the land, and the construction of buildings alters the natural processes of consolidation [22–24]. To date, only a few studies have examined the relationship between land subsidence and urban development in Beijing. Chen et al. [25] found a weak positive correlation between Landsat-derived urban indices and land subsidence between 2003–2010. Chen et al. [26] reported that the complexity of urban space utilisation may affect uneven settlement at five settlement centres. Both studies only analysed settlement funnel areas in which groundwater withdraw was the major control factor of land subsidence, and it is hard to attribute land subsidence to the urbanisation in these areas. Jiao et al. [27] studied the western Central Business District—a very small urban area—and found the skyscrapers with volumes over 3×10^5 m^3 showed a higher land subsidence rate than smaller buildings.

The aforementioned literature review shows that the impact of building construction on land subsidence have not been fully investigated in Beijing urban areas. How building construction impacts the subsidence processes at different spatial scales is not clear. The present study aims to fill this gap. The objective is to evaluate the relationship between building construction and land subsidence in Beijing urban areas at regional, block, and building scales using the PS-InSAR technique. Building characteristics, including the age of building construction and building volume, are considered. The Chaoyang District, where both urbanisation and land subsidence have developed rapidly over the past few decades, was selected as the study area. Five hydrogeological regions were partitioned over the study area, with each demonstrating a similar groundwater level and compressible layer thickness. Land subsidence data between 2003–2016 were derived based on 39 Envisat ASAR images and 55 TerraSAR-X images. For each hydrogeological region, the relationships between building characteristics, namely, the age of construction and building volume, with land subsidence, were evaluated at both the block scale and the single-building scale. A detailed description of the study area, datasets, and methodology is seen in Sections 2 and 3. Sections 4 and 5 present and discuss the results. The main conclusions are summarised in Section 6.

2. Study Area and Dataset

2.1. Study Area

Beijing (39°28′–40°05′N and 115°25′–117°30′E) is an international metropolitan city with rapid urban sprawl and a high intensity of human activities. In 2016, the population of Beijing reached 21,729 million, inhabiting an area of approximately 16,410 km^2, of which 1410 km^2 was built-up area. Beijing is affected by a monsoon-influenced semi-arid and semi-humid continental climate. The temporal and spatial distribution of annual precipitation occurs unevenly, and 60.4% of the precipitation is concentrated in the summer (July–September). Two thirds of the water sources for the Beijing municipality come from groundwater. Groundwater is mainly withdrawn from the shallow confined aquifer layer in single-layered structure areas, and the middle confined aquifer layer and deep confined aquifer layer in multi-layered structure zones [4]. Since 1950, the Beijing plain area has formed five major ground subsidence funnels in the Chaoyang, Tongzhou, Shunyi, Changping, and Daxing Districts.

In this paper, we chose the Dongbalizhuang–dajiaoting settlement funnel and its surrounding area in the Chaoyang District as the study area (205 km^2). The groundwater level at the second confined aquifer varies from -37 m to -10 m, and the thickness of the clay layer in the first 100 m depth ranges from less than 50 m to 70 m. The Nanyuan–Tongxian geological fault is located across the study area (Figure 1), and rapid urbanisation has occurred in this area since the 1980s. Currently, the development in the study area is typical of urban areas in Beijing, with complex and intensive urban land use with mixed low-rise and high-rise buildings, historic and modern buildings, and commercial and residential buildings. Urban villages with considerable numbers of single-storey houses still exist.

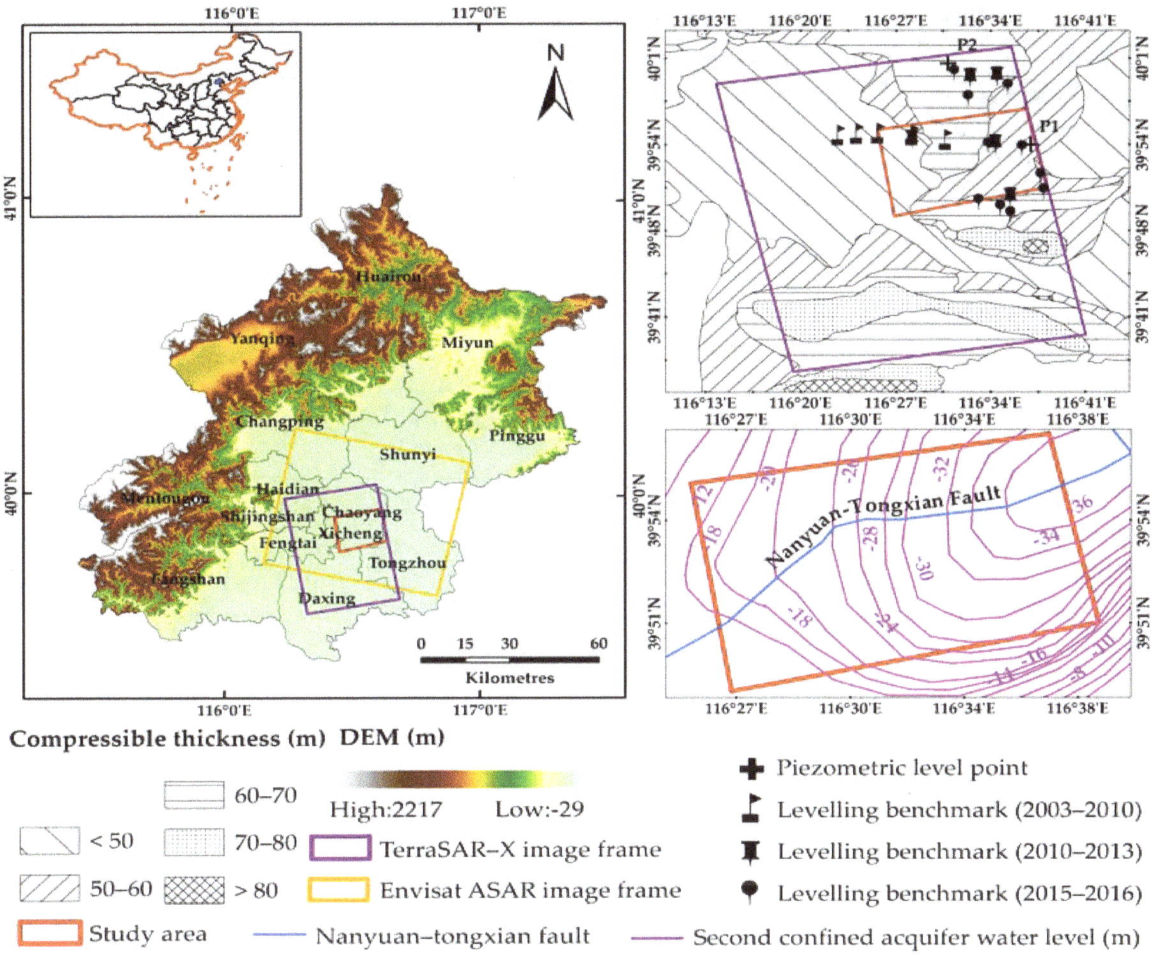

Figure 1. Study area and dataset coverage.

2.2. SAR Images and Validation Datasets

To determine the land subsidence evolution in the study area, two sets of SAR images comprising 39 descending single look complex (SLC) scenes collected by Envisat ASAR from 18 June 2003 to 25 August 2010 and 55 ascending scenes collected by the TerraSAR-X satellite with Stripmap mode from 13 April 2010 to 24 May 2016 were utilised in this study (Figure 1). The spatial resolutions of the Envisat ASAR and TerraSAR-X images were approximately 30 m and 3 m on the ground, respectively. Although high spatial resolution SAR datasets such as the TerraSAR-X imagery are preferable for identifying permanent scatterers in urban areas, this type of high-resolution satellite was not available before 2008. Thus, Envisat ASAR images were used to derive surface deformation between 2003–2010. The external Digital Elevation Model (DEM) that was applied for PS-InSAR processing to remove the topographic phase and flatten effects came from Shuttle Radar Topography Mission (SRTM) with a spatial resolution of 90 m (http://dds.cr.usgs.gov/srtm/).

Land surface deformation measurements collected from 12 levelling benchmarks from 2003 to 2010, five levelling benchmarks between 2010–2013, and 13 levelling benchmarks between 2015–2016 were used for validation (Figure 1). Due to the limited availability of long-term observations, none of the in situ datasets covered the whole time span from 2010 to 2016.

2.3. Building Properties

From Google Earth, high-resolution images, and Baidu Street Map (map.baidu.com), the block locations were visually interpreted and manually outlined in ArcGIS software. In our study, a "block" refers to a gated community that is usually surrounded by main streets or roads. In the study area, a total of 178 blocks were identified. Most of the blocks were residential communities, and among them, some communities had one or two commercial buildings. Apart from the residential communities, several school campuses and commercial business districts were also located in the area. Sixteen blocks were urban villages with similar low-rise buildings. It has been reported that many of these villages have private wells, which may affect the local groundwater level. Previous research has shown that the subsidence rate increases as the distance to the pumping well decreases [2]. Twenty-four blocks were located within a 50-m buffer area of the Nanyuan–Tongxian geological fault. To minimise the influence of local groundwater level variation and the fault on the uneven subsidence within the blocks, we eliminated these blocks from further analysis. As a result, 138 blocks were selected with a total of 6023 buildings. For each building, properties such as building height, base area, and construction age were acquired from real estate companies. Building volume was estimated as the product of the building base area and height. Volume was used to represent the load of the buildings.

3. Methodology

First, land surface deformation over the study area from 2003 to 2016 was derived from Envisat ASAR and TerraSAR-X datasets using the PS-InSAR technique. The results were evaluated by a comparison with levelling measurements. The study area was then partitioned into five hydrogeological regions. Each region had a similar groundwater level and compressible thickness of the clay layer. For each region, the spatiotemporal pattern of land subsidence was analysed. The relationship between land subsidence and building characteristics, including construction age and building load, was then evaluated at both the block scale and building scale.

3.1. Land Subsidence Monitoring Using the PSI Technique

The PSI method is capable of detecting points with strong and stable radiometric characteristics based on a stack of SAR data and deriving surface deformation information. In this study, the Stanford method for persistent scatterers (StaMPS) PSI method developed by Hooper et al. [28] was used to retrieve time series deformation from the Envisat ASAR images. First, the image acquired on 18 April 2007 was selected as the master image. The StaMPS selected initial persistent scatterer (PS) candidates with a small dispersion index value (lower than 0.4 in this study). The phase stability analysis was then performed for each PS candidate, and the probability that each pixel was a PS pixel was refined based on the stability indicator. For PS pixels, the wrapped phase was corrected, and unwrapping was then applied. The spatially correlated errors were eliminated with the aid of a high-pass filter in time, and a low-pass filter in space. Finally, the time series deformation along the line-of-sight (LOS) direction was derived.

The PSI method presented by Ferretti et al. [29,30] in SARPROZ software was used to obtain a deformation time series from the 55 TerraSAR-X images. First, the TerraSAR-X image acquired on 1 November 2013 was selected as the master image by considering a shorter spatial baseline and temporal baseline. Other images were co-registered to the master image. Second, a series of differential interferograms were constructed with the aid of SRTM DEM and precise orbital data. Third, persistent scatterer candidates (PSCs) were obtained with an amplitude difference dispersion index lower than 0.3. Then, multi-image grid phase unwrapping was conducted, and an atmospheric phase screen (APS) was estimated and removed. Afterwards, PS points with a temporal coherence index greater than 0.75 were selected. This ensures that the selected PS points have a high coherence and show phase stability over a long period of time. Finally, the displacement time series for each PS point along the LOS were derived by separating the phase components of the interferometric phase. Research has shown that use of the StaMPS and the PSI method in SARPROZ software yields consistent results for the same datasets [31,32], while we found that StaMPS was more computationally intensive. Therefore, we used SARPROZ software to process the high amount of TerraSAR-X datasets. Note that we used the same reference points with known zero deformation in the overlapped area of two data frames. The vertical deformation rate was then estimated from the LOS deformation rate by assuming that the horizontal movement of the land surface can be neglected [2,7].

3.2. Multi-Scale Analysis of the Relationship between Building Characteristics and Land Subsidence

In Beijing, lithology provides the geological background for land subsidence, and the spatial variation in the groundwater level is closely related to the regional distribution of subsidence rates. To minimise the impact of hydrogeological conditions on the subsidence-building relationship analysis, the study area was partitioned into five hydrogeological regions based on the compressible thickness of the clay layer and groundwater level at the second confined aquifer (Table 1 and Figure 2). Compressible layer thickness was classified into three classes, including <50 m, 50–60 m and 60–70 m. Groundwater level at the second confined aquifer was classified into four classes, including −37 to −32 m, −32 to −27 m, −27 to −22 m, and −22 to −17 m. Blocks were grouped based on the construction age. There were a total of 25 blocks built prior to 1998, 65 blocks built between 1998–2005, and 48 blocks built after 2005 (Table 1).

Table 1. Compressible layer thickness, groundwater level, and number of blocks within each region.

Region		I	II	III	IV	V
Compressible Layer Thickness (m)		50–60	50–60	60–70	60–70	<50
Groundwater Level (m) at Second Confined Aquifer		−37–32	−32––27	−32––27	−27––22	−22––17
Number of Blocks	Before 1998	7	0	3	7	8
	1998–2005	21	4	8	13	19
	After 2005	22	6	0	12	8
	Total	50	10	11	32	35

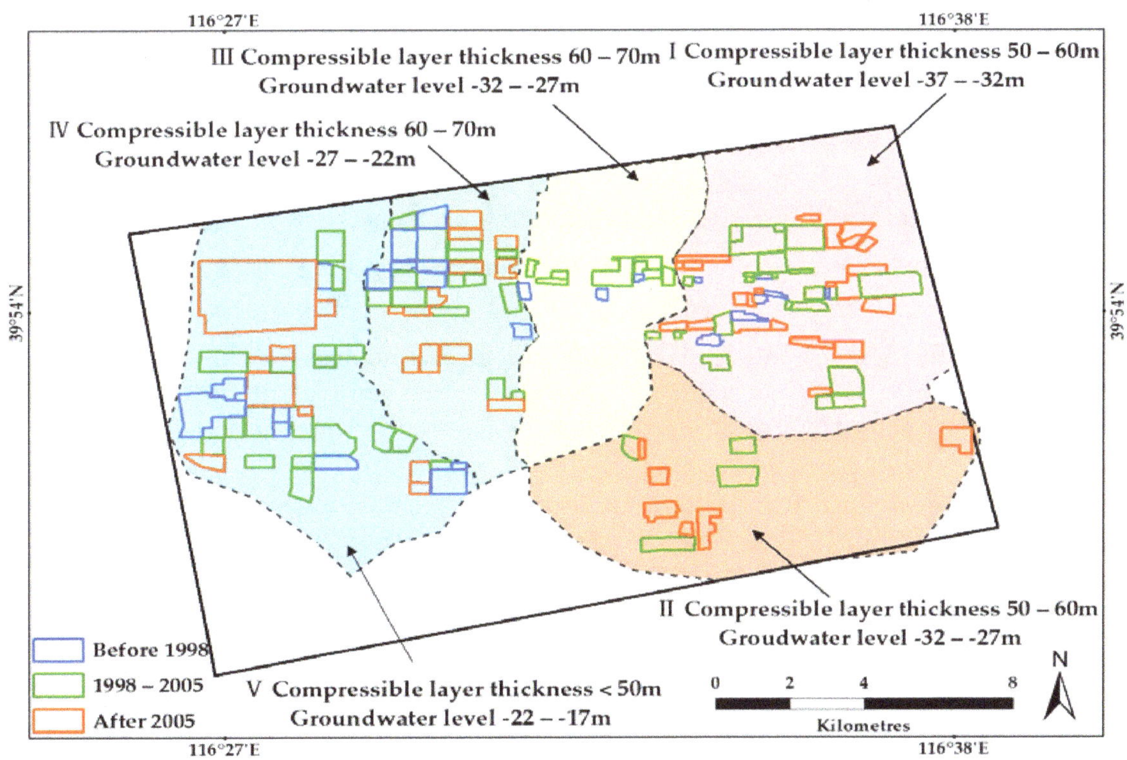

Figure 2. Location and extent of five regions.

For each region, the spatiotemporal pattern of the vertical deformation rate during the two observation periods was analysed. The spatial unevenness and temporal instability of the ground deformation within the blocks and their relationship with the block construction age and building volume variation was analysed. Further, the relationship between building-scale characteristics and settlement was analysed.

4. Results

4.1. Land Surface Deformation Derived from the PSI Techniques and Validation

A total of 167,690 pixels and 1,099,639 pixels were detected as PS points from the Envisat ASAR and TerraSAR-X datasets, respectively. As illustrated in Figure 3, the deformation rate ranged from −109.1 mm/year to +13.1 mm/year during 2003–2010 for the Envisat ASAR datasets and from −150.5 mm/year to +19 mm/year during 2010–2016 for the TerraSAR-X datasets. Figure 3a clearly shows five settlement funnels in the Chaoyang, Shunyi, and Changping districts in the Beijing plain. The two settlement funnels with the highest subsidence rates were located within the TerraSAR-X data frame (Figure 3b). At the Laiguangying settlement funnel, the maximum deformation rate reached −92 mm/year from 2003 to 2010, and −151 mm/year from 2010 to 2016. At the Dongbalizhuang-dajiaoting settlement funnel, the maximum deformation rate reached −109.1 mm/year and −141.5 mm/year from 2003 to 2010 and 2010 to 2016, respectively.

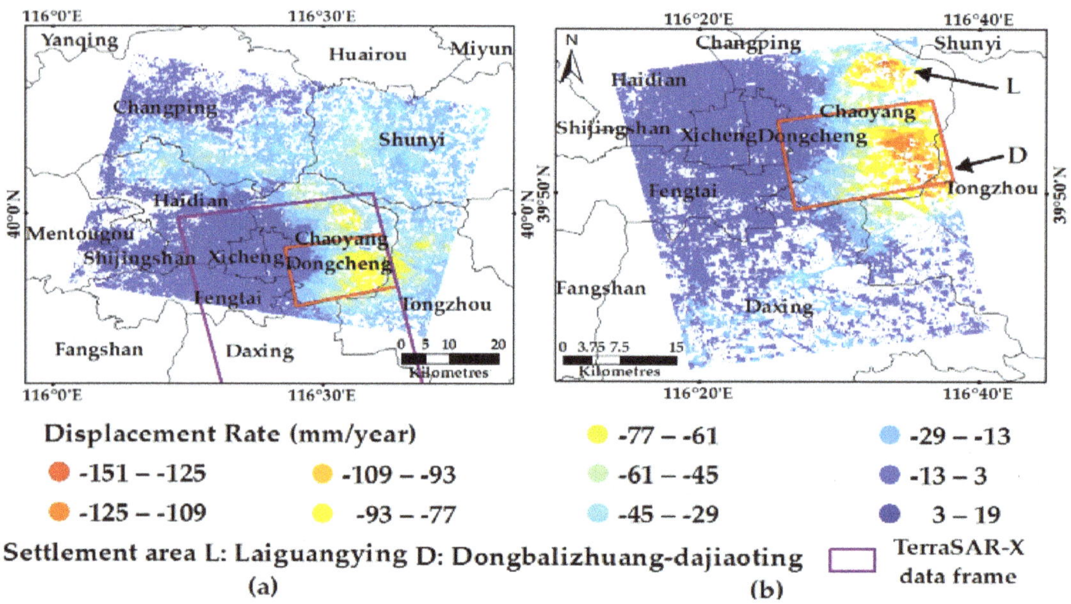

Figure 3. Average displacement rate derived from the (**a**) Envisat Advanced Synthetic Aperture Radar (ASAR) datasets during 2003–2010 and (**b**) TerraSAR-X datasets during 2010–2016. The red outline shows the boundary of the study area.

The deformation rates during the two periods were further assessed by 30 in situ levelling measurements collected from 2003 to 2010, 2010 to 2013, and 2015 to 2016. The PS pixels closest to the levelling benchmarks were selected. As there were no measurements available throughout the time period from 2010 to 2016, we calculated the average displacement rate during 2010–2013 and 2015–2016 from the cumulative displacement time series derived from the TerraSAR-X datasets. Figure 4 shows that the deformation rate measurements from the PSI techniques and levelling benchmarks are in good agreement. The average biases of the estimates from the PSI techniques are −1.57 mm/year (root mean square error (RMSE) = 2.45 mm/year, R^2 = 0.90) for 2003 to 2010, 4.53 mm/year (RMSE = 8.01 mm/year, R^2 = 0.98) for 2010 to 2013, and 5.00 mm/year (RMSE = 9.06 mm/year, R^2 = 0.86) for 2015 to 2016. The

StaMPS method applied to Envisat ASAR datasets and the SAR PROcessing tool by periZ (SARPROZ) method applied to TerraSAR-X datasets have different procedures; for example, StaMPS does not require any priori assumptions for the temporal nature of the deformation for PS selection. However, the validation results using levelling measurements indicate that reliable land deformation results were derived from both PSI methods. This supports the following analysis on the relationship between land subsidence and building characteristics.

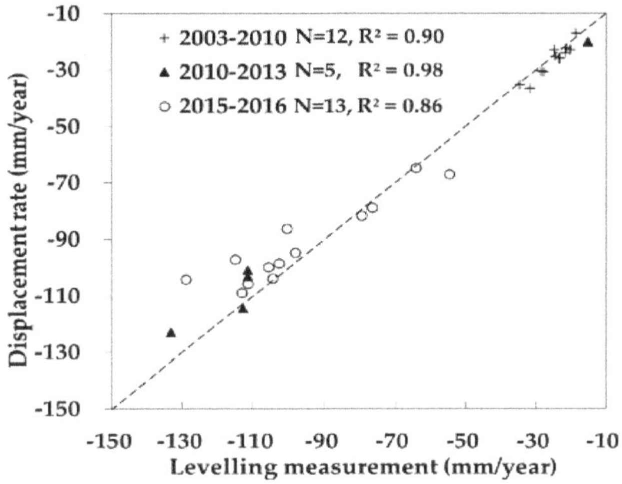

Figure 4. Comparison of the displacement rates derived from the permanent scatterer interferometric synthetic aperture radar (PS-InSAR) technique and levelling measurements.

4.2. Spatiotemporal Characteristics of Land Subsidence at Each Region

Tables 2 and 3 list the deformation rate statistics in each region and within the blocks during 2003–2010 and 2010–2016. There were a total of 14,729 PS pixels (90 pixels/km^2) and 163,072 PS pixels (939 pixels/km^2) detected by the Envisat ASAR and TerraSAR-X datasets, respectively, in the five regions. Note that the point density from the TerraSAR-X datasets was over 10 times that from the Envisat ASAR datasets. This is mainly due to the differences in spatial resolution of the two datasets. The StripMap mode of TerraSAR-X acquisition provides up to 3-m resolution, while Envisat ASAR acquisition provides 30-m resolution imagery. During both time periods, the mean deformation rate in region I (−75.2 mm/year and −93.1 mm/year), region II (−68.1 mm/year and −94.0 mm/year), and region III (−53.8 mm/year and −64.1 mm/year) were much higher than those in regions IV and V. Regions I, II, and III also had higher deformation rate standard deviations (SDs) (>15 mm/year) than regions IV and V, indicating a greater unevenness in the subsidence of regions I, II, and III. Thus, we considered regions I, II, III, i.e., the regions with most severe subsidence, as the subsidence centre.

Table 2. Deformation rate of persistent scatterer (PS) points during 2003–2010 within regions and blocks. SD: standard deviations.

Region	I	II	III	IV	V
Total Number of PS points	3288	2305	2103	2520	4513
Mean Deformation Rate of PS points (mm/year)	−75.2	−68.1	−53.8	−26.4	−13.8
SD of Deformation Rate of PS Points (mm/year)	16.7	11.0	11.9	10.6	8.5
Number of PS Points within Blocks	975	261	254	968	1727
Mean Deformation Rate within Blocks (mm/year)	−69.6	−64.8	−60.1	−25.9	−10.8
SD of Deformation Rate within Blocks (mm/year)	20.2	8.6	10.4	9.4	5.7

Table 3. Deformation rate of PS points during 2010–2016 within regions and blocks.

Region	I	II	III	IV	V
Total Number of PS Points	35,159	20,743	18,348	29,687	59,135
Mean Deformation Rate of PS Points (mm/year)	−93.1	−94.0	−64.1	−30.4	−13.6
SD of Deformation Rate of PS Points (mm/year)	22.3	16.4	18.4	13.4	10.7
Number of PS Points within Blocks	15,023	5710	2527	16,419	25,978
Mean Deformation Rate within Blocks (mm/year)	−89.0	−91.6	−82.0	−30.0	−11.2
SD of Deformation Rate within Blocks (mm/year)	25.2	12.4	12.9	13.2	8.9

We further selected the PS pixels within the blocks. The number of PS pixels within the blocks accounted for approximately 29% (Envisat ASAR) and 40% (TerraSAR-X) of the PS pixels in the whole region. However, the density of the PS pixels in the blocks was higher than the overall PS pixel density due to the existence of buildings (120 pixels/km^2 from Envisat ASAR and 1876 pixels/km^2 from TerraSAR-X). Although the Envisat ASAR PS pixel density is much lower than the TerraSAR-X PS pixel density, there is still an average of over 30 PS pixels within each block, enabling further statistical analysis within the blocks. For each region, the mean and SD of the deformation rate within the blocks had a similar pattern as those of the whole region, indicating that the subsidence rates in these blocks can represent the overall regional situation (Tables 2 and 3).

A comparison between Tables 2 and 3 shows that the settlement rate increased significantly from 2003–2010 to 2010–2016 in all of the regions and blocks except for in region V. The SD of the settlement increased in all of the regions and blocks. In the subsidence centre area (regions I, II, and III), the increments in the mean and SD of the subsidence rates were even more notable. Specifically, the mean settlement rate increased from 75.2 mm/year to 93.1 mm/year in region I, from 68.1 mm/year to 94.0 mm/year in region II, and from 53.8 mm/year to 64.1 mm/year in region III. Figure 5 also shows that the spatial distribution of the subsidence rate is similar during the two time periods at each region, while the magnitude of the subsidence rate during 2010–2016 is significantly higher than that during 2003–2010.

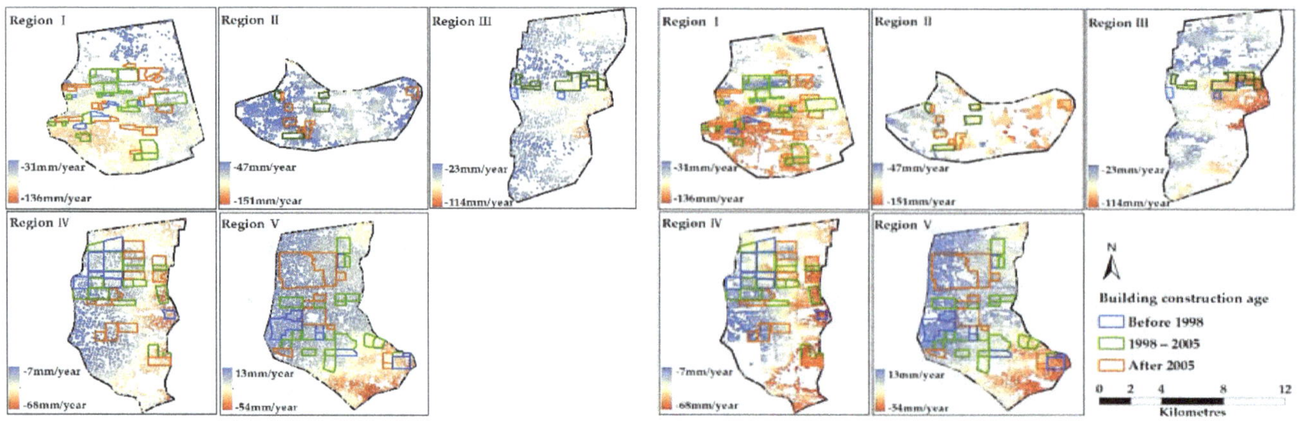

(a) (b)

Figure 5. Deformation rates of each region during (**a**) 2003–2010 and (**b**) 2010–2016.

4.3. Block-Scale Building Characteristics and Subsidence

From the two SAR datasets, the PS pixels within the blocks were extracted, and the average and range of the cumulative settlement were calculated for each block. In this study, we define the range, denoted as $R(s)$, as the difference between the 95th percentile and the 5th percentile of the cumulative settlement within a given block during 2003–2010 based on the ASAR datasets or during 2010–2016 based on the TerraSAR-X datasets. The 95th percentile and 5th percentile were used in order to eliminate the outliers of PS-InSAR-derived cumulative settlement. $R(s)$ was utilised to represent the spatial variation or unevenness of the land subsidence within each block. For each PS point derived from the ASAR dataset, we found the closest PS point derived from the TerraSAR-X dataset within a 5-m buffer area. The settlement velocity change, which was denoted as Δv, from 2003–2010 to 2010–2016 was then calculated for the pairs of PS points. The range of the velocity change $R(\Delta v)$, i.e., the difference between the 95th percentile and 5th percentile velocity change within each block, was calculated to represent the stability of the block settlement. The basic assumption was that blocks are relatively stable if there was no change in the displacement velocity during the two periods, or if the velocity change was similar across the block area; a high $R(\Delta v)$ within the block indicates poor stability.

Table 4 summarises the average cumulative settlement and velocity change within the blocks constructed before 1998, between 1998–2005, and after 2005. In the subsidence centre area (regions I–III), the average cumulative settlement ranged from 563.2 mm to 574.3 mm between 2003–2010, and from 552.9 mm to 612.8 mm during 2010–2016; the average velocity change ranged from 14.5 mm/year to 17.5 mm/year. In the areas far from the subsidence centre (regions IV and V), the average cumulative settlement ranged from 144.8 mm to 176.2 mm between 2003–2010, and from 126.7 mm to 134.6 mm between 2010–2016; the velocity change varied from 6.1 mm/year to 7.4 mm/year. Note that there were six blocks constructed after 2005 in region I that had no PS pixels detected in the Envisat ASAR dataset due to the lack of permanent scatterers; hence, we did not include these blocks in the statistical calculation based on the Envisat ASAR dataset. The blocks constructed during the three time periods did not have considerably different settlements, regardless of the dataset used (Envisat ASAR TerraSAR). The velocity changes were also similar.

Table 4. The average cumulative settlement ($\bar{s}$), average settlement velocity change ($\overline{\Delta v}$), average range of cumulative displacement ($\overline{R(s)}$), and average range of velocity change ($\overline{R(\Delta v)}$) of PS pixels within blocks.

Region	Construction age (N)	$\bar{s}$ (mm)		$\overline{\Delta v}$ (mm/year)	$\overline{R(s)}$ (mm)		$\overline{R(\Delta v)}$ (mm/year)
		2003–2010	2010–2016		2003–2010	2010–2016	
I–III	Before 1998 (10)	574.3	612.8	17.5	52.1	47.7	5.8
	1998–2005 (33)	596.9	552.9	14.5	86.2	93.2	11.0
	After 2005 (28) *	563.2 *	578.6	15.8	82.1 *	92.5	10.5
IV, V	Before 1998 (15)	144.8	130.2	7.4	63.3	64.5	7.0
	1998–2005 (32)	165.8	126.7	6.1	70.1	73.7	7.2
	After 2005 (20)	176.2	134.6	6.2	68.9	78.8	8.7

*: Among the 28 blocks constructed after 2005 in regions I–III, six blocks had no PS pixels detected by the Envisat ASAR dataset. Thus, they were not considered in the statistical calculation based on the Envisat ASAR dataset.

Table 4 compares the average $R(s)$ and $R(\Delta v)$ within the blocks constructed during the three time periods. The spatial unevenness of the subsidence in the newly constructed blocks was higher than that in the old blocks, as shown by both the Envisat ASAR and TerraSAR-X datasets. In the subsidence centre area, the blocks that were built prior to 1998 had an average unevenness of 52.1 mm between 2003–2010, and 47.7 mm between 2010–2016; the blocks built after 2005 had an average unevenness of 82.1 mm between 2003–2010 and 92.5 mm between 2010–2016. In addition, the newer blocks had a greater increase in displacement unevenness from 2003–2010 to 2010–2016. For example, in regions IV and V, the subsidence unevenness of the old blocks (constructed before 1998) increased from 63.3 mm to 64.5 mm, and that of the new blocks (constructed after 2005) increased from 68.9 mm to 78.8 mm. Figure 6 illustrates the percentage of blocks with each interval of $R(s)$. As illustrated in Figure 6, in the subsidence centre area, only 20% of the old blocks had a subsidence unevenness over 80 mm between 2003–2010; in contrast, the unevenly subsiding blocks ($R(s) > 80$ mm) accounted for 48.4% and 40.9% of the newer blocks, respectively. Between 2010–2016, the percentage of unevenly subsiding old blocks decreased to 10%, while for newer blocks, the percentages were 48.4% and 43%. Similarly, in regions IV and V, the percentage of unevenly subsiding old blocks was lower than that of the unevenly subsiding new blocks. In the study area, there were four groups of adjacent blocks with new blocks close to old blocks. The described pattern, that is, that newer blocks had greater spatial unevenness and temporal instability than older blocks, was more evident at the local scale (Figure 7). Figure 7e illustrates the settlement range for each block of each group. The figure clearly shows that blocks constructed between 1998–2005 and after 2005 had greater $R(s)$ values than the neighbouring old blocks.

In addition, Table 4 shows that newer blocks have greater $R(\Delta v)$ values than older blocks. In regions I, II, and III, none of the old blocks had $R(\Delta v)$ values greater than 15 mm/year, while 24.2% of the blocks that were constructed between 1998–2005, and 13.6% of the blocks constructed after 2005, had $R(\Delta v)$ values greater than 15 mm/year. Similarly, the old blocks had lower $R(\Delta v)$ values than the newer blocks in regions IV and V. The percentages of blocks with $R(\Delta v) > 15$ mm/year were 0%, 9.4%, and 5% for old blocks, blocks constructed between 1998–2005, and blocks constructed after 2005, respectively (Figure 8). As $R(\Delta v)$ represents the spatial variability of changes in subsidence velocity within the blocks, these results show that the new blocks were less stable than the old blocks.

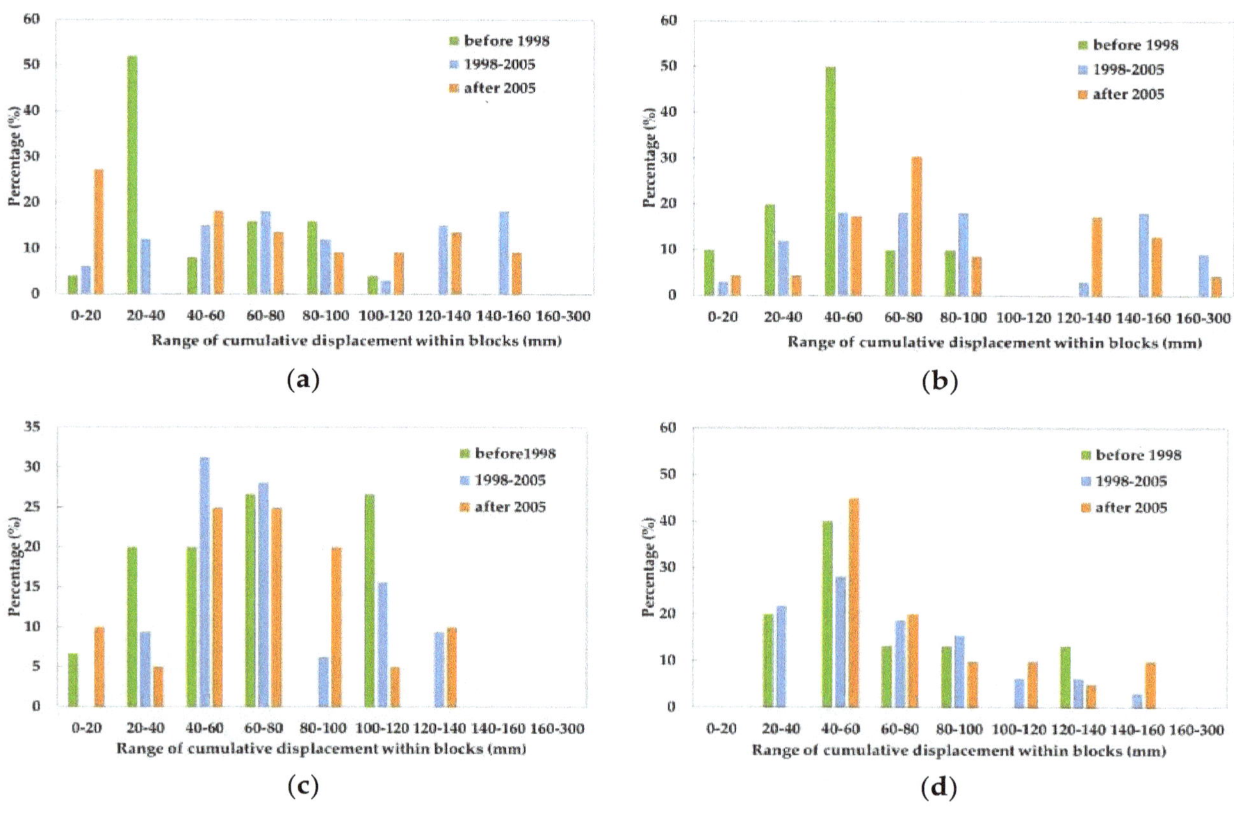

Figure 6. Percentage of blocks with different intervals of $R(s)$ within blocks in (**a**) regions I, II, and III derived from the ASAR datasets; (**b**) regions I, II, and III derived from the TerraSAR-X datasets; (**c**) regions IV and V derived from the ASAR datasets; and (**d**) regions IV and V derived from the TerraSAR-X datasets.

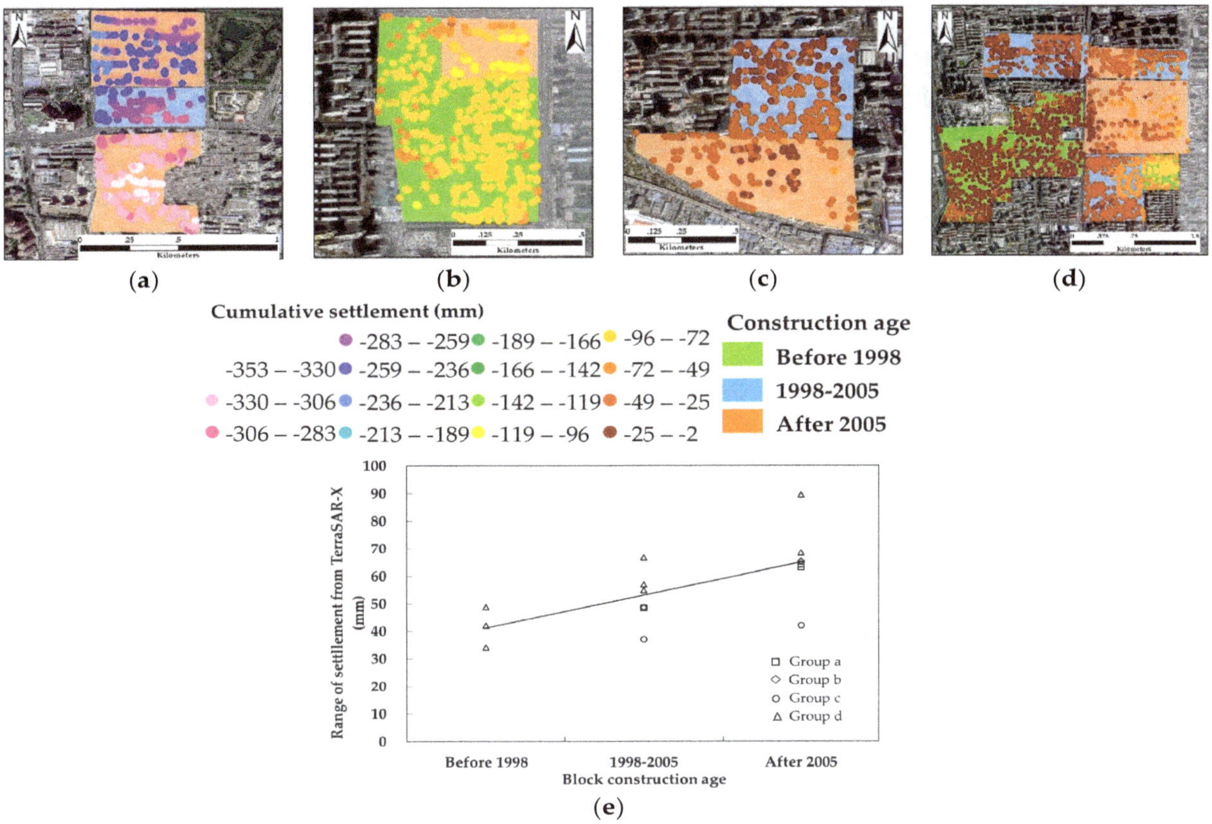

Figure 7. Four groups of neighbouring blocks ((**a**) group a, (**b**) group b, (**c**) group c, and (**d**) group d) and (**e**) the range of cumulative settlement derived from TerraSAR-X datasets for each group.

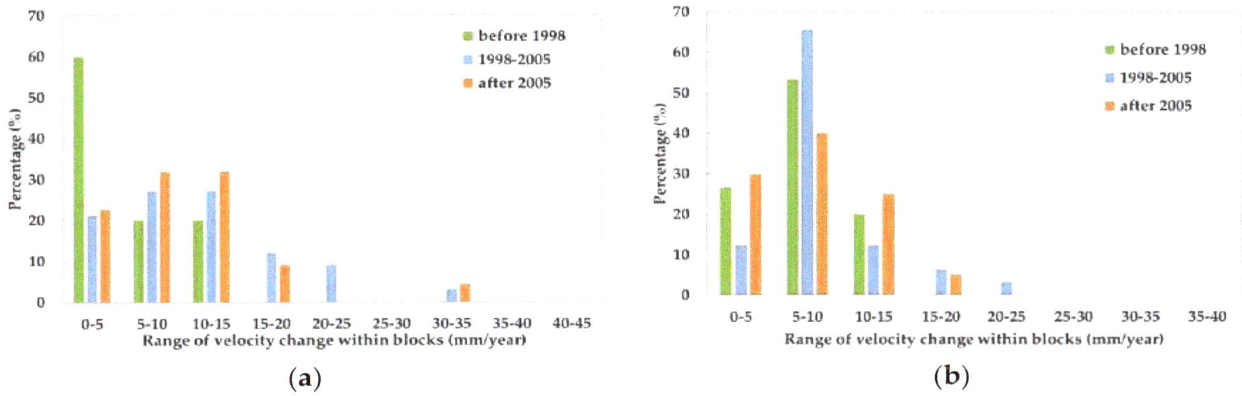

Figure 8. Percentage of blocks with different intervals of $R(\Delta v)$ within blocks in (**a**) regions I, II, and III, and (**b**) regions IV and V.

We further focussed on the 48 blocks constructed after 2005 (28 blocks in regions I, II, and III, and 20 blocks in regions IV and V) and analysed the relationship between subsidence and building volume for these blocks. The 48 blocks were selected because they had a greater spatial unevenness and temporal instability of land subsidence than the old blocks. Compared with the blocks that were constructed between 1998–2005, they also had a greater increase in $R(s)$ from 2003–2010 to 2010–2016; therefore, we speculated that the ground settlement within these blocks was more susceptible to the impact of building construction. Figure 9 shows the relationship between the range of cumulative displacement ($R(s)$) derived from the TerraSAR-X datasets and the range of building volume within the 48 new blocks in regions I, II, IV, and V. Note that there were no new blocks in region III. Except for in region I, $R(s)$ increased with the building load range, i.e., greater differences in building load corresponded to greater unevenness in the settlement within the blocks. For example, in region V, three blocks had building volume differences greater than 2×10^5 m^3 and average $R(s)$ values that reached 93 mm, which was a greater value than that of the blocks with small differences in building volume.

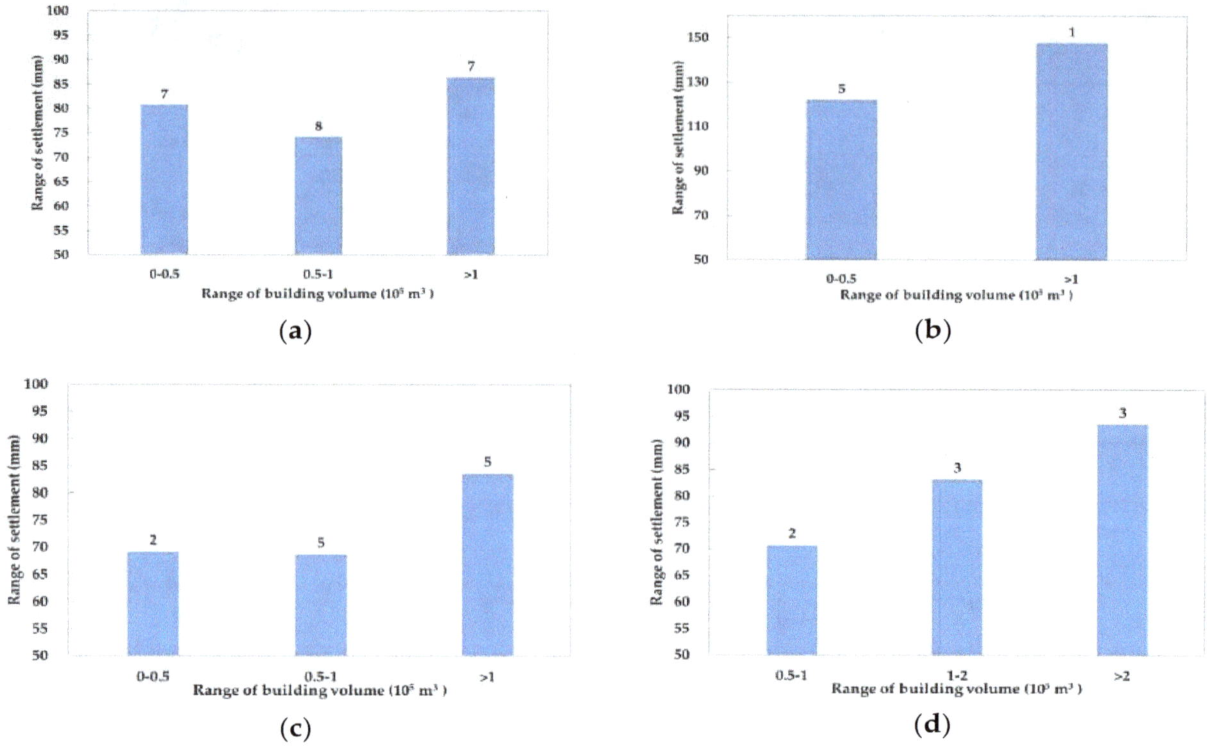

Figure 9. Average range of cumulative settlement ($\overline{R(s)}$) between 2010–2016 for blocks with different building volume ranges in (**a**) region I, (**b**) region II, (**c**) region IV, and (**d**) region V. The numbers above each bar denote the number of blocks.

4.4. Building-Scale Subsidence and Building Volume

A building-scale analysis was performed on the 16 new blocks (seven in region I, 1 in region II, 5 in region IV and 3 in region V) with building volume differences greater than 10^5 m^3 since they have relatively high subsidence unevenness. We calculated the cumulative displacement of each building from 2010 to 2016 using the co-registered PS points derived from the TerraSAR-X datasets, and analysed the relationship between building volume and displacement. Table 5 lists the coefficient, intercept, and R^2 values of the linear regression model with building volume (10^5 m^3) as the explanatory variable, and building cumulative displacement as the response variable. Despite low R^2 values (0.1–0.43), most of the blocks (13 out of 16) showed a negative relationship (negative coefficient) between building volume and displacement; that is, building settlement increased with the magnitude of building volume. For example, in the Dafangju block community in region II, the buildings in the west subsided by approximately 740 mm between 2010–2016, while those in the east only subsided by approximately 635 mm in the same period. The volume of the western buildings was approximately 3.3×10^5 m^3, and that of the eastern buildings was approximately 0.3×10^5 m^3 (Figure 10a). In the Jintaixianfeng block community, there was a building cluster in the southeast with a greater subsidence rate than that of the other buildings. An examination of the building cluster showed that the buildings were commercial high rises with a volume of approximately 1.5×10^5 m^3, while the other buildings had a volume of approximately 0.4×10^5 m^3 (Figure 10b). However, in the remaining three blocks, a similar pattern was not found. For example, in the Huamao block of region V, there were three high-rise buildings constructed between 2005–2008 with 28–36 floors above the ground and four basement floors comprising a commercial centre. However, the displacement rate of these skyscrapers was obviously lower than that of the surrounding low-rise buildings.

Table 5. Parameters of the linear regression model between building volume and building deformation. Deformation = $a \times$ volume $(10^5) + b$.

Region				I				II
a	−149.5	−36.0	−216.4	−17.9	0.35	−4.9	−1.8	−10.5
b	−335.4	−366.5	−319.6	−841.1	−840.4	−400.2	−665.1	−679.0
R^2	0.43	0.10	0.12	0.23	0.21	0.11	0.11	0.35
Region			IV				V	
a	11.4	−5.80	−9.4	−7.8	−16.5	4.5	−6.2	−10.0
b	−165.5	−113.9	−66.8	−280.9	−83.0	−69.9	−48.6	−45.7
R^2	0.12	0.10	0.15	0.24	0.13	0.24	0.08	0.33

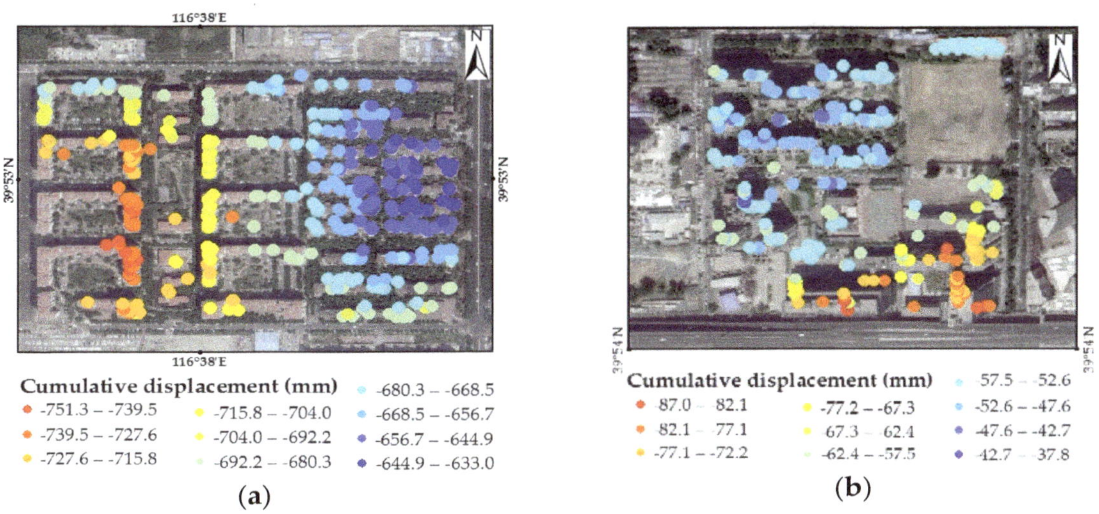

Figure 10. Cumulative displacement within the (**a**) Dafangju block and (**b**) Jintaixianfeng block.

5. Discussion

5.1. Causes of Land Subsidence at the Regional Scale

As illustrated in Tables 2 and 3 and Figure 5, land subsidence between 2003–2010 and 2010–2016 presented similar spatial patterns. The settlement funnels in the Chaoyang district were clearly present in both time periods, and the deformation rates in regions I, II, and III were significantly higher than those in regions IV and V. We overlaid the contour map of the average groundwater table in the second confined aquifer between 2003–2010 and 2010–2016 on the displacement map, and found that the spatial variation in the land subsidence rate was generally consistent with the groundwater level contours (Figure 11a,b). However, the distribution of the subsidence bowl did not exactly align with the groundwater depression cone, which may be affected by the compressibility of the soil structure and the geological fault zone [7]. The comparison between Figure 11a,b shows that the groundwater level decreased considerably from 2003–2010 to 2010–2016. Correspondingly, the subsidence bowls expanded, and the subsidence rate increased during the two time periods. By comparing the land surface displacement time series derived from the TerraSAR-X data with the confined groundwater level changes at two observation wells between 2005–2014 [33], we found that the land subsidence trend was closely related to temporal variations in the groundwater level (Figure 11c,d). When the groundwater level declined, the magnitude of displacement increased correspondingly. Intense groundwater withdraw is the predominant cause of subsidence in the Beijing plain area, since regional drops in piezometric levels could reduce pore pressure and increase effective stresses [4].

The thickness of the soft clay layer provides the lithology and structural background for land subsidence in the Beijing plain area. In our study area, the highest deformation rate occurred in regions I, II, and III, where the compressible clay layer thickness reached 50–70 m. Thicker compressible soil tends to have a relatively high deformation rate when piezometric levels decrease. For each region, the mean and SD of the deformation rate within the blocks had similar patterns as those of the whole region, indicating that the subsidence rate of these blocks can represent the overall regional situation (Table 1). The spatial–temporal deformation rate pattern revealed that regional-scale land subsidence was mainly caused by drops in the groundwater level and mainly controlled by structures, including the compressible thickness of the clay layer and faults.

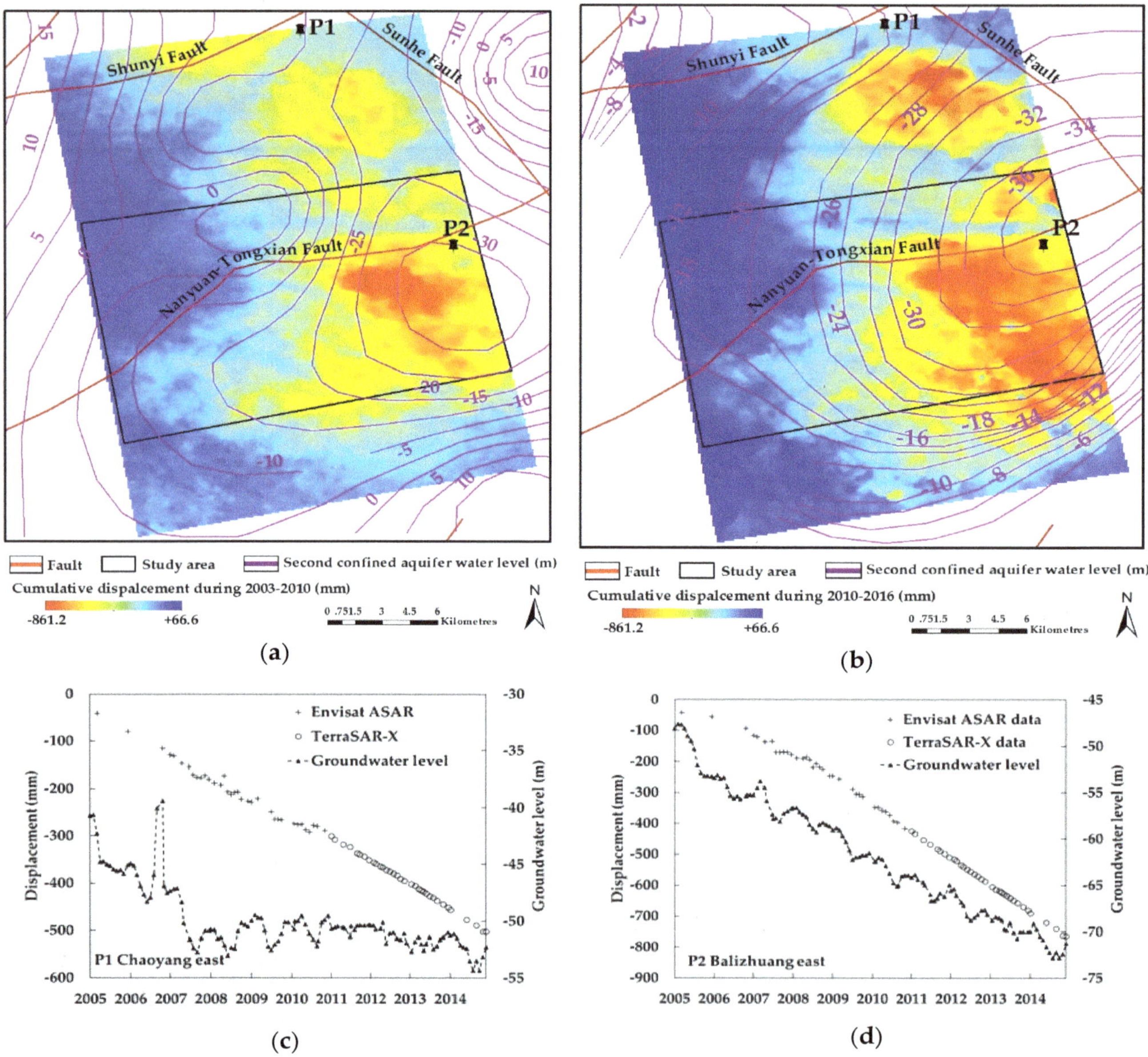

Figure 11. (**a**) Compressible thickness and groundwater level at the second confined aquifer superimposed onto an interpolated cumulative land deformation map. (**b**) Cumulative displacement derived by PS-InSAR and groundwater level at (**c**) the eastern Chaoyang and (**d**) eastern Balizhuang observation wells.

5.2. *Effects of Building Characteristics on Block-Scale Subsidence Unevenness*

The spatiotemporal evolution of land subsidence in the study area was mainly controlled by the aforementioned hydrogeological factors at the regional scale. Therefore, no substantial differences in cumulative settlement ($\bar{s}$) and the subsidence rate change ($\overline{\Delta v}$) were discovered among the three types of blocks classified by the age of construction in either the subsidence centre area (regions I, II, and III) or the area far from the subsidence centre (regions IV and V). However, we did find that newer blocks had a higher spatial unevenness of subsidence than older blocks. In addition, the increase in displacement unevenness from 2003–2010 to 2010–2016 and the spatial variation of the velocity change in the new blocks were higher than those in the old blocks. In contrast, the old blocks were more stable than the new blocks. These observations were the most obvious in the four groups of blocks with new blocks and old blocks near each other (Figure 7). At such a local scale, the lithology features and groundwater conditions were similar; thus, the contribution of groundwater variations to displacement unevenness was limited. On the other hand, the construction of buildings could lead to significant load changes, thus resulting in uneven subsidence within blocks. As the old blocks had been built more than 10 years before the TerraSAR-X data acquisition time, the primary consolidation processes may had completed before this period; thus, the old blocks were more stable than the nearby new blocks.

Solari et al. [11] reported a positive relationship between the building construction age and settlement rate in small urban areas in Pisa, Italy. Dong et al. [34] also presented that the construction of urban infrastructure induces local-scale uneven subsidence in Shanghai, China. Stramondo et al. [10] detected some local areas affected by subsidence rates greater than 10 mm/year in Rome, and revealed the effect of building construction on land subsidence. Pratesi et al. [12] reported that the subsidence of newly constructed building displayed a sharp slope of displacement velocities during the first year of completion of construction in Florence, Italy. In these studies, regional subsidence showed homogeneous spatial patterns in urban areas, while localised high deformation rates were detected in some newly developed buildings or districts. Unlike these studies, the urban area in our study showed great spatial variation in deformation associated with variations in the groundwater level and geological condition. Even within the same region, spatial subsidence gradients were visible. Therefore, it is difficult to attribute the difference in the deformation rate of the new and old blocks to the contribution of building construction, because these differences may also be caused by groundwater variations. Nonetheless, a comparison of the settlement unevenness across the blocks, instead of the settlement itself, could minimise the impact of groundwater in such a small area. From our results, the higher spatial unevenness and temporal instability within the new blocks confirmed that the effect of building construction on subsidence could be revealed at a local scale.

For the new blocks that were constructed after 2005, the deformation unevenness was found to be related to the variations in building loads within the blocks. In the study area, most of the residential buildings in the blocks had a construction period less than five years. The intense building construction in a small area disturbed the balance of stresses in the overlying strata. Greater building load variation indicated greater differences in building structures or founding techniques, and thus differential settlement [10]. Skyscrapers may produce greater downward pressure, which consolidates the soft clay layer, than low-rise buildings, and thereby may cause greater subsidence.

5.3. *Effects of Building Volume on Building-Scale Subsidence*

The magnitude of the imposed building load has been demonstrated as a factor promoting consolidation and high displacement rates [11]. In our study area, the analysis of the 16 blocks showed that building settlement increased with building volume in most blocks. Similarly, Jiao et al. [18] reported that buildings with a volume of over 3×10^5 m^3 had higher subsidence rates than small-volume buildings in the western CBD area. We also found that the R^2 of the positive relationship was relatively low, and a few blocks, such as those in the Huamao Centre, even showed an opposite trend. A possible explanation for this is that the load contribution to ground settlement also depended on the type of foundation and corresponding depth [10]. For example, the three high-rise buildings in the Huamao Centre had four basement floors that each had a depth of at least 18 m. The foundation piles of these buildings may reach down to the substrate. The use of long piles or super long piles in high or super high buildings can effectively reduce settlement, and the increase in the building load does not cause the same proportion of ground settlement. Nonetheless, pile foundation construction may cause ground deformation in surrounding areas as a result of the stress transfer effect [34], and thus affect the spatial unevenness of the subsidence within a block.

6. Conclusions

In this study, we analysed the relationship between ground subsidence and building characteristics in the Beijing Chaoyang and Tongzhou districts using the PS-InSAR technique. First, 39 Envisat ASAR images between 2003–2010 and 55 TerraSAR-X images between 2010 to 2016 were used to obtain surface deformation. The results showed that the ground deformation rate in the Beijing urban area ranged from −109 mm/year to +13.1 mm/year from 2003 to 2010, and from −151 mm/year to +19 mm/year from 2010 to 2016. There were two obvious subsidence bowls in the eastern part of the Chaoyang District. The displacement rates that were estimated by the PS-InSAR datasets agreed well with the levelling observations, and the average biases were 1.57 mm/year at 12 levelling benchmarks from 2003 to 2010, 4.53 mm/year at five levelling benchmarks from 2010–2013, and 5.01 mm/year at 13 levelling benchmarks from 2015–2016. The study area was partitioned into five hydrogeological regions with similar groundwater levels at the second confined aquifer and compressible layer thicknesses. A total of 138 blocks with 6023 buildings were selected and analysed in the five regions. Based on the land subsidence monitoring results, we analysed the land subsidence pattern and its relationship with different building characteristics at regional, block, and building scales, and the following conclusions were drawn:

(1) At the regional scale, the spatiotemporal evolution of land subsidence was mainly controlled by declines in the groundwater level, compressible layer thickness, and geological faults. The spatial pattern of the land subsidence rate distribution was consistent with groundwater level contours during the two time periods, and the highest deformation rate occurred in regions I, II, and III, where the compressible clay layer thickness reached 50–70 m. Geological faults also affected the subsidence unevenness at a regional scale. The mean and SD of ground displacement increased significantly from 2003–2010 to 2010–2016 in almost all of the regions. For each region, the mean and SD of ground displacement within the blocks showed a similar spatiotemporal pattern as that within the whole region.

(2) At the block scale, we analysed the relationship between the age of block construction and the deformation at the subsidence centre area (regions I, II, and III) and the area far from the subsidence centre (regions IV and V). Interestingly, we found that newly constructed blocks (constructed between 1998–2005 and after 2005) had a considerably higher spatial unevenness of ground settlement than the old blocks (constructed before 1998), especially during the time period of 2010–2016, as shown by the TerraSAR-X dataset. This pattern was more obvious for the block cluster with adjacent new and old blocks. The temporal instability of the deformation within the new blocks was also greater than that within the old blocks. For the new buildings, we found that subsidence unevenness was related to the variation in building volume within the block. Greater variations in building volume corresponded to greater subsidence unevenness. The block-scale results indicated that intense building construction within a small area could disturb the balance of stresses in the overlying strata, and thus cause differential settlement.

(3) At the building scale, an analysis of 16 new blocks with a building volume range over 10^5 m^3 demonstrated a weak positive relationship between single-building settlement and building volume in 13 blocks. However, in the remaining three blocks, we found the settlement rates of some high-rise buildings were lower than those of low-rise buildings. Single-building settlement can be caused by the combined effects of load magnitude, foundation structure, and foundation depth.

In summary, the impact of building construction on land subsidence was difficult to determine at the regional scale in a Beijing urban area, as hydrogeological conditions are the main drivers of subsidence, as reported in previous research. At the block scale and single-building scale, we found block construction age and building volume could affect the spatial unevenness and instability of subsidence. As the urban area in Beijing is characterised by complex land use and varied building properties, the impact of building construction on subsidence might be a combination of effects from these factors. Nevertheless, our conclusion implies that building construction contributed to the spatial unevenness of ground displacement at the local scale, and attention should be paid to those uneven settlement blocks.

Author Contributions: Q.Y. performed the experiment, analysed the data and wrote the manuscript. Y.K. provided crucial guidance and support during the research, and revised the manuscript. D.Z. contributed to the data validation and analysis. B.C. and H.G. provided important suggestions on the research direction. M.L. provided support for the map drawing. L.Z. and X.L. made important suggestions for data processing.

Acknowledgments: This work was supported by the National Natural Science Foundation of China under Grant [41401493], by the 2015 Beijing Nova Program under Grant [xx2015B060] and by the Beijing Natural Science Foundation under Grant [5172002]. We also thank the reviewers for their valuable suggestions.

References

1. Zhu, L.; Gong, H.L.; Li, X.J.; Wang, R.; Chen, B.B.; Dai, Z.X.; Teatini, P. Land subsidence due to groundwater withdrawal in the northern Beijing plain, China. *Eng. Geol.* **2015**, *193*, 243–255. [CrossRef]
2. Chen, M.; Tomás, R.; Li, Z.H.; Motagh, M.; Li, T.; Hu, L.Y.; Gong, H.L.; Li, X.J.; Yu, J.; Gong, X.L. Imaging Land Subsidence Induced by Groundwater Extraction in Beijing (China) Using Satellite Radar Interferometry. *Remote Sens.* **2016**, *8*, 468. [CrossRef]
3. Chen, B.B.; Gong, H.L.; Li, X.J.; Lei, K.C.; Zhu, L.; Gao, M.L.; Zhou, C.F. Characterization and causes of land subsidence in Beijing, China. *Int. J. Remote Sens.* **2017**, *38*, 808–826. [CrossRef]
4. Lei, K.C.; Luo, Y.; Chen, B.B.; Guo, G.X.; Zhou, Y. Distribution characteristics and influence factors of land subsidence in Beijing area. *Geol. China* **2016**, *6*, 2216–2228.
5. Zhou, C.F.; Gong, H.L.; Chen, B.B.; Zhu, F.; Duan, G.Y.; Gao, M.L.; Lu, W. Land subsidence under different land use in the eastern Beijing plain, China 2005–2013 revealed by InSAR timeseries analysis. *GISci. Remote Sens.* **2016**, *53*, 671–688. [CrossRef]
6. Zhou, C.; Gong, H.; Chen, B.; Li, J.; Gao, M.; Zhu, F.; Chen, W.; Liang, Y. InSAR Time-Series Analysis of Land Subsidence under Different Land Use Types in the Eastern Beijing Plain, China. *Remote Sens.* **2017**, *9*, 380. [CrossRef]
7. Gao, M.L.; Gong, H.L.; Chen, B.B.; Zhou, C.F.; Chen, W.F.; Liang, Y.; Shi, M.; Si, Y. InSAR time-series investigation of long-term ground displacement at Beijing Capital International Airport, China. *Tectonophysics* **2016**, *691*, 271–281. [CrossRef]
8. Teatini, P.; Tosi, L.; Strozzi, T.; Carbognin, L.; Cecconi, G.; Rosselli, R.; Libardo, S. Resolving land subsidence within the Venice Lagoon by persistent scatterer SAR interferometry. *Phys. Chem. Earth* **2012**, *40–41*, 72–79. [CrossRef]
9. Deng, Z.; Ke, Y.H.; Gong, H.L.; Li, X.J.; Li, Z.H. Land subsidence prediction in Beijing based on PS-InSAR technique and improved Grey-Markov model. *GISci. Remote Sens.* **2017**, *54*, 797–818. [CrossRef]
10. Stramondo, S.; Bozzano, F.; Marra, F.; Wegmuller, U.; Cinti, F.R.; Moro, M.; Saroli, M. Subsidence induced by urbanisation in the city of Rome detected by advanced InSAR technique and geotechnical investigations. *Remote Sens. Environ.* **2008**, *112*, 3160–3172. [CrossRef]

11. Solari, L.; Ciampalini, A.; Raspini, F.; Bianchini, S.; Moretti, S. PSInSAR Analysis in the Pisa Urban Area (Italy): A Case Study of Subsidence Related to Stratigraphical Factors and Urbanization. *Remote Sens.* **2016**, *8*, 120. [CrossRef]
12. Pratesi, F.; Tapete, D.; Del Ventisette, C.; Moretti, S. Mapping interactions between geology, subsurface resource exploitation and urban development in transforming cities using InSAR Persistent Scatterers: Two decades of change in Florence, Italy. *Appl. Geogr.* **2016**, *77*, 20–37. [CrossRef]
13. Chen, G.; Zhang, Y.; Zeng, R.Q.; Yang, Z.K.; Chen, X.; Zhao, F.M.; Meng, X.M. Detection of land subsidence associated with land creation and rapid urbanization in the Chinese Loess Plateau using Time Series InSAR: A case study of Lanzhou New District. *Remote Sens.* **2018**. [CrossRef]
14. Tosi, L.; Strozzi, T.; Da Lio, C.; Teatini, P. Regional and local land subsidence at the Venice coastland by TerraSAR-X PSI. *Proc. Int. Assoc. Hydrol. Sci.* **2015**, *372*, 199–205. [CrossRef]
15. Tosi, L.; Da Lio, C.; Strozzi, T.; Teatini, P. Combining L- and X-Band SAR Interferometry to Assess Ground Displacements in Heterogeneous Coastal Environments: The Po River Delta and Venice Lagoon, Italy. *Remote Sens.* **2016**, *8*, 308. [CrossRef]
16. Maghsoudi, Y.; van der Meer, F.; Hecker, C.; Perissin, D.; Saepuloh, A. Using PS-InSAR to detect surface deformation in geothermal areas of West Java in Indonesia. *Int. J. Appl. Earth Obs. Geoinform.* **2018**, *64*, 386–396. [CrossRef]
17. Liao, M.S.; Pei, Y.Y.; Wang, H.M.; Fang, Z.L.; Wei, L.H. Subsidence Monitoring in Shanghai Using the PSInSAR Technique. *Shanghai Land Resour.* **2012**, *33*, 5–10.
18. Qin, X.Q.; Yang, M.S.; Wang, H.M.; Yang, T.L.; Lin, J.X.; Liao, M.S. Application of High-resolution PS-InSAR in Deformation Characteristic Probe of Urban Rail Transit. *Acta Geod. Geophys. Sin.* **2016**, *45*, 713–721.
19. Chen, W.F.; Gong, H.L.; Chen, B.B.; Liu, K.S.; Gao, M.L.; Zhou, C.F. Spatiotemporal evolution of land subsidence around a subway using InSAR time-series and the entropy method. *GISci. Remote Sens.* **2017**, *54*, 78–94. [CrossRef]
20. Solari, L.; Ciampalini, A.; Raspini, F.; Bianchini, S.; Zinno, I.; Bonano, M.; Manunta, M.; Moretti, S.; Casagli, N. Combined Use of C- and X-Band SAR Data for Subsidence Monitoring in an Urban Area. *Geosciences* **2017**, *7*, 21. [CrossRef]
21. Tapete, D.; Morelli, S.; Fanti, R.; Casagli, N. Localising deformation along the elevation of linear structures: An experiment with spaceborne InSAR and RTK GPS on the Roman Aqueducts in Rome, Italy. *Appl. Geogr.* **2015**, *58*, 65–83. [CrossRef]
22. Tang, Y.; Cui, Z.; Wang, J.; Yan, L.; Yan, X. Application of grey theory-based model to prediction of land subsidence due to engineering environment in Shanghai. *Environ. Geol.* **2008**, *55*, 583–593. [CrossRef]
23. Cui, Z.; Tang, Y.; Yan, X. Centrifuge modeling of land subsidence caused by the high-rise building group in the soft soil area. *Environ. Earth Sci.* **2010**, *59*, 1819–1826. [CrossRef]
24. Xu, Y.; Ma, L.; Du, Y.; Shen, S. Analysis of urbanisation-induced land subsidence in Shanghai. *Nat. Hazards* **2012**, *63*, 1255–1267. [CrossRef]
25. Chen, B.B.; Gong, H.L.; Li, X.J.; Lei, K.C.; Ke, Y.H.; Duan, G.Y.; Zhou, C.F. Spatial correlation between land subsidence and urbanization in Beijing, China. *Nat. Hazards* **2015**, *75*, 2637–2652. [CrossRef]
26. Chen, B.B.; Gong, H.L.; Li, X.J.; Lei, K.C.; Gao, M.L.; Zhou, C.F.; Ke, Y.H. Spatial–temporal evolution patterns of land subsidence with different situation of space utilization. *Nat. Hazards* **2015**, *77*, 1765–1783. [CrossRef]
27. Jiao, S.; Yu, J.; Milas, A.S.; Li, X.; Liu, L. Assessing the Impact of Building Volume on Land Subsidence in the Central Business District of Beijing with SAR Tomography. *Can. J. Remote Sens.* **2017**, *43*, 177–193. [CrossRef]
28. Hooper, A.; Zebker, H.; Segall, P.; Kampes, B. A new method for measuring deformation on volcanoes and other natural terrains using InSAR persistent scatterers. *Geophys. Res. Lett.* **2004**, *31*. [CrossRef]
29. Ferretti, A.; Prati, C.; Rocca, F. Nonlinear subsidence rate estimation using permanent scatterers in differential SAR interferometry. *IEEE Trans. Geosci. Remote Sens.* **2000**, *38*, 2202–2212. [CrossRef]
30. Perissin, D.; Rocca, F. High-accuracy urban DEM using permanent scatterers. *IEEE Trans. Geosci. Remote Sens.* **2006**, *44*, 3338–3347. [CrossRef]
31. Ruiz-Armenteros, A.M.; Bakon, M.; Lazecky, M.; Delgado, J.M.; Sousa, J.J.; Perissin, D.; Caro-Cuenca, M. Multi-Temporal InSAR Processing Comparison in Presence of High Topography. *Procedia Comput. Sci.* **2016**, *100*, 1181–1190. [CrossRef]

32. Çomut, F.C.; Ustun, A.; Lazecky, M.; Aref, M.M. Multi band InSAR analysis of subsidence development based on the long period time series. *Int. Arch. Photogramm. Remote Sens. Spat. Inf. Sci.* **2015**, *40*, 115–121. [CrossRef]
33. Wu, A. *Chinese Institute of Geological Environment Monitoring. China Groundwater Level Yearbook for Geo-Environmental Monitoring*; China Land Press: Beijing, China, 2014; pp. 154–196. ISBN 9787802463813.
34. Dong, S.; Smsonov, S.; Yin, H.; Ye, S.; Cao, Y. Time-series analysis of subsidence associated with rapid urbanization in Shanghai, China measured with SBAS InSAR method. *Environ. Earth Sci.* **2014**, *72*, 677–691. [CrossRef]

4

Detection and Monitoring of Forest Fires using Himawari-8 Geostationary Satellite Data in South Korea

Eunna Jang [1,†], Yoojin Kang [1,†], Jungho Im [1,*], Dong-Won Lee [2], Jongmin Yoon [2] and Sang-Kyun Kim [2]

1 School of Urban and Environmental Engineering, Ulsan National Institute of Science and Technology (UNIST), Ulsan 44919, Korea; enjang@unist.ac.kr (E.J.); kangyj@unist.ac.kr (Y.K.)
2 Environmental Satellite Centre, Climate and Air Quality Research Department, National Institute of Environmental Research, Incheon 22689, Korea; ex12@korea.kr (D.-W.L.); objyoon@korea.kr (J.Y.); nierkum@korea.kr (S.-K.K.)
* Correspondence: ersgis@unist.ac.kr.
† The first two authors equally contributed to the paper.

Abstract: Geostationary satellite remote sensing systems are a useful tool for forest fire detection and monitoring because of their high temporal resolution over large areas. In this study, we propose a combined 3-step forest fire detection algorithm (i.e., thresholding, machine learning-based modeling, and post processing) using Himawari-8 geostationary satellite data over South Korea. This threshold-based algorithm filtered the forest fire candidate pixels using adaptive threshold values considering the diurnal cycle and seasonality of forest fires while allowing a high rate of false alarms. The random forest (RF) machine learning model then effectively removed the false alarms from the results of the threshold-based algorithm (overall accuracy ~99.16%, probability of detection (POD) ~93.08%, probability of false detection (POFD) ~0.07%, and 96% reduction of the false alarmed pixels for validation), and the remaining false alarms were removed through post-processing using the forest map. The proposed algorithm was compared to the two existing methods. The proposed algorithm (POD ~ 93%) successfully detected most forest fires, while the others missed many small-scale forest fires (POD ~ 50–60%). More than half of the detected forest fires were detected within 10 min, which is a promising result when the operational real-time monitoring of forest fires using more advanced geostationary satellite sensor data (i.e., with higher spatial and temporal resolutions) is used for rapid response and management of forest fires.

Keywords: forest fire; Himawari-8; threshold-based algorithm; machine learning

---

1. Introduction

Forest fires can have a significant impact on terrestrial ecosystems and the atmosphere, as well as on society in general. In order for a site to recover from a forest fire, a lot of time and effort are required. According to the 2015 forest standard statistics, forest areas in South Korea cover 6,335,000 ha, accounting for 63.2% of the national land. This forest-to-land ratio of South Korea is the fourth largest among the Organization for Economic Co-operation and Development (OECD) countries [1]. Since forests in South Korea are densely distributed, a forest fire can easily spread outwards, resulting in huge amounts of damage. The forest growing stock of South Korea is 146 m^3/ha, which is higher than the average of OECD countries (131 m^3/ha) [1]. Approximately 36.9% of forest in South Korea are coniferous, and their growing stock reaches 172.7 m^3/ha. Since coniferous forests have a large amount of branches and leaves, those under the canopy dry easily. Thus, when a forest fire occurs, it can easily develop into a large one if early extinguishment fails, resulting in huge amounts of damage [2]. In 2017 in South Korea, the total area damaged by forest fires was 1,480 ha with the amount of damage totaling 80,150,000,000 KRW (71,594,462 USD), while the number of casualties was 16 [3]. Most forest fires that occur in South Korea are caused by anthropogenic factors and are thus unpredictable and hard to control. To minimize forest fire damage, South Korea has been conducting forest fire monitoring through tower systems and closed-circuit television (CCTV) [2]. An alternative to such field monitoring is satellite-based monitoring, which can cover vast areas including inaccessible regions with fine temporal resolution [4]. Satellite data have been widely used in forest fire management, such as pre-fire condition management, forest fire hot spot detection, smoke detection, and burn severity mapping [5]. Various satellite sensors have been used for forest fire detection, such as polar-orbiting satellite sensors (Moderate Resolution Imaging Spectroradiometer (MODIS), Advanced Very High Resolution Radiometer (AVHRR), the Landsat series, and the Visible Infrared Imaging Radiometer Suite (VIIRS)), and geostationary satellite sensor systems (Geostationary Operational Environmental Satellite (GOES), Spinning Enhanced Visible and Infrared Imager (SEVIRI), Communication, Ocean and Meteorological Satellite (COMS), and Himawari-8).

Fires are typically detected through their high surface temperature, which is easily distinguishable in mid-infrared and thermal remote sensing data [5]. One of the most widely used methods for detecting forest fires is a simple threshold-based algorithm, which distinguishes fire pixels based on given empirical threshold values applied to band radiance, brightness temperature (BT), or the band ratio of specific wavelengths. However, this method produces a relatively high number of false alarms and often misses fires because of the varied characteristics of forests, topography, and climate between different regions [4]. Contextual algorithms, which were developed from the threshold-based algorithm, use local maxima and other multispectral criteria based on the difference between fire pixels and the background temperature [6–15]. Furthermore, the modeling of the fire pixel diurnal temperature cycle (DTC), which shows a diurnal variation of the brightness temperature of the pixel, has been also used [16–19]. Other ways to detect forest fires include using artificial neural networks (ANN)-based modeling [20] and hierarchical object-based image analysis (OBIA), which classifies active fire pixels using a ruleset based on image-specific object values [21]. Most existing forest fire detection and monitoring algorithms along with their related products have been developed and tested over Europe and the US. For example, MODIS active fire data is one of the most widely used products for fire management in many countries [4]. However, these algorithms often produce a very high false alarm rate in East Asia including South Korea. Collection 5 MODIS fire products struggle to detect small fires because of the overly high global thresholds of regions such as East Asia. Although the Collection 6 MODIS fire products slightly increased the probability of detection (POD) ~1% [9], it is still not enough to detect small forest fires in South Korea. When we calculated the accuracy of the Collection 6 MODIS active fire products (M*D14) in South Korea from March to May (i.e., dry season) in 2017 (Appendix A), only 22 of the 145 forest fires were detected (POD ~ 15.2%). In addition, 266 forest fires were falsely alarmed among the 288 MODIS-detected forest fires (false alarm rate (FAR) ~ 92.4%). Consequently, the algorithms are not good at detecting small forest fires, which frequently occur over rugged terrain in South Korea. Thus, there is a strong need to develop a novel forest fire detection and monitoring algorithm suitable for South Korea.

Several studies have been conducted to develop forest fire detection algorithms focusing on fires in South Korea. [22] developed an algorithm for detecting missed sub-pixel scale forest fires in MODIS active fire data using a spectral mixed analysis. While it showed a POD ~ 70% and a FAR ~ 40%, it was only tested with data in April for 2004 and 2005. [2] developed a forest fire detection model for South Korea using the COMS Meteorological Imager (MI) data, which modifies the MODIS algorithm [8] based on the spectral characteristics of MI. This algorithm was able to detect small-scale forest fires with damaged areas ~1 ha at 15-min intervals, but resulted in a relatively high mis-detection rate. Another forest fire detection method has been proposed, which is based on the negative relationship between vegetation density and land surface temperature with a contextual approach using MODIS [23]. This method improves the previous MODIS contextual forest fire detection algorithm, but still has a low temporal resolution.

In this study, we used Himawari-8 geostationary satellite data to detect and monitor forest fires in South Korea. It is suitable for continuous forest fire monitoring and early detection because of its high temporal resolution (<= 10 min), even though its spatial resolution is not as good as polar-orbiting satellite sensor systems. Early detection and monitoring of forest fires are crucial to reduce damage and save human lives and property. Recently, some researchers have used Himawari-8 to detect forest fires over Asia and Australia using its multiple spectral bands and high temporal resolution [10,13,15,19]. These studies were based on contextual algorithms targeting large forest fires in East Asia and Australia, which were not evaluated for detecting small fires in South Korea.

There is no official definition for small-scale forest fires in South Korea. In this study, we defined small-scale forest fires considering the damaged areas and the spatial resolution of input Himawari-8 satellite data (4 km^2 = 400 ha). Forest fires damaging areas measuring less than 8 ha (2% of the Himawari-8 pixel size) were defined as 'small-scale' in this study. It should be noted that the real-time detection of such small forest fires (<8 ha) using Himawari-8 data is possible because of the spread of fire-induced heat and gaseous materials to the much larger surrounding area of a fire than the damaging area recorded by the expert after the fire. Although small-scale forest fires are dominant in South Korea, they often have a significant effect on people, infrastructure, and the environment. Since the population density of South Korea is very high and many fires occur near farmhouses and roads, even small-scale forest fires can result in costly damages. Thus, the detection and monitoring of small-scale forest fires is crucial in South Korea. For example, on 19 April 2018, a small forest fire (the damaged area ~3 ha) that occurred in Yangyang, Gangwon-do, required not only 387 firefighters and 41 units of equipment to be mobilized, but also 9 evacuation helicopters to be dispatched to the scene to extinguish and monitor the forest fire [24].

The forest fire detection algorithm proposed in this study consists of three steps: a threshold-based algorithm, machine learning modeling, and post processing. First, we developed a threshold-based algorithm optimized for detecting small forest fires in South Korea with the tradeoff of a relatively high false alarm rate from Himawari-8 data. The proposed threshold-based algorithm adopted a thresholding approach adaptive to corresponding satellite imagery to detect small-scale forest fires considering the diurnal cycle and seasonality. Then, machine learning and post processing approaches were applied to the potential fire pixels to effectively reduce false alarms. Existing threshold-based forest fire detection algorithms often miss small forest fires, resulting in too many false alarms due to the fixed thresholds. Our proposed approach is focused on increasing the detection of small-scale forest fires and significantly reducing false alarms.

The objectives of this study were to (1) develop a machine learning-combined approach for detecting small to large-scale forest fires in South Korea, (2) examine the feasibility of early detection of forest fires based on the approach, and (3) monitor forest fires using Himawari-8 satellite data at high temporal resolution. This study can provide a basis for the geostationary satellite-based operational monitoring of forest fires in South Korea.

2. Data

2.1. Study Area

South Korea has an area of 10,030,000 ha, with forests covering 6,335,000 ha (about 63.2% of the total area) [1,25]. South Korea has suffered forest fires every year especially during the spring and fall seasons because of the large number of visitors to forests in May and October and the high frequency of agricultural incinerations in Spring [3,26]. It belongs to the mid-latitude cold temperate region and has a continental climate. When compared to other regions with similar latitudes, the range of annual temperature is large. The average lowest monthly temperature is −6 to 3 °C, the average highest temperature is 23 to 26 °C, and the annual average rainfall is 1000–1900 mm [27]. Approximately

200–800 forest fires have occurred annually in South Korea and many of them were small-scale fires. For example, 94% of the areas damaged by forest fires in 2017 were smaller than 1 ha [3]. Over the last 10 years, more than 60% of forest fires have occurred in the spring season (March to May). More than 80% of the forest fires that occurred in 2017 were caused by humans, for reasons including the carelessness of hikers, and agricultural and waste incineration [3]. The study period is from July 2015 to December 2017 when Himawari-8 satellite data are available.

2.2. Forest Fire Reference Data

In situ forest fire occurrence data provided by the Korea Forest Service were used as reference data in this study (Appendix B). Each forest fire case contains information about the starting and extinguishing date/time, location (specific address), damaged area, and cause. When a forest fire occurs in a region, the public officials in charge of the region confirm and report the fire in detail. Damaged areas are calculated by trained forest fire experts based on visual observations, actual measurements using Global Positioning System (GPS) survey equipment, aerial photographs, and/or topographic maps with a scale of 1:25,000 [28]. Small forest fires damaging less than 0.7 ha of land were not considered in this study because most of them did not show little spectral difference in the Himawari-8 time-series data based on visual inspection of the images. It should be noted, though that pixel radiance is affected by not only a fire, but also many other factors. Among the 114 forest fires that resulted in damaged areas of over 0.7 ha during the study period, 64 cases that were clearly distinguishable from the satellite data without being blocked by clouds were selected as reference data, resulting in 2165 fire pixels and 18,085 non-fire pixels between 2015 and 2017. Note that the non-fire pixels were randomly extracted from the forested areas from the images after excluding fire and cloud pixels.

2.3. Himawari-8 AHI Satellite Data

Himawari-8, launched in October 2014, is the geostationary satellite sensor system operated by the Japan Meteorological Agency (JMA), the latest line of Multifunctional Transport Satellite (MTSAT) series. The Advanced Himawari Imager (AHI) sensor onboard Himawari-8 collects data every 10 min as full disk images in 16 bands from visible to infrared wavelengths at a 500 m–2 km resolution, covering from East Asia to Australia. From a monitoring perspective, geostationary satellite data with a very high temporal resolution may be a better option than polar-orbiting satellite data even though its spatial resolution is typically not as good. Tables 1 and 2 summarize Himawari-8 derived input variables used in the threshold-based algorithm and machine learning modeling, respectively.

Table 1. Himawari-8 AHI-derived input variables used in the threshold-based algorithm.

Himawari-8 AHI	Band Number	Central Wavelength (μm)	Spatial Resolution (km)
Input variables	5	1.61	2
	7	3.85	
	14	11.20	
	Band 5/Band 7		
	Band 7 brightness temperature—Band 14 brightness temperature		

Table 2. Himawari-8 AHI bands and variables used in machine learning modeling in this study (Ch is the radiance of each band and BT is the brightness temperature of each band).

Himawari-8 AHI	Band Number	Bandwidth (μm)	Central Wavelength (μm)	Spatial Resolution (km)
	4	0.85–0.87	0.86	1
	5	1.60–1.62	1.61	2
	6	2.25–2.27	2.26	
	7	3.74–3.96	3.85	
	8	6.06–6.43	6.25	
	9	6.89–7.01	6.95	
	10	7.26–7.43	7.35	
	11	8.44–8.76	8.60	
	12	9.54–9.72	9.63	
	13	10.30–10.60	10.45	
	14	11.10–11.30	11.20	
	15	12.20–12.50	12.35	
	16	13.20–13.40	13.30	
Input variables	Ch07	BT07	BT13-BT14	BT07/BT14
	Ch04-Ch07	BT07-BT11	BT13-BT15	BT07/BT15
	Ch05-Ch07	BT07-BT12	BT07/BT09	BT07/BT16
	Ch06-Ch07	BT07-BT13	BT07/BT10	BT09/BT16
	Ch07-Ch12	BT07-BT14	BT07/BT11	BT13/BT15
	Ch07-Ch15	BT07-BT15	BT07/BT12	
	Ch12-Ch15	BT12-BT16	BT07/BT13	

2.4. Land Cover Data and Forest Map

Land cover data obtained from the Ministry of Environment of South Korea was used to identify forest areas (Figure 1). The land cover data were produced using Landsat TM images collected in 2010 and the overall accuracy is reported as 75% [29]. It has 7 classes—built-up, agricultural land, forest, grassland, wetland, barren land and water—at 30 m resolution. The land cover map was upscaled to 2 km corresponding to the spatial resolution of the input AHI data using a majority filtering. Considering many forest fires occurred along roads or agricultural land in the boundaries of forests, one pixel (2 km)-buffered areas from the forest pixels were used as the forest mask.

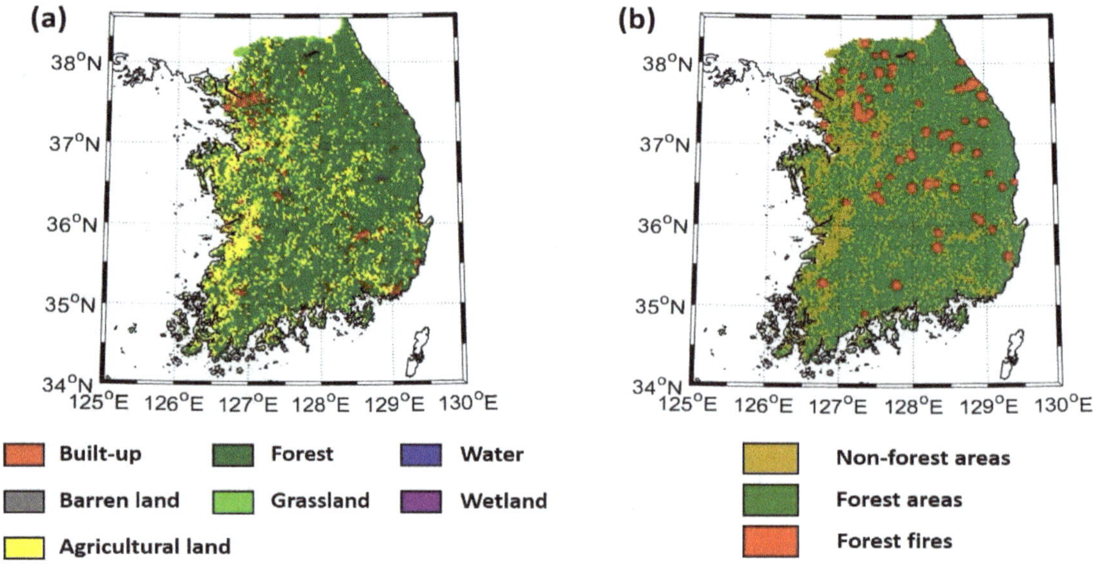

Figure 1. The study area of this research (South Korea) and (**a**) land cover map from the Ministry of Environment of South Korea and (**b**) forest region map with forest fires occurred during the study period.

3. Methodology

3.1. Forest Fire Detection Aalgorithm

The forest fire detection model proposed in this study consists of 3 steps. Figure 2 shows the process flow diagram of the proposed approach. The first step is to identify the candidate pixels of forest fires using infrared bands based on a threshold-based algorithm. The proposed threshold-based algorithm uses multi-temporal analysis to consider the stationary heat sources and varied thermal signals from the surface due to the diurnal cycle of forest fires. This first step tries to focus on identifying potential fire pixels, regardless of a high false alarm rate. The following machine learning and post-processing approaches, then, try to reduce false alarms effectively from the results of the first step.

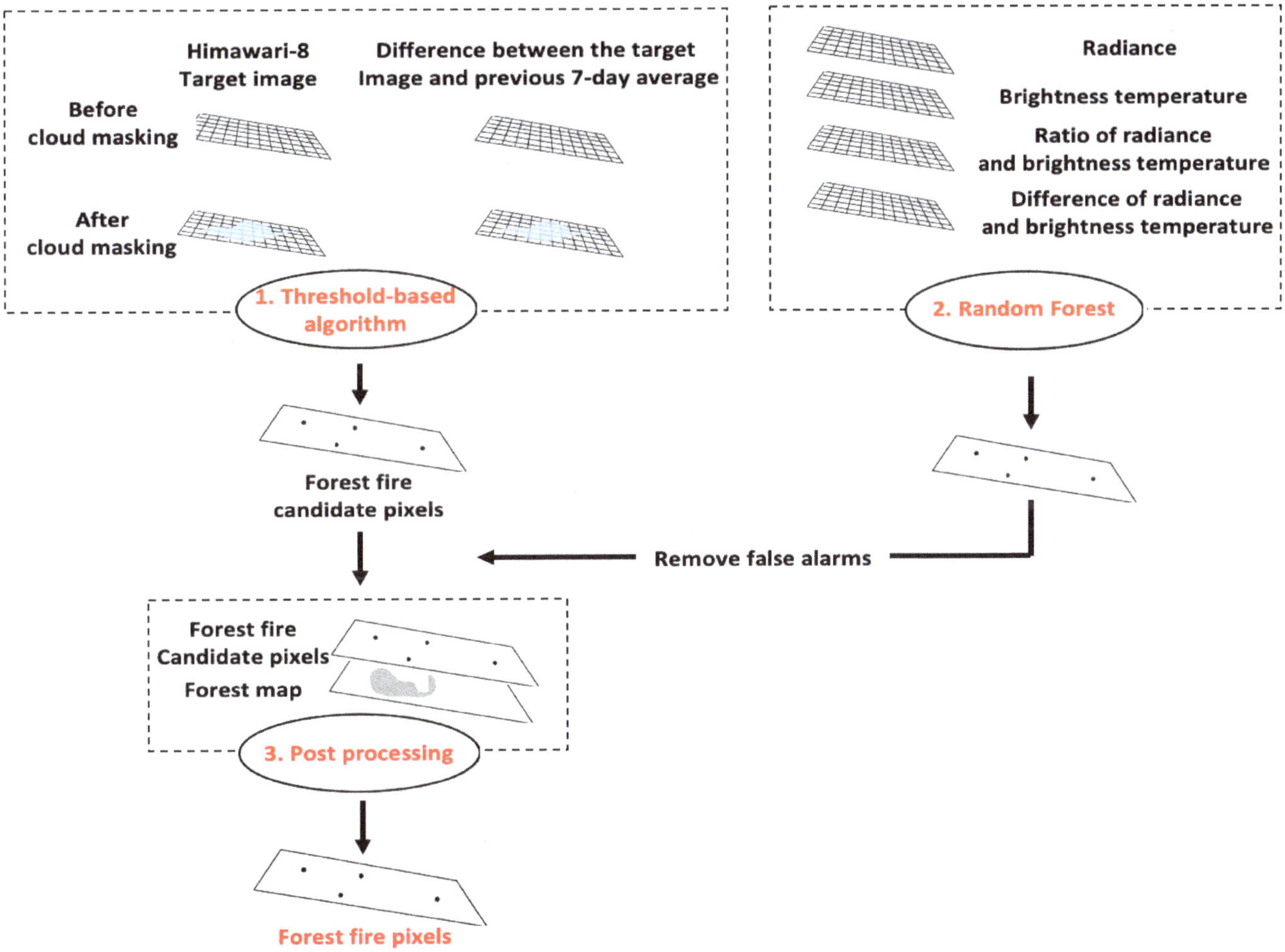

Figure 2. The process flow of detecting forest fire pixels based on the threshold-based algorithm, machine learning, and post-processing approaches.

3.2. Threshold-Based Algorithm

To increase the probability of detection of small-scale forest fires, the threshold-based algorithm of this study modified the existing threshold-based forest fire detection algorithms, which used 4 and 11 μm bands [9,13], considering the characteristics of forest fires in South Korea. Since average temperatures of active fires range from 800 K to 1200 K, the fires are detectable in the mid-infrared and thermal bands with high intensities [5]. Himawari-8 AHI band 5 (1.61 μm), 7 (3.85 μm) and 14 (11.2 μm) data were used in the threshold-based algorithm in this study (Table 1).

Figure 3 summarizes the proposed threshold-based algorithm with multiple AHI channels and their time series. In the first condition of the threshold-based algorithm, the band 7 radiance was used to distinguish forest fire candidate pixels. The MIR band (i.e., band 7) is effective in observing radiative emissions from objects radiating at temperatures similar to those of forest fires [13]. Thus, it has been used in most existing fire detection algorithms [10,11,13,15,30]. Many factors such as land cover type, topographic characteristics, time of day, and day of the year affect the threshold [4]. Unlike with existing algorithms, the threshold in the proposed algorithm was not fixed to better identify small-scale forest fires. Instead, the top 7% value in the forest region buffered by one-pixel in South Korea for each image was assigned as an adaptive threshold through multiple empirical tests. In order to identify heat sources other than forest fires, a multi-temporal component was considered in the threshold-based algorithm. The multi-temporal component uses the difference between the radiance of the target image and the averaged radiance for 7 days before the target image. In this way, it can effectively remove stationary heat sources (e.g., industrial facilities), radiometrically bright objects (e.g., hot and reflective rooftops such as solar cells on the roof) and other unique structures such as solar farms, which can be classified as potential forest fire pixels. The other step in the threshold-based algorithm is cloud masking. The cloud pixels usually have a negative effect on the multi-temporal analysis and are classified as forest fires, because clouds have high albedo or reflectance in visible and near infrared bands [15]. As Himawari-8 has not provided a publicly available cloud mask product yet, cloud pixels were defined by the cloud masking algorithm developed by [31]. When the operational Himawari-8 cloud mask product is available in the future, the proposed algorithm will be able to use the product to more effectively remove clouds from images.

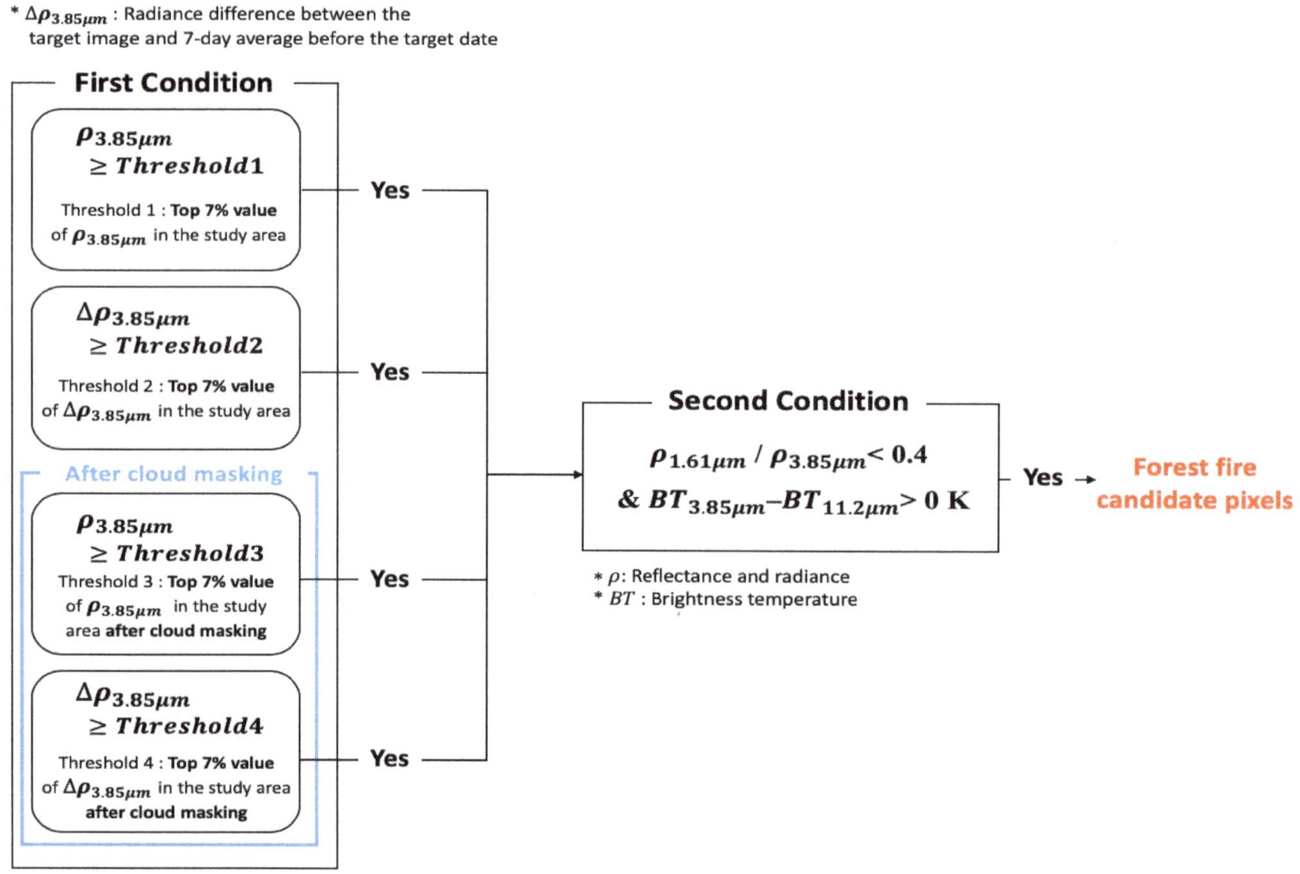

Figure 3. The threshold-based algorithm proposed in this study.

Secondly, the pixels which were classified as potential forest fire pixels in the first condition were checked against another series of empirically selected parameters and thresholds that reflect forest fire characteristics (Figure 3). Shortwave infrared (1.58–1.64 μm) bands are used in cloud, sun glint and water distinction in the existing fire detection algorithms [11]. The ratio of band 5 reflectance to band 7 radiance was the most distinct parameter among several ratios and differences of bands based on the reference data (forest fire vs. non-forest fire pixels). An optimum threshold for the ratio was determined through empirical testing of multiple thresholds. Another parameter is the difference between BTs of bands 7 and 14. The large difference between the BTs in the shortwave (3–4 μm) and longwave (~11 μm) bands is related to fires [32]. Thus, it has been widely used in other fire detection algorithms [9,15,33,34]. The threshold value of this parameter was also defined by empirical tests using the reference data.

3.3. Random Forest

RF is widely used in various remote sensing applications for both classification and regression [35–39]. RF is based on Classification and Regression Tree (CART) methodology [40], which is a rule-based decision tree. RF adopts two randomization strategies to produce many independent CARTs: a random selection of training samples for each tree, and a random selection of input variables at each node of a tree [41–43]. Final output from RF is achieved through an ensemble of individual CARTs. This ensemble approach can mitigate overfitting and the sensitivity to training data configurations, which are major limitations of CART [44–46]. Using many independent decision trees, RF makes a final decision by (weighted) averaging and majority voting approaches for regression and classification, respectively. RF also provides useful information on the contribution of input variables to the model, which is based on relative variable importance using out-of-bag (OOB) data [47–49]. OOB errors are the differences between the actual value and the decision value that is estimated using data not used in training.

In this study, the 64 forest fire reference cases between 2015 and 2017 (2165 fire pixels and 18,085 non-fire pixels) were divided into two groups considering their damaged area, time and location: 50 fire cases (80%; 1775 fire pixels and 15,043 non-fire pixels) to develop an RF model and the remaining 14 cases (20%; 390 fire pixels and 3042 non-fire pixels) to validate the model.

First, a total of 191 input variables—band radiance, BT, band ratios, BT differences and BT ratios of bands 4–16—(Appendix C) were used as the input parameters of the RF model. We used a simple feature selection based on the relative variable importance provided by RF through iterative testing with different sets of input variables. Finally, 26 parameters were selected (Table 2), which were used to develop the RF model for effectively removing false alarms of forest fire detection.

3.4. Post Processing

In order to further refine forest fire detection results, additional post processing was applied. The post processing was designed to effectively remove salt-and-pepper noise and fires from non-forest areas. We applied a buffer to the forest boundary to effectively detect almost all forest fires that occurred near roads or areas between agricultural land and forests (Section 2.4), but the buffered area inevitably contains non-forest regions which results in salt-and-pepper noise (mostly fires from agricultural land or hot spots in urban areas). We used the forest map (refer to the Ministry of Environment of South Korea) in the post processing. If more than three out of the eight surrounding pixels of a pixel classified as forest fire by the RF model were forest, they were then considered to be forest fire pixels. Otherwise, the pixels were removed as non-forest area fires (e.g., agricultural fires).

3.5. Accuracy Assessment

The performance of the proposed approach was evaluated using the probability of detection (POD; $\frac{Correctly\ detected\ forest\ fire\ pixels}{Correctly\ detected\ forest\ fire\ pixels + Mis\text{-}detected\ forest\ fire\ pixels} \times 100$), the probability of false detection (POFD; $\frac{Miss\ detected\ forest\ fire\ pixels}{Correctly\ detected\ non\text{-}forest\ fire\ pixels + Mis\text{-}detected\ forest\ fire\ pixels} \times 100$) and the overall accuracy (OA; $\frac{Correctly\ detected\ forest\ fire\ and\ non\text{-}fire\ pixels}{Total\ pixels} \times 100$).

The accuracy of the proposed algorithm was further compared to those of the two existing forest fire detection algorithms and the Collection 6 MODIS fire products (M*D14; [9]). One is the COMS algorithm, which was proposed by [2]. The COMS algorithm is threshold-based, and is based on the MODIS wildfire detection algorithm using COMS MI sensor data. It classified the forest fire candidate pixels by the two thresholds and surrounding statistical values using the 3.7 and 10.8 μm wavelength bands of COMS MI. The other algorithm is an AHI Fire Surveillance Algorithm (AHI-FSA), which was developed by [13]. The AHI-FSA algorithm is based on three different wavelengths (RED-band 3; NIR-band 4; MIR-band 7) of Himawari-8 AHI sensor to detect burnt areas and smoke.

4. Results and Discussion

4.1. Forest Fire Detectioin

The forest fire detection algorithm was evaluated using 14 reference forest fire cases among a total of 64 reference cases (Section 2.2). More than 90% of the forest fires (i.e., 46 out of 50 in calibration cases, and 14 out of 14 in validation cases) were detected by the threshold-based algorithm, and an additional 5 forest fires were detected which were not included in the reference data due to their small damage areas. Although the threshold-based algorithm detected most forest fires, it resulted in a high rate of false alarms.

The final RF model was constructed using the 26 input variables (Table 2), which were selected based on variable importance identified by RF among over 191 variables (Appendix C). When using both BT differences and ratios, the RF model produced higher accuracy (OA = 98.75%, POD = 89.74%, and POFD = 0.10%) than using either one of the sets (OA = 96.44%, POD = 70%, and POFD = 0.16% when using only BT differences; OA = 98.60%, POD = 88.97%, and POFD = 0.16% when using only BT ratios). Many of the selected input variables were related to band 7 (MIR band), which was used in the threshold-based algorithm. This corresponds to the literature in that the MIR band (i.e., band 7) is sensitive to forest fire temperature [13]. BT differences and ratios between band 7 and thermal bands were considered important variables. While the peak radiation at thermal wavelengths (8–12 μm) is related to a normal environmental temperature, hot temperature by forest fires can be detected at a shorter wavelength than the Earth's surface, especially 3–4 μm (band 7) [5]. The large BT difference between the shortwave (3–4 μm) and thermal bands can be observed in fire pixels, and thus the BT difference has been used in other fire detection algorithms [9,15,33,34]. High radiance values of band 5 are related to near the center of fires [12], and 2.2 μm (band 6) wavelength is sensitive to hot targets [50]. The NIR (0.846–885 μm; band 4) and the shortwave infrared (1.58–1.64 μm; band 5) regions are used to discriminate cloud, sun glint and water in the fire detection algorithm [9,11]. The reflectance values of band 4 are used to remove highly reflective surface and sun glint characteristics from non-fire pixels [30]. These wavelengths (bands 4–6) are also used in existing wildfire detection algorithms [11,12,50].

Figure 4 summarizes the relative variable importance of the selected 26 input variables provided by the RF model. The difference between bands 5 and 7 was identified as the most contributing variable to the model, followed by the difference between bands 6 and 7, and that between bands 4 and 7. The BT ratios and differences between bands 13 (10.45 μm) and 15 (12.35 μm) were also identified as contributing variables. These variables are known to be effective for separating active fires from fire-free background [11,30]. The usefulness of Himawari-8 AHI sensor data for forest fire detection is largely unassessed because of the relatively young age of the sensor and the minimal existing published work [17]. Thus, it is desirable to test various variable combinations to find an optimum set of Himawari-8 derived input variables for forest fire detection.

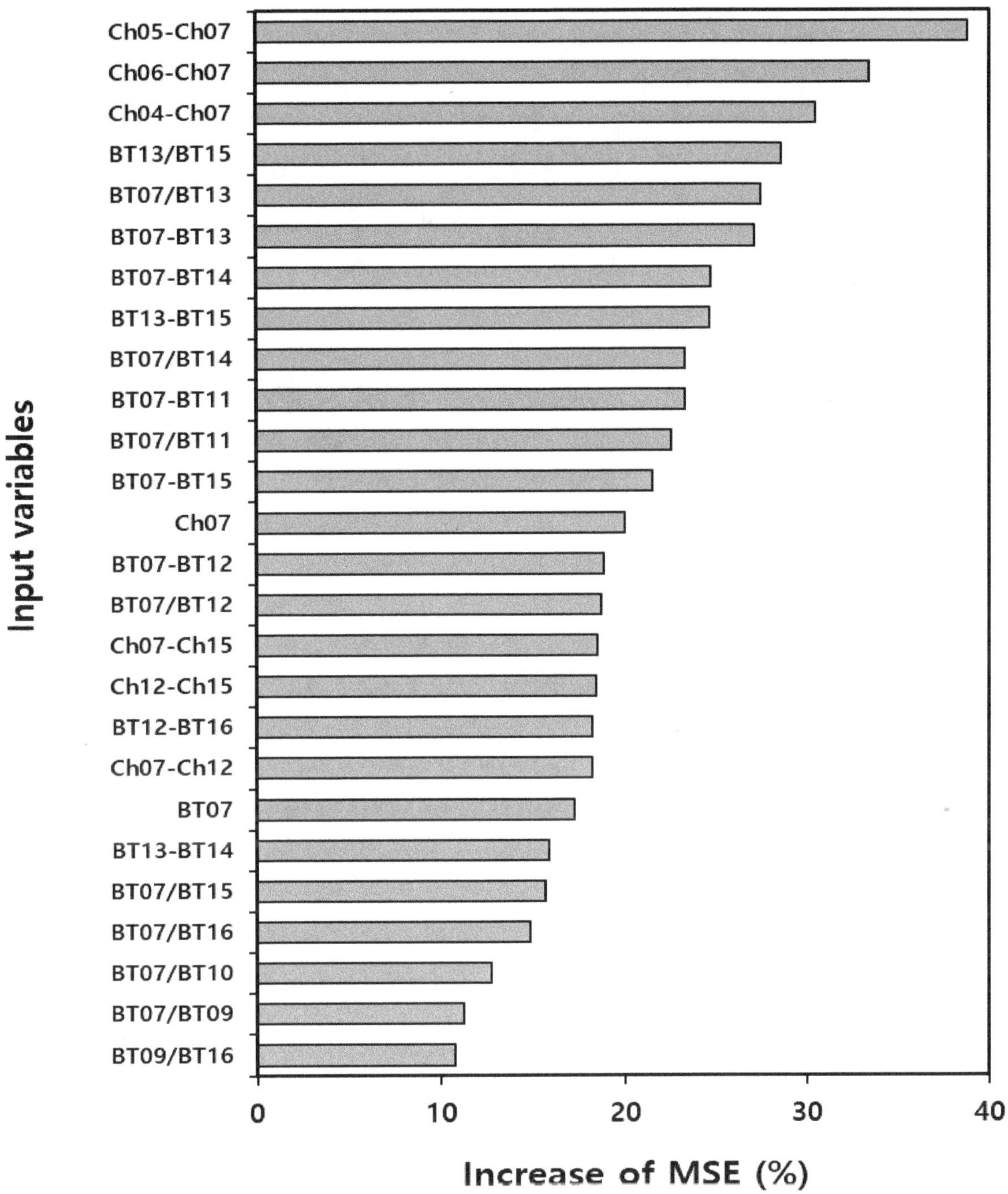

Figure 4. Variable importance for removal of false alarms using the RF model. Increase of mean squared error (MSE) was calculated using out of bag (OOB) data. More detailed information about the increase of MSE%) is given in Section 3.3.

Table 3 shows the accuracy assessment results of the RF model using the calibration and validation data. The RF model resulted in 100% training accuracy. The validation accuracy was also high (OA = 99.16%, POD = 93.08%, and POFD = 0.07%) with 27 forest fire pixels (7% of reference forest fire pixels) being classified as non-fire pixels. From this result, one of the 14 forest fires (validation cases) detected by the threshold algorithm were removed after the RF model was combined. Figure 5 shows that the RF model effectively removed false alarms for the validation forest fire cases (Figure 5b,f) when compared to the results of the threshold-based algorithm (Figure 5a,e,i,m). However, a few false alarms (Figure 5j,n) still remained. For the validation cases, about 96% of the false alarmed pixels from the thresholding results were successfully removed by RF.

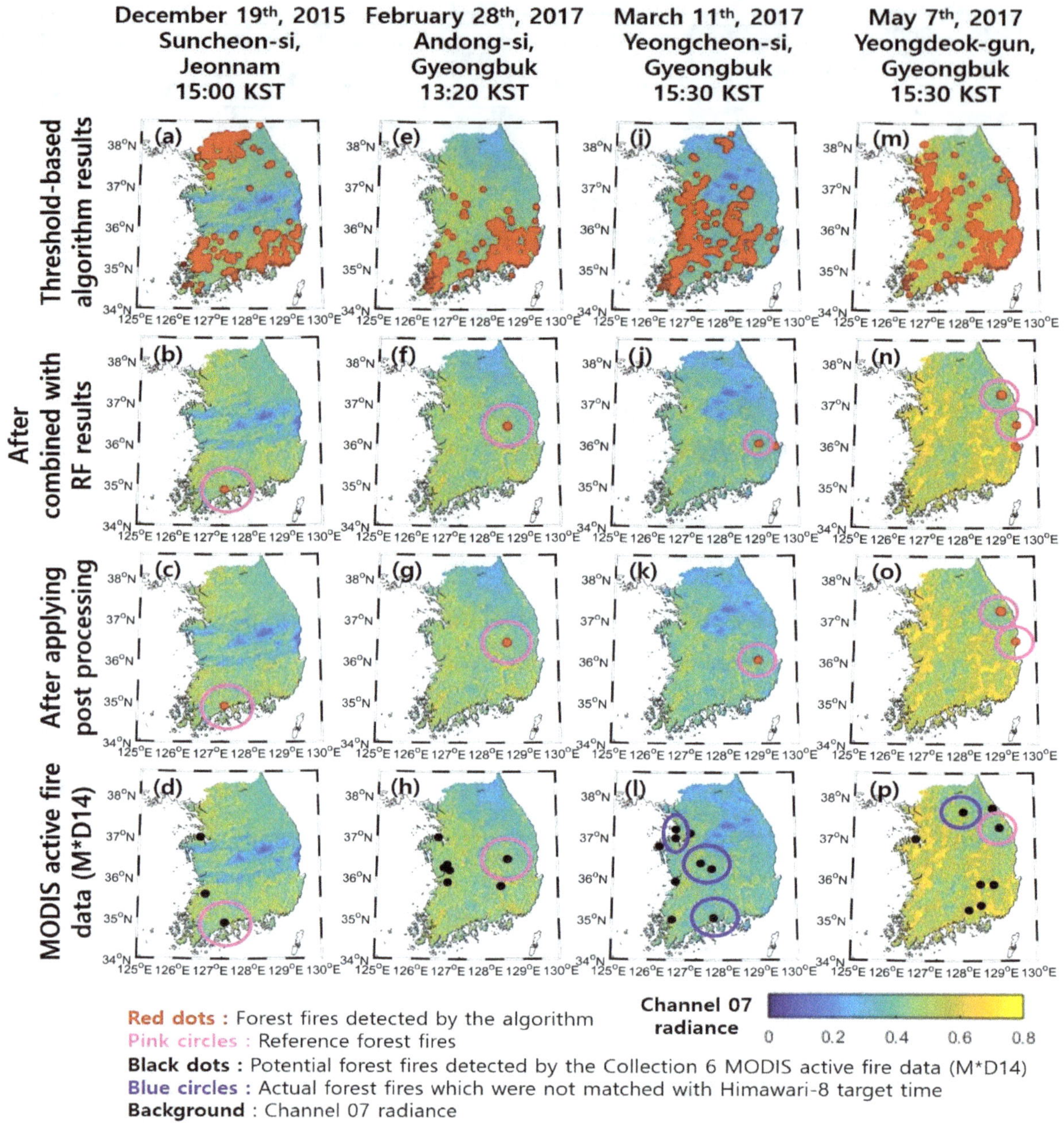

Figure 5. The maps of detected forest fires after each step: the threshold-based algorithm (**a**,**e**,**i**,**m**), RF modeling (**b**,**f**,**j**,**n**), and post processing (**c**,**g**,**k**,**o**) and the MODIS active fire data (M*D14; **d**,**h**,**l**,**p**). (1) the forest fire (Suncheon-si, Jeonnam; from 13:13 to 16:00 (KST) on 19th October 2015) occurred by shaman rituals with the damaged area ~1 ha (**a**,**b**,**c**,**d**); (2) the forest fire (Andong-si, Gyeongbuk; from 13:10 to 15:20 (KST) on 28th February 2017) occurred due to an unknown reason with the damaged area ~0.8 ha (**e**,**f**,**g**,**h**); (3) forest fire (Yeongcheon-si, Gyeongbuk; from 15:13 to 18:30 (KST) on 11th March 2017) occurred due to the incineration of agricultural waste with the damaged area ~5.2 ha (**i**,**j**,**k**,**l**); and (4) the forest fire (upper; Gangneung-si, Gangwon; from 15:32 on 6th to 6:34 (KST) on 9th May 2017) occurred for an unknown reason with the damaged area ~252 ha and the forest fire (lower; Yeongdeok-gun, Gyeongbuk; from 14:45 on 7th to 7:00 (KST) on 8th May 2017) occurred due to a cigarette with the damaged area ~5.9 ha (**m**,**n**,**o**,**p**). The red dots are the potential forest fire pixels detected by the proposed algorithm, pink circles show actual forest fire cases from the reference data, black dots are the potential forest fire pixels detected by the Collection 6 MODIS active fire data, blue circles show actual forest fire cases which were not matched with Himawari-8 target time, and the band 7 radiance of Himawari-8 AHI of each image is used as a background image. These five forest fire cases of four dates come from the validation data, which were not used in training of the RF model.

Table 3. Accuracy assessment results of best combinations of RF based on variable importance. (OA = Overall accuracy (%); POD = Probability of detection (%)).

		Reference			
		Fire	No fire	Sum	
Calibration	Fire	1775	0	1775	OA = 100% POD = 100% POFD = 0%
	Non-fire	0	15,043	15,043	
	Sum	1775	15,043	16,818	
Validation	Fire	363	2	365	OA = 99.16% POD = 93.08% POFD = 0.07%
	Non-fire	27	3040	3067	
	Sum	390	3042	3432	

Finally, after the post processing using the forest region map, 13 of the 14 forest fires (validation cases) were detected using the 3-step forest fire detection algorithm, and an additional 5 small-scale forest fires (damaged areas were 0.02–0.3 ha), which were not included in the reference data, were detected. About 64% of the remaining false alarms were further removed by incorporating the post processing with the results of the threshold-based and RF approaches. Among 50 calibration forest fire cases, two forest fires were removed when the post-processing was applied. The location of these two fire cases were very close to the dense urban areas, and thus, removed due to the coarse resolution of the AHI images. In addition, since the forest map used in this study is not 100% accurate, there might be false alarms or mis-detection of forest fires caused by using the map. Nonetheless, the post processing based on the forest map resulted in an increase in POD and a decrease in false alarms. If more accurate and higher resolution forest data can be used, the performance of the proposed approach could be further enhanced.

The final results of forest fire detection were compared to two existing algorithms (refer to Section 3.5; Table 4). Among 14 validation forest fires, 13 forest fires were detected using the proposed approach, while 7 and 8 forest fires were detected by the COMS algorithm and the AHI-FSA algorithm, respectively (Table 4). Among the 12 validation small-scale forest fires (damaged area < 8 ha), 11 forest fires were detected using the proposed 3-step algorithm, while 5 and 6 forest fires were detected by the two existing algorithms respectively. The POD of the proposed 3-step algorithm was higher than the two existing algorithms. Two of the five additional small-scale forest fires detected by the proposed approach were also detected by the AHI-FSA algorithm. However, none of them was detected by the COMS algorithm.

This implies that the proposed approach works well for small-scale fires when compared to the existing algorithms. The same Himawari-8 AHI sensor was used to detect the same forest fire cases, but the algorithm proposed in this study detected the forest fires better than the AHI-FSA algorithm. The final results of the proposed 3-step algorithm were also compared to the Collection 6 MODIS fire products (Figure 5d,h,l,p). Among 14 validation data, 6 forest fires (5 small-scale forest fires) were detected by MODIS. Among 8 forest fires which were not detected by MODIS, 3 forest fires were not detected because MODIS didn't pass at the time of forest fires. This implies that the use of geostationary satellite data has great potential in the real time monitoring of forest fires.

Table 4. The number of detected forest fire and the detection rate of the proposed 3-step algorithm, the COMS algorithm, and the AHI-FSA algorithm.

		3-Step Algorithm	COMS Algorithm	AHI-FSA Algorithm
Validation forest fires (14)	The number of detected forest fire	13	7	8
	Detection rate	93%	50%	57%
	Average damaged area	13.29 ha	22 ha	20.14 ha
Small-scale validation forest fires (12)	The number of detected forest fire	11	5	6
	Detection rate	92%	42%	50%

We further applied the 3-step algorithm to the Himawari-8 data collected from January to February in 2018 (i.e., more recent than the research period used in the study), and the results were compared to the Collection 6 MODIS active fire data (Table S1). Among new 18 reference forest fires, 12 forest fires were detected using the proposed approach (the detection rate was 66.7%), while 6 forest fires were detected by the Collection 6 MODIS active fire data (the detection rate was 33.3%). Among detected 12 forest fires, 9 forest fire cases were clearly detected without false alarms. Four of the 6 undetected forest fires were detected by the threshold-based algorithm, but they were excluded when the RF model was applied. The relatively lower detection rate of the proposed model when compared to its results for the previous years can be explained by the fact that only a few training samples from January and February were used to train the model. The detection rate can increase when the RF model is improved with more training data. Please note that forest fires in other months in 2018 were not tested because Himawari-8 time-series data were not always available to the public. The proposed 3-step algorithm was also applied to the East Asia and it detected reference fires well (pink circle in Figure S1) and compared with the Collection 6 MODIS active fire data (M*D14). We got the information about the reference forest fires in China from the website of the China Forest Fire Management [51]. The first forest fire (Figure S1b,c; [52]) was detected by both 3-step algorithm (Himawari-8 target time was 17:50 (UTC)) and MODIS/Aqua active fire data (passing time was 17:55 (UTC)). The second forest fire (Figure S1d,e; [53]) was detected by 3-step algorithm (Himawari-8 target time was 18:30 (UTC)) but not by MODIS/Aqua active fire data (passing time was 18:30 (UTC)). Other forest fires detected by the 3-step algorithm and MODIS data have no reference data, and thus, their accuracy is unknown.

4.2. Monitoring of Forest Fires

Since time series data with a 10 min interval (Himawari-8 AHI sensor) were used, the lead time on how early the proposed algorithm detected fires was examined. Among the 52 forest fires detected by the proposed approach, shows the number of forest fires with respect to initial detection time, and 25 forest fires were detected within 10 min after fires occurred and 39 forest fires were detected within 30 min. The Samcheok forest fire, which is the largest forest fire from the reference data (i.e., the damaged area was 765.12 ha and the duration was longer than 3 days), and other forest fires (with a damaged area range from 0.8 to 252 ha) were detected within 10 min. This shows that detection is generally possible within a short period of time after a forest fire has broken out. The average initial detection time using the proposed approach was about 24 min (median value was 20 min). The averaged initial detection time increased due to several forest fires with detection times of more than 30 min. There was no significant correlation between the initial detection time and the starting time, location, and size of the forest fires. When we carefully examined the high resolution Google Earth images before and after the forest fires, many late-detected (with the initial detection time of longer than 30 min) forest fires had little difference between the before and after images, which implies that the forest fires occurred mostly under the canopy and did not show significant difference in remotely sensed images at the canopy level during the initial period of fires. It should be noted that Himawari-8 can scan the focus area including Japan and Korea about every 2 min, and thus, there is a greater chance of reducing the initial detection time in the future using more dense time series data.

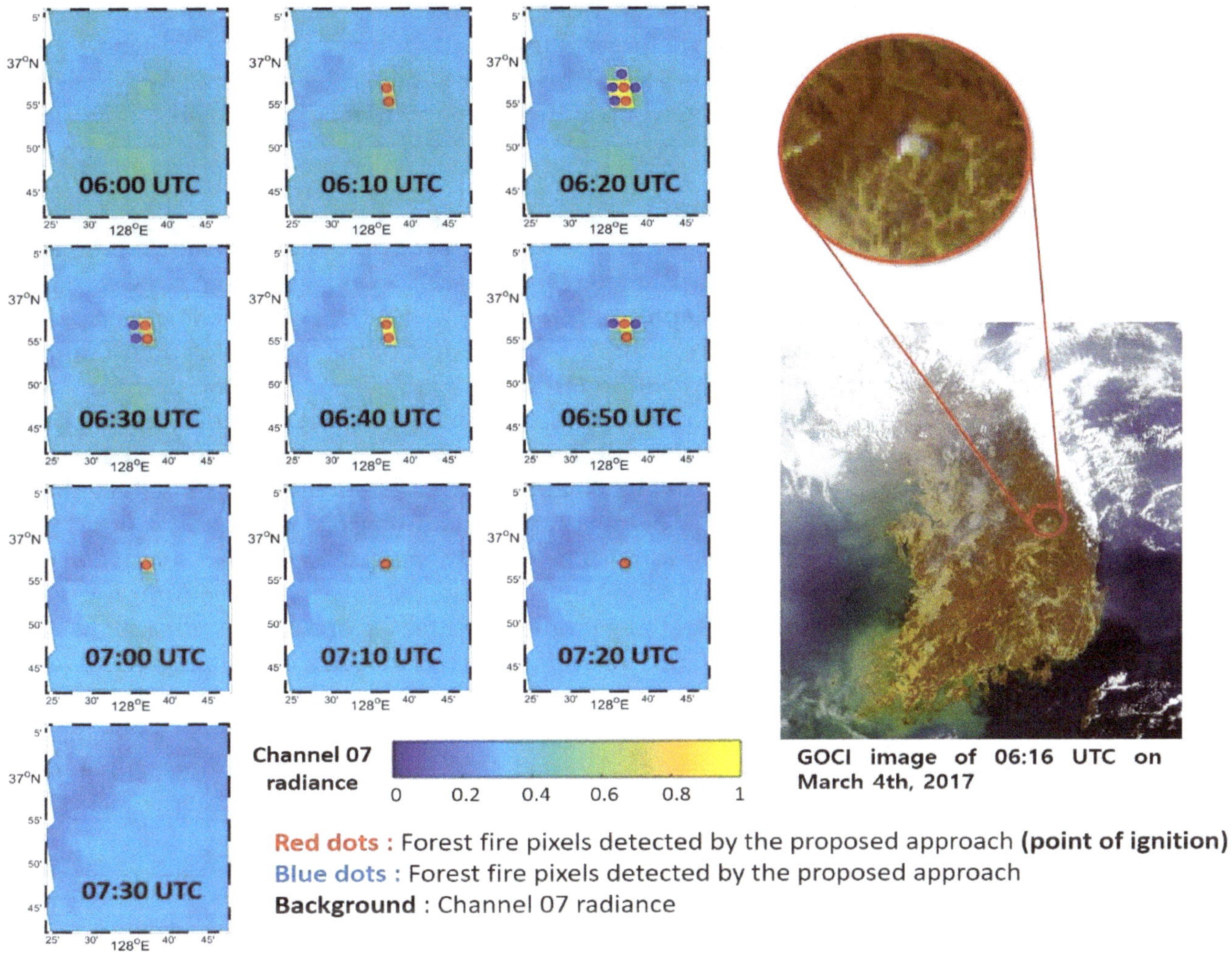

Figure 6. Forest fire in Yeongju-si, Gyeongbuk and Geostationary Ocean Color Imager (GOCI) image on the 4th March 2017. The fire monitoring results from 6:00 to 7:20 UTC by the proposed algorithm (red and blue dots) are shown with the band 7 radiance of Himawari-8 AHI as the background.

4.3. Novelty and Limitations

This study proposed an integrated approach for the detection of small to large-scale forest fires in South Korea. The proposed algorithm detected forest fires in South Korea better than the other two existing algorithms, especially for small-scale forest fires. The proposed approach consists of three steps. In the threshold-based algorithm, as the first step, an adaptive thresholding approach was adopted for each image considering the diurnal cycle and seasonality, unlike the existing threshold-based algorithm. While the first step resulted in very high POD and false alarms, the following RF model and post processing effectively removed the false alarms (Figure 5). In particular, the post processing using the forest map and the filtering approach was very useful for South Korea, which has a complex and rugged terrain with small patches of land cover. This study showed promising results that more advanced geostationary satellite sensor systems with higher spatial (<1 km) and temporal (~few minutes) resolutions can be used to monitor even small forest fires, i.e., less than 1 ha. Since high temporal resolution geostationary satellite data are used in the proposed approach, the early detection and spreading direction of fires can be identified by the monitoring results (Figure 6), which can be used to provide appropriate information for rapid response. [54] calculated a wildfire spread rate and burned area using Himawari-8 satellite data and active fire data developed by [15]. They defined the burned area and fire center using active fire data. This method demonstrates Himawari-8 data is useful for computing the fire spread rate. The burned area mapping and fire spread rate calculation can be combined with active fire data produced by our proposed 3-step algorithm to better manage forest fires. When approximately two-minute interval images by Himawari-8 or Geo-Kompsat-2A (GK-2A) satellite sensor systems (rapid scan mode; GK2 satellite successfully launched on 5 December 2018) are available, the proposed approach can contribute to faster initial detection and monitoring of forest fires.

However, there are some limitations in this study. First, clouds are always problematic when optical sensor data are used, as is the case in the proposed approach. Second, adaptive threshold values in the threshold-based algorithm might not work well for very small fires when large forest fires occur at the same time. Local tuning of the adaptive thresholding should be conducted before applying the proposed approach to different areas. Providing a certain range of the thresholds might improve the performance of forest fire detection. Third, although the RF model showed good performance in reducing false alarms, it did not detect a few forest fires. Such mis-detection by the RF model can be improved by using additional forest fire and non-forest fire samples for training because the RF model is an empirical model, and requires new training when applied to different areas. Although the locally optimized algorithm has the disadvantage of being a time-consuming process, it can produce high accuracy in the target study area. Fourth, two forest fires were not detected and were removed by the post processing because they were not included in the forest map. Considering the complex terrain and patched land cover in the small size of South Korea, a more precise forest map with a higher level of accuracy can mitigate such a problem. Finally, very small forest fires (i.e., damaged areas less than 0.7 ha) are hard to detect due to the limitation of spatial resolution of input geostationary satellite data. Higher spatial resolution (e.g., 500 m) thermal data from geostationary satellite sensor systems than Himawari-8 may further improve the detection of very small forest fires especially from an operational forest fire monitoring perspective in South Korea where small-scale forest fires frequently occur.

5. Conclusions

In this study, a combined 3-step algorithm (threshold-based algorithm, RF model, and post processing) was proposed to detect and monitor forest fires in South Korea using Himawari-8 geostationary satellite data. Existing forest fire detection algorithms using satellite data are not used in the operational monitoring system in South Korea due to the high rate of false alarms, mis-detection of small-scale forest fires, and the low temporal resolution of satellite data. This proposed 3-step algorithm using geostationary satellite data provides a basis for use in the operational forest fire monitoring system. The early detection and spreading direction of fires using high temporal resolution of geostationary satellite data enables efficient rapid response. The active fire data resulting from the 3-step algorithm can be used to calculate the size of the burned area and fire spread rates. Such information is of great help for efficient forest fire monitoring, extinguishment, and recovery management. Although the 3-step algorithm proposed in this study is locally optimized, it is necessary to effectively detect and monitor forest fires in a study area such as South Korea, where the environmental characteristics are unique in terms of land cover, topography, and climate. This algorithm can be extended to the rest of East Asia after refining all three steps (i.e., tuning adaptive threshold values, RF modeling with additional samples, and post-processing using a fine resolution forest map of East Asia).

More than half of the detected forest fires were detected within 10 min, which is a promising result when the operational real-time monitoring of forest fires using more advanced geostationary satellite sensor data is considered for the rapid response and management of forest fires. The algorithm proposed in this study can be optimized and used for the Geo-Kompsat-2 Advanced Meteorological Imager (AMI), a new geostationary meteorological satellite, which was successfully launched on 4th December 2018 by the Korean Meteorological Administration. The satellite has similar specifications to the Himawari-8 AHI sensor, and provides data every 10 min in full disk, and approximately every 2 min in the focusing area around the Korean peninsula, which can be useful for continuous forest fire detection and monitoring.

Author Contributions: E.J. and Y.K. equally contributed to the paper. They led manuscript writing and contributed to the data analysis and research design. J.I. supervised this study, contributed to the research design and manuscript writing, and served as the corresponding author. D.-W.L., J.Y., and S.-K.K. contributed to the discussion of the results.

Acknowledgments: This study was supported by grants from the Space Technology Development Program and the Basic Science Research Program through the National Research Foundation of Korea (NRF) funded by the Ministry of Science, ICT, & Future Planning and the Ministry of Education of Korea, respectively (NRF-2017M1A3A3A02015981; NRF-2017R1D1A1B03028129); from the National Institute of Environmental Research (NIER), funded by the Ministry of Environment (MOE) of the Republic of Korea (NIER-2017-01-02-063). This work was also supported by the Development of Geostationary Meteorological Satellite Ground Segment (NMSC-2014-01) program, funded by the National Meteorological Satellite Centre (NMSC) of the Korea Meteorological Administration (KMA).

Appendix A

List of Collection 6 MODIS fire products in South Korea from March to May in 2017.

Date	Actual Fires	Detected Fires by MODIS	Correctly Detected by MODIS	Falsely Detected by MODIS	Miss Detected Fires
2nd March	0	1	0	1	0
4th March	1	14	0	14	1
5th March	0	4	0	4	0
6th March	0	2	0	2	0
7th March	0	13	0	13	0
8th March	1	0	0	0	1
9th March	2	6	1	5	1
10th March	8	4	0	4	8
11th March	7	12	5	7	2
12th March	3	1	0	1	3
13th March	2	7	0	7	2
14th March	3	11	0	11	3
15th March	4	2	1	1	3
16th March	5	6	0	6	5
17th March	4	4	0	4	4
18th March	2	4	1	3	1
19th March	10	4	1	3	9
21st March	1	4	0	4	1
22nd March	1	3	0	3	1
23rd March	2	1	0	1	2
26th March	1	0	0	0	1
27th March	1	2	0	2	1
28th March	1	0	0	0	1
29th March	3	0	0	0	3
30th March	3	7	0	7	3

Date	Actual Fires	Detected Fires by MODIS	Correctly Detected by MODIS	Falsely Detected by MODIS	Miss Detected Fires
1st April	1	1	0	1	1
3rd April	1	13	0	13	1
4th April	6	2	0	2	6
5th April	1	0	0	0	1
6th April	0	1	0	1	0
7th April	1	3	0	3	1
8th April	1	4	0	4	1
9th April	0	2	0	2	0
10th April	3	3	0	3	3
11th April	0	1	0	1	0
12th April	2	4	1	3	1
13th April	4	1	0	1	4
15th April	1	2	0	2	1
19th April	0	14	0	14	0
21th April	0	1	0	1	0
22th April	1	1	0	1	1
23th April	3	2	0	2	3
24th April	1	8	1	7	0
26th April	3	4	0	4	3
27th April	2	0	0	0	2
28th April	3	16	1	15	2
29th April	3	2	0	2	3
30th April	6	7	1	6	5
1st May	1	3	0	3	1
2nd May	1	4	0	4	1
3rd May	5	7	1	6	4
4th May	1	0	0	0	1
5th May	1	3	0	3	1
6th May	6	2	2	0	4
7th May	4	10	2	8	2
8th May	3	2	0	2	3
9th May	1	0	0	0	1
11th May	0	1	0	1	0
14th May	0	14	0	14	0
15th May	0	1	0	1	0
17th May	0	4	0	4	0
18th May	0	1	0	1	0
19th May	1	5	1	4	0
20th May	1	2	1	1	0

Date	Actual Fires	Detected Fires by MODIS	Correctly Detected by MODIS	Falsely Detected by MODIS	Miss Detected Fires
21th May	3	3	1	2	2
23th May	0	2	0	2	0
24th May	0	2	0	2	0
25th May	1	1	0	1	1
26th May	2	4	0	4	2
27th May	2	2	0	2	2
28th May	1	10	1	9	0
29th May	1	0	0	0	1
30th May	2	1	0	1	2
Total	**145**	**288**	**22**	**266**	**123**

Appendix B

List of forest fires used as reference data in this study provided by the Korea Forest Service.

Location	Ignition Date	Ignition Time (UTC)	Extinguished Date	Extinguished Time (UTC)	Cause	Damaged Area (ha)
64 reference forest fires						
Yeongok-myeon, Gangneung-si, Gangwon-do	17th October 2015	0:20	17th October 2015	6:00	Unknown cause	0.8
Byeollyang-myeon, Suncheon-si, Jeollanam-do	19th October 2015	4:20	19th October 2015	6:10	Shaman rituals	1
Dong-myeon, Chuncheon-si, Gangwon-do	4th February 2016	4:40	4th February 2016	6:50	Other	1
Ucheon-myeon, Hoengseong-gun, Gangneung-si, Gangwon-do	5th February 2016	7:00	5th February 2016	8:50	Waste incineration	0.8
Buseok-myeon, Yeongju-si, Gyeongsangbuk-do	7th February 2016	6:30	7th February 2016	7:20	Agricultural Waste Incineration	1.5
Jungbu-myeon, Gwangju-si, Gyeonggi-do	26th February 2016	1:00	26th February 2016	3:20	Arson	2.7
Geumgwang-myeon, Anseong-si, Gyeonggi-do	16th March 2016	6:50	16th March 2016	8:00	Waste incineration	2
Yeongyang-eup, Yeongyang-gun, Gyeongsangbuk-do	27th March 2016	7:20	27th March 2016	8:10	Agricultural Waste Incineration	0.7
Gimhwa-eup, Cheorwon-gun, Gangwon-do	28th March 2016	4:10	28th March 2016	5:40	Climber accidental fire	2
Namdong-gu, Incheon Metropolitan City	29th March 2016	21:00	29th March 2016	22:30	The others	1
Hwado-eup, Namyangju-si, Gyeonggi-do	30th March 2016	3:50	30th March 2016	6:30	Agricultural Waste Incineration	0.8

Location	Ignition Date	Ignition Time (UTC)	Extinguished Date	Extinguished Time (UTC)	Cause	Damaged Area (ha)
Oeseo-myeon, Sangju-si, Gyeongsangbuk-do	30th March 2016	5:50	31th March 2016	9:40	Paddy field incineration	92.6
Sanae-myeon, Hwacheon-gun, Gangwon-do	31th March 2016	4:00	31th March 2016	5:00	The others	1.5
Jangheung-myeon, Yangju-si, Gyeonggi-do	31th March 2016	5:30	31th March 2016	9:30	The others	8.3
Nam-myeon, Yanggu-gun, Gangwon-do	1st April 2016	3:50	1st April 2016	5:50	The others	14.4
Gonjiam-eup, Gwangju-si, Gyeonggi-do	1st April 2016	2:20	1st April 2016	5:30	Paddy field incineration	2.6
Seolseong-myeon, Icheon-si, Gyeonggi-do	1st April 2016	4:00	1st April 2016	6:40	Waste incineration	1
Kim Satgat myeon, Yeongwol-gun, Gangwon-do	2nd April 2016	6:30	2nd April 2016	7:50	The others	1
Seo-myeon, Hongcheon-gun, Gangwon-do	2nd April 2016	5:20	2nd April 2016	7:50	Work place accidental fire	3.9
Gapyeong-eup, Gapyeong-gun, Gyeonggi-do	2nd April 2016	6:00	2nd April 2016	9:00	The others	7
Opo-eup, Gwangju-si, Gyeonggi-do	2nd April 2016	4:20	3rd April 2016	7:50	Ancestral tomb visitor accidental fire	2
Chowol-eup, Gwangju-si, Gyeonggi-do	2nd April 2016	5:50	2nd April 2016	8:00	The others	1
Dong-gu, Daejeon Metropolitan City	2nd April 2016	6:00	3rd April 2016	8:00	The others	4.8
Mosan-dong, Jecheon-si, Chungcheongbuk-do	2nd April 2016	5:40	2nd April 2016	8:00	Ancestral tomb visitor accidental fire	4.7
Suanbo-myeon, Chungju-si, Chungcheongbuk-do	5th April 2016	6:10	6th April 2016	9:40	Waste incineration	53.8
Nam-myeon, Jeongson-Gun, Gangwon-do	14th May 2016	6:20	14th May 2016	7:50	Work place accidental fire	2
Yeongchun-myeon, Danyang-gun, Chungcheongbuk-do	22th May 2016	3:00	23th May 2016	12:20	Wild edible greens collector accidental fire	13
Dongi-myeon, Okcheon-gun, Chungcheongbuk-do	22th May 2016	4:40	22th May 2016	10:20	The others	1
Jinbu-myeon, Pyeongchang-gun, Gangwon-do	30th May 2016	4:50	22th May 2016	6:50	Waste incineration	1
Jipum-myeon, Yeongdeok-gun, Gyeongsangbuk-do	4th February 2017	4:10	4th February 2017	7:10	The others	0.98
Iljik-myeon, Andong-si, Gyeongsangbuk-do	28th February 2017	4:10	4th February 2017	6:20	The others	0.8

Location	Ignition Date	Ignition Time (UTC)	Extinguished Date	Extinguished Time (UTC)	Cause	Damaged Area (ha)
Buseok-myeon, Yeongju-si, Gyeongsangbuk-do	4th March 2017	6:00	4th March 2017	7:20	Agricultural Waste Incineration	2
Jangseong-eup, Jangseong-gun, Jeollanam-do	6th March 2017	8:00	4th March 2017	8:00	The others	1
Okgye-myeon, Gangneung-si, Gangwon-do	9th March 2017	1:30	10th March 2017	13:30	The others	160.41
Saengyeon-dong, Dongducheon-si, Gyeonggi-do	11th March 2017	1:30	11th March 2017	4:00	Waste incineration	0.72
Hwanam-myeon, Yeongcheon-si, Gyeongsangbuk-do	11th March 2017	6:20	11th March 2017	7:40	Paddy field incineration	5.2
Wolgot-myeon, Gimpo-si, Gyeonggi-do	18th March 2017	6:30	18th March 2017	7:20	Paddy field incineration	3
Seojong-myeon, Yangpyeong-gun, Gyeonggi-do	18th March 2017	8:00	18th March 2017	8:50	Waste incineration	2
Hanam-myeon, Hwacheon-gun, Gangwon-do	19th March 2017	2:00	19th March 2017	5:50	Agricultural Waste Incineration	1.5
Buk-myeon, Gapyeong-gun, Gyeonggi-do	19th March 2017	5:40	19th March 2017	7:30	Agricultural Waste Incineration	2
Baekseok-eup, Yangju-si, Gyeonggi-do	19th March 2017	4:10	19th March 2017	6:30	Climber accidental fire	0.9
Beopjeon-myeon, Bonghwa-gun, Gyeongsangbuk-do	22th March 2017	7:10	22th March 2017	7:10	The others	2.2
Dain-myeon, Uiseong-gun, Gyeongsangbuk-do	23th March 2017	5:30	23th March 2017	6:00	Paddy field incineration	1.5
Namyang, Hwaseong-si, Gyeonggi-do	3rd April 2017	5:50	3rd April 2017	8:00	Waste incineration	2.5
Noseong-myeon, Nonsan-si, Chungcheongnam-do	3rd April 2017	7:30	3rd April 2017	9:10	The others	0.8
Buk-myeon, Gapyeong-gun, Gyeonggi-do	23th April 2017	3:40	23th April 2017	7:30	Climber accidental fire	1.5
Goesan-eup, Goesan-gun, Chungcheongbuk-do	26th April 2017	8:20	26th April 2017	13:10	The others	2
Gonjiam-eup, Gwangju-si, Gyeonggi-do	28th April 2017	2:20	28th April 2017	6:50	The others	1
Jojong-myeon, Gapyeong-gun, Gyeonggi-do	29th April 2017	5:10	29th April 2017	7:20	Climber accidental fire	2
Dogye-eup, Samcheok-si, Gangwon-do	6th May 2017	2:50	9th May 2017	13:30	The others	765.12
Seongsan-myeon, Gangneung-si, Gangwon-do	6th May 2017	6:40	9th May 2017	17:30	The others	252

Location	Ignition Date	Ignition Time (UTC)	Extinguished Date	Extinguished Time (UTC)	Cause	Damaged Area (ha)
Tongjin-eup, Gimpo-si, Gyeonggi-do	6th May 2017	6:50	6th May 2017	7:50	The others	1
Sabeol-myeon, Sangju-si, Gyeongsangbuk-do	6th May 2017	5:10	8th May 2017	13:30	Agricultural Waste Incineration	86
Gaeun-eup, Mungyeong-si, Gyeongsangbuk-do	6th May 2017	7:30	6th May 2017	9:30	Agricultural Waste Incineration	1.5
Yeonghae-myeo, Yeongdeok-gun, Gyeongsangbuk-do	7th May 2017	5:50	7th May 2017	9:00	Cigarette accidental fire	5.9
Seonnam-myeon Seongju-gun, Gyeongsangbuk-do	4th June 2017	3:10	4th June 2017	11:50	Waste incineration	2
Munui-myeon, Sangdang-gu, Cheongju-si, Chungcheongbuk-do	11th June 2017	14:30	11th June 2017	17:50	The others	3.12
Miwon-myeon, Sangdang-gu, Cheongju-si, Chungcheongbuk-do	14th June 2017	12:10	14th June 2017	15:10	The others	0.7
Hwanam-myeon, Yeongcheon-si, Gyeongsangbuk-do	23th November 2017	20:40	23th November 2017	23:50	The others	0.8
Hyeonbuk-myeon, Yangyang-gun, Gangwon-do	4th December 2017	10:40	4th December 2017	12:10	House fire spread	1.86
Sicheon-myeon, Sancheong-gun, Gyeongsangnam-do	5th December 2017	21:30	5th December 2017	3:50	The others	5
Buk-gu, Ulsan Metropolitan City	12th December 2017	14:50	12th December 2017	23:40	The others	18
Gogyeong-myeon, Yeongcheon-si, Gyeongsangbuk-do	16th December 2017	8:30	16th December 2017	10:50	The others	1.89
Gaejin-myeon, Goryeong-gun, Gyeongsangbuk-do	19th December 2017	5:00	19th December 2017	7:30	Climber accidental fire	1.5
5 additionally detected forest fires						
Bibong-myeon, Wanju-gun, Jeollabuk-do	16th March 2016	6:20	16th March 2016	8:30	Agricultural Waste Incineration	0.2
Dosan-myeon, Andong-si, Gyeongsangbuk-do	30th March 2016	8:47	30th March 2016	10:00	Paddy field incineration	0.02
Sari-myeon, Goesan-gun, Chungcheongbuk-do	1st April 2016	5:10	1st April 2016	7:45	The others	0.3
Sosu-myeon, Goesan-gun, Chungcheongbuk-do	5th April 2016	7:20	5th April 2016	8:50	Waste incineration	0.1
Hyeonsan-myeon, Haenam-gun, Jeollanam-do	19th March 2017	3:55	19th March 2017	5:25	Agricultural Waste Incineration	0.03

Appendix C

List of 191 input variables—band radiance, BT, band ratios, BT differences and BT ratios of bands 4–16—which were used for identifying input data to the RF model.

Band radiance (13)	Ch04	Ch05	Ch06	Ch07
	Ch08	Ch09	Ch10	Ch11
	Ch12	Ch13	Ch14	Ch15
	Ch16			
Band ratios (78)	Ch04/Ch05	Ch04/Ch06	Ch04/Ch07	Ch04/Ch08
	Ch04/Ch09	Ch04/Ch10	Ch04/Ch11	Ch04/Ch12
	Ch04/Ch13	Ch04/Ch14	Ch04/Ch15	Ch04/Ch16
	Ch05/Ch06	Ch05/Ch07	Ch05/Ch08	Ch05/Ch09
	Ch05/Ch10	Ch05/Ch11	Ch05/Ch12	Ch05/Ch13
	Ch05/Ch14	Ch05/Ch15	Ch05/Ch16	
	Ch06/Ch07	Ch06/Ch08	Ch06/Ch09	Ch06/Ch10
	Ch06/Ch11	Ch06/Ch12	Ch06/Ch13	Ch06/Ch14
	Ch06/Ch15	Ch06/Ch16		
	Ch07/Ch08	Ch07/Ch09	Ch07/Ch10	Ch07/Ch11
	Ch07/Ch12	Ch07/Ch13	Ch07/Ch14	Ch07/Ch15
	Ch07/Ch16			
	Ch08/Ch09	Ch08/Ch10	Ch08/Ch11	Ch08/Ch12
	Ch08/Ch13	Ch08/Ch14	Ch08/Ch15	Ch08/Ch16
	Ch09/Ch10	Ch09/Ch11	Ch09/Ch12	Ch09/Ch13
	Ch09/Ch14	Ch09/Ch15	Ch09/Ch16	
	Ch10/Ch11	Ch10/Ch12	Ch10/Ch13	Ch10/Ch14
	Ch10/Ch15	Ch10/Ch16		
	Ch11/Ch12	Ch11/Ch13	Ch11/Ch14	Ch11/Ch15
	Ch11/Ch16			
	Ch12/Ch13	Ch12/Ch14	Ch12/Ch15	Ch12/Ch16
	Ch13/Ch14	Ch13/Ch15	Ch13/Ch16	
	Ch14/Ch15	Ch14/Ch16		
	Ch15/Ch16			
BT (10)	BT07	BT08	BT09	BT10
	BT11	BT12	BT13	BT14
	BT15	BT16		
BT differences (45)	BT07-BT08	BT07-BT09	BT07-BT10	BT07-BT11
	BT07-BT12	BT07-BT13	BT07-BT14	BT07-BT15
	BT07-BT16			
	BT08-BT09	BT08-BT10	BT08-BT11	BT08-BT12
	BT08-BT13	BT08-BT14	BT08-BT15	BT08-BT16
	BT09-BT10	BT09-BT11	BT09-BT12	BT09-BT13
	BT09-BT14	BT09-BT15	BT09-BT16	
	BT10-BT11	BT10-BT12	BT10-BT13	BT10-BT14
	BT10-BT15	BT10-BT16		
	BT11-BT12	BT11-BT13	BT11-BT14	BT11-BT15
	BT11-BT16			
	BT12-BT13	BT12-BT14	BT12-BT15	BT12-BT16
	BT13-BT14	BT13-BT15	BT13-BT16	
	BT14-BT15	BT14-BT16		
	BT15-BT16			

	BT07/BT08	BT07/BT09	BT07/BT10	BT07/BT11
	BT07/BT12	BT07/BT13	BT07/BT14	BT07/BT15
	BT07/BT16			
	BT08/BT09	BT08/BT10	BT08/BT11	BT08/BT12
	BT08/BT13	BT08/BT14	BT08/BT15	BT08/BT16
	BT09/BT10	BT09/BT11	BT09/BT12	BT09/BT13
	BT09/BT14	BT09/BT15	BT09/BT16	
BT ratios (45)	BT10/BT11	BT10/BT12	BT10/BT13	BT10/BT14
	BT10/BT15	BT10/BT16		
	BT11/BT12	BT11/BT13	BT11/BT14	BT11/BT15
	BT11/BT16			
	BT12/BT13	BT12/BT14	BT12/BT15	BT12/BT16
	BT13/BT14	BT13/BT15	BT13/BT16	
	BT14/BT15	BT14/BT16		
	BT15/BT16			

References

1. Shin, W. *2015 Forest Standard Statistics*; Ryu, G., Ed.; Korea Forest Service: Daejeon Metropolitan City, Korea, 2016.
2. Kim, G. A Study on Wildfire Detection Using Geostationary Meteorological Satellite. Master's Thesis, Pukyoung National University, Busan Metropolitan City, Korea, 2015.
3. Lee, J.; Park, D. *2017 Statistical Yearbook of Forest Fire*; Kim, J., Lee, S., Nam, M., Eds.; Korea Forest Service: Daejeon Metropolitan City, Korea, 2018.
4. Leblon, B.; San-Miguel-Ayanz, J.; Bourgeau-Chavez, L.; Kong, M. Remote sensing of wildfires. In *Land Surface Remote Sensing*; Elsevier: Amsterdam, The Netherlands, 2016; pp. 55–95.
5. Leblon, B.; Bourgeau-Chavez, L.; San-Miguel-Ayanz, J. Use of remote sensing in wildfire management. In *Sustainable Development-Authoritative and Leading Edge Content for Environmental Management*; InTech: London, UK, 2012.
6. Di Biase, V.; Laneve, G. Geostationary sensor based forest fire detection and monitoring: An improved version of the SFIDE algorithm. *Remote Sens.* **2018**, *10*, 741. [CrossRef]
7. Filizzola, C.; Corrado, R.; Marchese, F.; Mazzeo, G.; Paciello, R.; Pergola, N.; Tramutoli, V. RST-FIRES, an exportable algorithm for early-fire detection and monitoring: Description, implementation, and field validation in the case of the MSG-SEVIRI sensor. *Remote Sens. Environ.* **2017**, *192*, e2–e25. [CrossRef]
8. Giglio, L.; Descloitres, J.; Justice, C.O.; Kaufman, Y.J. An enhanced contextual fire detection algorithm for MODIS. *Remote Sens. Environ.* **2003**, *87*, 273–282. [CrossRef]
9. Giglio, L.; Schroeder, W.; Justice, C.O. The collection 6 MODIS active fire detection algorithm and fire products. *Remote Sens. Environ.* **2016**, *178*, 31–41. [CrossRef] [PubMed]
10. Na, L.; Zhang, J.; Bao, Y.; Bao, Y.; Na, R.; Tong, S.; Si, A. Himawari-8 satellite based dynamic monitoring of grassland fire in China-Mongolia border regions. *Sensors* **2018**, *18*, 276. [CrossRef] [PubMed]
11. Schroeder, W.; Oliva, P.; Giglio, L.; Csiszar, I.A. The new VIIRS 375 m active fire detection data product: Algorithm description and initial assessment. *Remote Sens. Environ.* **2014**, *143*, 85–96. [CrossRef]
12. Schroeder, W.; Oliva, P.; Giglio, L.; Quayle, B.; Lorenz, E.; Morelli, F. Active fire detection using Landsat-8/OLI data. *Remote Sens. Environ.* **2016**, *185*, 210–220. [CrossRef]
13. Wickramasinghe, C.H.; Jones, S.; Reinke, K.; Wallace, L. Development of a multi-spatial resolution approach to the surveillance of active fire lines using Himawari-8. *Remote Sens.* **2016**, *8*, 932. [CrossRef]
14. Wickramasinghe, C.; Wallace, L.; Reinke, K.; Jones, S. Implementation of a new algorithm resulting in improvements in accuracy and resolution of SEVIRI hotspot products. *Remote Sens. Lett.* **2018**, *9*, 877–885. [CrossRef]
15. Xu, G.; Zhong, X. Real-time wildfire detection and tracking in Australia using geostationary satellite: Himawari-8. *Remote Sens. Lett.* **2017**, *8*, 1052–1061. [CrossRef]
16. Hally, B.; Wallace, L.; Reinke, K.; Jones, S. Assessment of the utility of the advanced Himawari imager to detect active fire over Australia. *Int. Arch. Photogramm. Remote Sens. Spat. Inf. Sci.* **2016**, *41*, 65–71. [CrossRef]

17. Hally, B.; Wallace, L.; Reinke, K.; Jones, S.; Skidmore, A. Advances in active fire detection using a multi-temporal method for next-generation geostationary satellite data. *Int. J. Digit. Earth* **2018**, 1–16. [CrossRef]
18. Roberts, G.; Wooster, M. Development of a multi-temporal Kalman filter approach to geostationary active fire detection & fire radiative power (FRP) estimation. *Remote Sens. Environ.* **2014**, *152*, 392–412.
19. Xie, Z.; Song, W.; Ba, R.; Li, X.; Xia, L. A spatiotemporal contextual model for forest fire detection using Himawari-8 satellite data. *Remote Sens.* **2018**, *10*, 1992. [CrossRef]
20. Miller, J.; Borne, K.; Thomas, B.; Huang, Z.; Chi, Y. Automated wildfire detection through Artificial Neural Networks. In *Remote Sensing and Modeling Applications to Wildland Fires*; Springer: New York, NY, USA, 2013; pp. 293–304.
21. Atwood, E.C.; Englhart, S.; Lorenz, E.; Halle, W.; Wiedemann, W.; Siegert, F. Detection and characterization of low temperature peat fires during the 2015 fire catastrophe in Indonesia using a new high-sensitivity fire monitoring satellite sensor (FireBird). *PLoS ONE* **2016**, *11*, e0159410. [CrossRef]
22. Kim, S. Development of an Algorithm for Detecting Sub-Pixel Scale Forest Fires Using MODIS Data. Ph.D. Thesis, Inha University, Incheon Metropolitan City, Korea, 2009.
23. Huh, Y.; Lee, J. Enhanced contextual forest fire detection with prediction interval analysis of surface temperature using vegetation amount. *Int. J. Remote Sens.* **2017**, *38*, 3375–3393. [CrossRef]
24. Seoul Broadcasting System (SBS) News Website. Available online: http://news.sbs.co.kr/news/endPage.do?news_id=N1004722267&plink=ORI&cooper=NAVER (accessed on 9 January 2019).
25. Korean Statistical Information Service Home Page. Available online: http://kosis.kr/statHtml/statHtml.do?orgId=101&tblId=DT_2KAA101&conn_path=I2 (accessed on 10 January 2019).
26. Lee, C. *2018 National Park Standard Statistics*; Ryu, G., Ed.; Korea National Park Service: Gangwon-do, Korea, 2018.
27. Korea Meteorological Administration. Available online: http://www.weather.go.kr/weather/climate/average_south.jsp (accessed on 27 August 2018).
28. Korea Forest Service Home Page. Available online: http://www.forest.go.kr/newkfsweb/html/HtmlPage.do?pg=/policy/policy_0401.html&mn=KFS_38_05_04 (accessed on 17 January 2019).
29. Environmental Geographic Information Service Home Page. Available online: http://www.index.go.kr/search/search.jsp (accessed on 9 January 2019).
30. Peterson, D.; Wang, J.; Ichoku, C.; Hyer, E.; Ambrosia, V. A sub-pixel-based calculation of fire radiative power from MODIS observations: 1: Algorithm development and initial assessment. *Remote Sens. Environ.* **2013**, *129*, 262–279. [CrossRef]
31. Lee, S.; Han, H.; Im, J.; Jang, E.; Lee, M.-I. Detection of deterministic and probabilistic convection initiation using Himawari-8 Advanced Himawari Imager data. *Atmos. Meas. Tech.* **2017**, *10*, 1859–1874. [CrossRef]
32. Koltunov, A.; Ustin, S.L.; Quayle, B.; Schwind, B.; Ambrosia, V.G.; Li, W. The development and first validation of the GOES Early Fire Detection (GOES-EFD) algorithm. *Remote Sens. Environ.* **2016**, *184*, 436–453. [CrossRef]
33. Amraoui, M.; DaCamara, C.; Pereira, J. Detection and monitoring of African vegetation fires using MSG-SEVIRI imagery. *Remote Sens. Environ.* **2010**, *114*, 1038–1052. [CrossRef]
34. Polivka, T.N.; Wang, J.; Ellison, L.T.; Hyer, E.J.; Ichoku, C.M. Improving nocturnal fire detection with the VIIRS day–night band. *IEEE Trans. Geosci. Remote Sens.* **2016**, *54*, 5503–5519. [CrossRef]
35. Forkuor, G.; Dimobe, K.; Serme, I.; Tondoh, J.E. Landsat-8 vs. Sentinel-2: Examining the added value of Sentinel-2's red-edge bands to land-use and land-cover mapping in Burkina Faso. *GIScі. Remote Sens.* **2018**, *55*, 331–354. [CrossRef]
36. Jang, E.; Im, J.; Park, G.-H.; Park, Y.-G. Estimation of fugacity of carbon dioxide in the East Sea using in situ measurements and Geostationary Ocean Color Imager satellite data. *Remote Sens.* **2017**, *9*, 821. [CrossRef]
37. Liu, T.; Abd-Elrahman, A.; Morton, J.; Wilhelm, V.L. Comparing fully convolutional networks, random forest, support vector machine, and patch-based deep convolutional neural networks for object-based wetland mapping using images from small unmanned aircraft system. *GIScі. Remote Sens.* **2018**, *55*, 243–264. [CrossRef]
38. Richardson, H.J.; Hill, D.J.; Denesiuk, D.R.; Fraser, L.H. A comparison of geographic datasets and field measurements to model soil carbon using random forests and stepwise regressions (British Columbia, Canada). *GIScі. Remote Sens.* **2017**, *54*, 573–591. [CrossRef]

39. Zhang, C.; Smith, M.; Fang, C. Evaluation of Goddard's lidar, hyperspectral, and thermal data products for mapping urban land-cover types. *GISci. Remote Sens.* **2018**, *55*, 90–109. [CrossRef]
40. Breiman, L. Random forests. *Mach. Learn.* **2001**, *45*, 5–32. [CrossRef]
41. Guo, Z.; Du, S. Mining parameter information for building extraction and change detection with very high-resolution imagery and GIS data. *GISci. Remote Sens.* **2017**, *54*, 38–63. [CrossRef]
42. Park, S.; Im, J.; Park, S.; Yoo, C.; Han, H.; Rhee, J. Classification and mapping of paddy rice by combining Landsat and SAR time series data. *Remote Sens.* **2018**, *10*, 447. [CrossRef]
43. Yoo, C.; Im, J.; Park, S.; Quackenbush, L.J. Estimation of daily maximum and minimum air temperatures in urban landscapes using MODIS time series satellite data. *ISPRS J. Photogramm. Remote Sens.* **2018**, *137*, 149–162. [CrossRef]
44. Park, S.; Im, J.; Park, S.; Rhee, J. Drought monitoring using high resolution soil moisture through multi-sensor satellite data fusion over the Korean peninsula. *Agric. For. Meteorol.* **2017**, *237*, 257–269. [CrossRef]
45. Sonobe, R.; Yamaya, Y.; Tani, H.; Wang, X.; Kobayashi, N.; Mochizuki, K.-I. Assessing the suitability of data from Sentinel-1A and 2A for crop classification. *GISci. Remote Sens.* **2017**, *54*, 918–938. [CrossRef]
46. Zhang, D.; Liu, X.; Wu, X.; Yao, Y.; Wu, X.; Chen, Y. Multiple intra-urban land use simulations and driving factors analysis: A case study in Huicheng, China. *GISci. Remote Sens.* **2018**, 1–27. [CrossRef]
47. Georganos, S.; Grippa, T.; Vanhuysse, S.; Lennert, M.; Shimoni, M.; Kalogirou, S.; Wolff, E. Less is more: Optimizing classification performance through feature selection in a very-high-resolution remote sensing object-based urban application. *GISci. Remote Sens.* **2018**, *55*, 221–242. [CrossRef]
48. Rhee, J.; Im, J. Meteorological drought forecasting for ungauged areas based on machine learning: Using long-range climate forecast and remote sensing data. *Agric. For. Meteorol.* **2017**, *237*, 105–122. [CrossRef]
49. Sim, S.; Im, J.; Park, S.; Park, H.; Ahn, M.H.; Chan, P.W. Icing detection over East Asia from geostationary satellite data using machine learning approaches. *Remote Sens.* **2018**, *10*, 631. [CrossRef]
50. Murphy, S.W.; de Souza Filho, C.R.; Wright, R.; Sabatino, G.; Pabon, R.C. HOTMAP: Global hot target detection at moderate spatial resolution. *Remote Sens. Environ.* **2016**, *177*, 78–88. [CrossRef]
51. China Forest Fire Management Home Page. Available online: http://www.slfh.gov.cn (accessed on 10 January 2019).
52. China Forest Fire Management Home Page. Available online: http://www.slfh.gov.cn/Item/24197.aspx (accessed on 10 January 2019).
53. China Forest Fire Management Home Page. Available online: http://www.slfh.gov.cn/Item/25469.aspx (accessed on 10 January 2019).
54. Liu, X.; He, B.; Quan, X.; Yebra, M.; Qiu, S.; Yin, C.; Liao, Z.; Zhang, H. Near real-time extracting wildfire spread rate from Himawari-8 satellite data. *Remote Sens.* **2018**, *10*, 1654. [CrossRef]

5

Machine Learning Approaches for Detecting Tropical Cyclone Formation using Satellite Data

Minsang Kim [1,2], Myung-Sook Park [3], Jungho Im [1], Seonyoung Park [4] and Myong-In Lee [1,*]

1 School of Urban and Environmental Engineering, Ulsan National Institute of Science and Technology, Ulsan 44919, Korea; mskinn@korea.kr (M.K.); ersgis@unist.ac.kr (J.I.)
2 Satellite Analysis Division, National Meteorological Satellite Center, Jincheon-gun, Chungcheongbuk-do 27803, Korea
3 Korea Ocean Satellite Center, Korea Institute of Ocean Science and Technology, Busan 49111, Korea; mspark@kiost.ac.kr
4 Satellite Operation and Application Center, Korea Aerospace Research Institute, Daejeon 34133, Korea; sypark90@kari.re.kr
* Correspondence: milee@unist.ac.kr.

Abstract: This study compared detection skill for tropical cyclone (TC) formation using models based on three different machine learning (ML) algorithms-decision trees (DT), random forest (RF), and support vector machines (SVM)-and a model based on Linear Discriminant Analysis (LDA). Eight predictors were derived from WindSat satellite measurements of ocean surface wind and precipitation over the western North Pacific for 2005–2009. All of the ML approaches performed better with significantly higher hit rates ranging from 94 to 96% compared with LDA performance (~77%), although false alarm rate by MLs is slightly higher (21–28%) than that by LDA (~13%). Besides, MLs could detect TC formation at the time as early as 26–30 h before the first time diagnosed as tropical depression by the JTWC best track, which was also 5 to 9 h earlier than that by LDA. The skill differences across MLs were relatively smaller than difference between MLs and LDA. Large yearly variation in forecast lead time was common in all models due to the limitation in sampling from orbiting satellite. This study highlights that ML approaches provide an improved skill for detecting TC formation compared with conventional linear approaches.

Keywords: tropical cyclone formation; WindSat; machine learning

---

1. Introduction

A tropical cyclone (TC) can lead to tremendous economic losses and casualties when it makes a landfall [1,2]. Since any tropical disturbance with a sufficient magnitude has a potential to be developed abruptly into a TC over the warm ocean, it is highly desirable to have an accurate forecast system for TC formation for a timely warning to the public. There are several approaches to predict TC formation. One is to use numerical weather prediction (NWP) models, which has been significantly improved in the past years due to the advance in modeling techniques and physics parameterizations, more available satellite data for the better initialization, and enhanced computing resource. For example, Halperin et al. [3] investigated the forecast skill for TC genesis in the North Atlantic from 2004 to 2011 by five NWP models, and showed that their conditional probability of hit ranged from 26% to 44%. Nevertheless, it still remains challenging for most NWP models to predict whether a tropical disturbance will develop to a TC or just decay as a non-developer [4,5].

An alternative is to use various types of statistical models that have been developed to predict the TC formation based on large-scale meteorological conditions identified as important processes for TC formation. Schumacher et al. [6] produced an estimation of 24-h probability of TC formation in each $5° \times 5°$ sub-region of various ocean basins. They combined large-scale environmental parameters derived from the National Center for Environmental Prediction (NCEP)-National Center for Atmospheric Research (NCAR) reanalysis [7], and multiple geostationary satellites. Similarly, Hennon and Hobgood [8] considered eight large-scale predictors such as latitude, daily genesis potential, maximum potential intensity, moisture divergence, precipitable water, pressure tendency, surface vorticity tendency, and 700-hPa vorticity tendency from 6-hourly global NCEP-NCAR reanalysis [7]. The foregoing studies [6,8] commonly used a Linear Discriminant Analysis (LDA) classifier in the prediction of TC formation. Fan [9] applied a multi-linear regression method from large-scale environmental factors based on NCEP-NCAR reanalysis data for detecting the Atlantic TC formation.

While the above-mentioned statistical methods require a linear statistical relationship between predictors (e.g., large-scale environmental parameters) and predictand variables (e.g., tropical cyclone formation possibility), TC formation involves complicated multi-scale interactions from large-scale environmental to mesoscale convective processes [10]. Hennon et al. [11] pointed out limitations in the use of the LDA or linear regression approaches due to the assumption of the linear relationship between predictors and predictands. On the other hand, machine learning (ML) approaches do not require any assumption in contrast to the conventional statistical techniques based on linear models [12]. There are several ML approaches widely used such as decision trees (DT), random forest (RF), and support vector machines (SVM). While DT is a simple classifier that recursively partitions data into subsets based on tree-like decision rules, RF uses a bootstrapping method to make an ensemble of classification trees [13]. SVM, which is a non-parametric statistical learning technique, builds a hyperplane to separate the dataset into a discrete, predefined number of classes. The hyperplane is adjusted for minimizing misclassifications during a training procedure [14]. Very recently, the ML approach has been applied to the model for classifying TC formation [15,16]. Zhang et al. [16] utilized DT to classify developing and non-developing tropical disturbances using several predictors (e.g., maximum 800-hPa relative vorticity, sea surface temperature, precipitation rate, vertically averaged divergence, and air temperature at 300 hPa) derived from the Navy Operational Global Atmospheric Prediction System analysis data [17]. They showed that the accuracy in the forecast of TC formation prior to 24 h was about 84.6%. Park et al. [15] also applied the DT technique for detecting TC formation. They trained DT rules using system-representative parameters such as the symmetry of low-level circulation pattern, intensity, and the organization indices that derived from the sea surface wind data from WindSat [18].

The dependency of ML approaches has been tested in various remote sensing applications. For example, Han et al. [19] developed a convective initiation algorithm from the Communication, Ocean, and Meteorological Satellite Meteorological Imager [20] based on the three ML approaches—DT, RF, and SVM. Their results showed that RF produced a slightly higher hit rate (HR) than DT with a comparable false alarm rate (FAR), while SVM resulted in relatively poor performance with much higher FAR compared with the other two techniques. On the other hand, Sesnie et al. [21] compared SVM with RF to classify 11 Costa Rican tropical rainforest type using the Landsat thematic mapper bands and the normalized difference vegetation index, where they found that SVM performed better with higher accuracy than RF. This suggests that there is no single best ML technique applicable to all cases, and the performance of the ML algorithm depends not only on the technique but also on the type of application and input data.

This study is an extension from the study of Park et al. [15]. Although it demonstrated a skill level practically useful in detecting TC formation, it was limited by using a single ML algorithm (i.e., DT) and could not evaluate the room for further improvement when the other algorithms were tested. In this study, three ML approaches have been tested to examine the dependence on the technique. An identical dataset of predictors is applied to the ML algorithms and calibrated independently, and then validated for comparing the detection skill. In addition, the performance is also compared with that by LDA with a quantitative skill assessment. TC formation processes are complicatedly involved with instability in the atmospheric and oceanic dynamics and thermodynamic processes that are not necessarily linear [22]. This study hypothesizes that ML, which can account for nonlinear relationships among the predictors [23,24], should be more suitable than LDA for the application to tropical cyclogenesis forecast.

The following section will describe data, methods for deriving predictors, and constructing the models. Section 3 compares the performance of four TC formation detection models (i.e., LDA and 3 ML-based models). A summary of the study is given in Section 4.

2. Data and Methods

2.1. Data and Preprocessing

To characterize tropical disturbances at their formation stage of TC, this study used all-weather wind speed and direction at 10 meter above the ocean surface and rain rate measured by the WindSat polarimetric microwave radiometer (http://www.remss.com/missions/windsat) onboard the Department of Defense Coriolis satellite developed by the Naval Research Laboratory, the Naval Center for Space Technology for the U.S. Navy, and the National Polar-orbiting Operational Environmental Satellite System (NPOESS) Integrated Program Office (IPO). The satellite was launched on 6 January 2003 and has been operating in the present time. Ocean surface wind vector product covers a wide area with an average swath width of 950 km. Retrieving ocean surface wind signal under rainy conditions has been a long-standing challenge for passive microwave radiometers due to the contamination by rain. The WindSat radiometer uses fully-polarimetric channels at 10.7, 18.7, and 37.0 GHz and dual-polarimetric channels at 6.8, and 23.8 GHz. Using multiple channels, particularly C-band (4 to 8 GHz) and X-band (8 to 12 GHz) where the atmospheric attenuation is relatively small, allows developing an algorithm for retrieving wind speed in hurricanes even under heavy rain to a reasonable degree of accuracy. The wind speed retrieval accuracy for tropical cyclones ranges from 2.0 m s^{-1} in light rain to 4.0 m s^{-1} in heavy rain. It also shows no degradation of wind speed signal at wind speeds up to 35 m s^{-1}, well above the magnitude of the tropical depression. The liquid water precipitation is derived by using 18.7, 23.8, and 37.0 GHz channels. The horizontal resolution of the WindSat measurement is 39 km $\times$ 71 km for 6.8 GHz, 25 km $\times$ 38 km for 10.7 GHz, 16 km $\times$ 27 km for 18.7 GHz, 20 km $\times$ 30 km km for 23.8 GHz, and 8 km $\times$ 13 km for 37.0 GHz, respectively. The dataset has been re-processed to $0.25^\circ \times 0.25^\circ$ gridded data for the current analysis.

This study uses 1,325 WindSat overpass images including approximately 630 tropical disturbances in the western North Pacific during the years of 2005–2009. The satellite observations are collocated with the system track data from the Joint Typhoon Warning Center (JTWC) best track, and Tropical Cloud Cluster [25]. A tropical disturbance is classified as a developing (DEV) disturbance when the maximum sustained wind (MSW) speed will be larger than 13 m s^{-1} later to be denoted as a tropical depression (TD) (announced by JTWC). In addition, the image needs to cover sufficiently large area near the center of disturbance (at least 60% of non-missing data within 4-degree radius of circle) at least once before TD stage and the data need to be sampled again no later than 72 h after TD stage. The remaining disturbances are defined as non-developing disturbances (non-DEV). In the case of non-DEV disturbance, all available satellite observations are collected.

Park et al. [15] used eight specific predictors that quantitatively describe dynamic and hydrological characteristics related to TC formation, and determine either DEVs or non-DEVs based on the DT-based rules. This study uses the same predictors, and a brief description of them is given here. Two indices, "wind_ave" and "rain_ave", are aimed to represent the intensity of low-level wind and rain rate near the tropical disturbance center, respectively. To quantify the degree of symmetry in the low-level circulation for developing cyclones, two indices are designed by calculating circular variance (CV), "wind_cv_fix" and "wind_cv_mv", in which the former calculated CV over a full domain ($16^\circ \times 16^\circ$ Lat./Lon.) centered at the target disturbance for characterizing synoptic-scale circulation, and the latter over a smaller subdomain ($4^\circ \times 4^\circ$ Lat./Lon.) moving around the disturbance for characterizing mesoscale circulation. The remaining four indices include "wind_ci", "wind_pladj", "rain_ci", and "rain_pladj", which represent the clumpiness index (CI) and the percentage of like adjacencies (PLADJ) for wind and rain rate, respectively. These two measures are defined from Fragstats statistics [26]. Both PLADJ and CI quantify the degree of organization of strong wind and heavy rainfall areas. Table 1 summarizes the eight predictors used in this study.

Table 1. Description of predictors that quantitatively describe dynamic and hydrological characteristics related to tropical cyclone (TC) formation.

Predictor	Description
wind_ave	Average of wind speed over the disturbance
wind_cv_fix	Circular variance (CV) of wind to measure the degree of symmetry in the circulation within a $16° \times 16°$ fixed widnow
wind_cv_mv	The maximum CV value found by moving a small ($4° \times 4°$) window over a larger area
wind_ci	Clumpiness Index (CI) of the wind speed over 15 m s^{-1} in a large area
wind_pladj	Percentage of Like ADJacency (PLADJ) of wind speed over 15 m s^{-1} in a large area
rain_ave	Average of rain rates near the center of the disturbance
rain_ci	Clumpiness Index (CI) of the rain rate exceeding 5 mm h^{-1} in a large area
rain_paldj	Percentage of Like ADJacency (PLADJ) of the rain rate exceeding 5 mm h^{-1} in a large area

2.2. *Sampling*

Training a prediction model should not be significantly influenced by insufficient samples. To examine the sample dependence, this study performed the k-fold cross validation, where all available WindSat data archived for five years were divided into the training and the validating sub-datasets [27]. Verification of the model fitted with different samples can help estimate the uncertainty and the sample dependence of the statistical model in a quantitative manner as well as evaluate the year-to-year variability of the prediction skill. Three-year data randomly selected out of five years were used for training, and the remaining two-year data were used for validation. This resulted in 10 different training-validation datasets to be tested. In the original dataset, the number of non-DEV was about three times larger than that of DEV. As the training with the ML approaches such as DT using unequal samples between DEV and non-DEV may cause a bias in the model [28], non-DEVs were resampled randomly as much as DEVs for training. Resampling was repeated for 100 times for each 10 different k-folding datasets, leading to 1000 datasets total for model development and validation for each TC formation detection model.

2.3. *Model Construction*

2.3.1. Linear Discriminant Analysis (LDA)

LDA has been widely used in a variety of classification studies [7,29,30]. LDA is a statistical technique to classify objects into two groups based on a set of predictors with specific threshold values. LDA projection has a known caveat that the classification ability is decreasing for the non-linear relationship between predictors and the predictand [31]. To construct a model based on LDA, multiple regression models were performed in this study given the eight potential predictors. The predictors were tested following procedures to avoid overfitting problems. All the possible combinations of predictors were tested in terms of the validating statistics using HR, FAR, and Peirce Skill Score [32] (hereafter referred to as PSS) and then the predictors of less significance were eliminated one by one.

2.3.2. Decision Trees (DT)

DT has been introduced in many recent remote sensing studies for both classification and regression [15,33–37], in which the data sample is subject to the partitioning into subdivisions repeatedly based on decision rules, resembling branches in a tree [38]. The advantage of the DT is to enable easy interpretation and physical insights to the classification rules as it provides visible if-then rules with the relative importance of predictors. This study used the C5.0 program developed by RuleQuest Research, Inc [39]. Unlike the LDA, DT (the other two ML approaches as well) selects an optimal combination of predictors empirically by itself through training. To evaluate the degree of fitting for 1000 different training datasets, overall accuracy is calculated which represents the ratio of samples correctly classified.

2.3.3. Random Forests (RF)

RF is an ensemble approach based on classification and regression trees (CART), which was originally designed to improve the well-known problem of DT that is sensitive to training data configuration [40]. RF has been widely used in remote sensing applications for both classification and regression tasks [41–44]. RF adopts two randomization processes to develop numerous independent decision trees and the final decision is made by majority voting (or weighted voting) strategy [19,45]. RF also provides the relative importance of a specific variable, as represented quantitatively by the mean decrease of accuracy when the variable is permuted to random values. The critical predictor has a higher variable importance than others. This study used the R software [46] to fit the data into the RF algorithm.

As the RF model contains a random process in the training, each round of training was likely to produce no identical result even with the same data. Each set of training dataset were repeated five times to produce the RF model for obtaining reliable results.

2.3.4. Support Vector Machines (SVM)

SVM is one of the widely-used machine learning models in remote sensing applications in recent years [47–51], which finds an optimal hyperplane to classify data [52]. SVM utilizes a kernel function to transform the data dimension into a higher one in order to identify an optimal hyperplane effectively [53]. Among the linear, polynomial, and radial basis functions, the radial basis function was turned out to be the best as a kernel function in our test. The kernel and penalty parameters used in the SVM model were automatically adjusted during the data training process to ensure the best performance in detecting tropical cyclone formation. Before data training, each predictor variable was linearly scaled to the range from 0 to 1 in order to consider the difference in magnitude across the variables. This procedure also helps reduce computational time in optimizing SVM model parameters. Compared with other machine learning algorithms, SVM does not provide the information on the relative importance of predictors instantaneously. As an alternative, the F-score test is applied to identify the major discriminating features that characterize the tropical cyclone formation based on the SVM method. The F score, also called the F1 score or F measure, is a measure of a test's accuracy, which is defined as the weighted harmonic mean of the test's precision and recall [54]. From the 2×2 contingency table of observed versus the model-classified number of DEVs and non-DEVs, precision is the number of correct DEV forecasts divided by all forecast DEVs, and recall is the correct DEV forecasts divided by all observed DEVs. The SVM model can make different results according to the kernel optimization procedure. Each sample was repeatedly trained five times to produce reliable output from the SVM model. This study used the library for the SVM software package, LIBSVM [55] version 3.22 available at http://www.csie.ntu.edu.tw/~{}cjlin/libsvm.

2.4. *Verification Methods*

The performance of the trained models is compared using HR, FAR, and PSS from the 2×2 contingency table of observed versus the model-classified number of DEVs and non-DEVs. HR is the number of model-classified DEVs divided by the number of observed DEVs. FAR is defined as the number of the model-classified DEVs divided by the number of observed non-DEVs. The PSS score [32] is based on the proportion correct as the basic accuracy measure and it is defined as the relative improvement over the reference forecast. The perfect score receives PSS =1, forecasts equivalent to the reference forecast receive zero scores, and forecasts worse than the reference forecasts receive negative scores. The reference accuracy measure is the proportion correct that would be achieved by unbiased random forecasts. It can be easily shown that the PSS is simply the difference between HR and FAR, and it provides a combined measure of the forecast skill. The detection lead time by model for each observed DEV was calculated, which was defined as a time difference between the first TC formation detection and the time when it actually developed to TD according to the JTWC best track. Negative (positive) lead time means that the model detection is made before (after) the actual TC formation.

3. Results

3.1. Model Calibration

Using eight predictors with the 1000 datasets, each TC formation model was trained independently using LDA and the three ML approaches. In calibrating the LDA model, all the possible combinations of potential predictors were tested in advance. Figure 1 compares the performance for various combinations of predictors and the number of predictors. HR increases significantly by reducing the number of predictors (Figure 1a). By contrast, FAR increased gradually in general. However, the lowest (best) values were found when the number of input predictors is 3 or 4 (Figure 1b). As seen in Figure 1c, PSS results indicated that the number of three predictors with the use of wind_ave, wind_cv_fix, and wind_pladj showed the best performance among the all other sets. Table 2 shows the average and standard deviation of coefficients for each predictor after applying the model for 1000 different training datasets. This model in Table 2 was applied to the validation dataset.

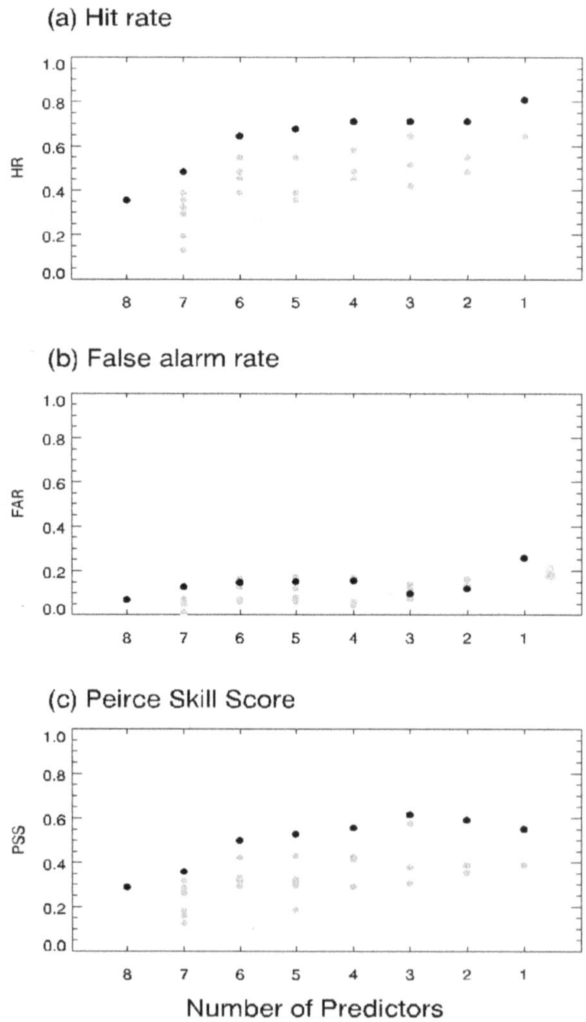

Figure 1. (**a**) Hit rate (HR), (**b**) false alarm rate (FAR), and (**c**) the Peirce skill score (PSS) compared for different selections of potential predictors in the Linear Discriminant Analysis (LDA) model. Each gray dot shows the performance of a different combinations of predictors and the black dot represents the highest performance in terms of PSS.

Table 2. Average and standard deviation of coefficients in the linear discriminant analysis equation using for training sample datasets.

LDA = R_1 × wind_ave + R_2 × wind_cvfix + R_3 × wind_pladj − constant				
	R_1	R_2	R_3	constant
Average	0.321	5.223	0.048	8.776
Standard deviation	4.517	0.806	0.124	1.215

The training results for 1000 different training datasets for the DT model showed the accuracy (i.e., the ratio of samples correctly classified) ranging from 72.9% to 91.1%. In each training, DT selected a different set of predictors. Nevertheless, the wind_cv_fix was included in all tests and wind_ave was included 83% of the tests. On the other hand, rain_ave and rain_ci were selected least as 33% and 30%, respectively. Although the major advantage of DT is to display its rules, each algorithm contains different combination of predictors and thresholds. Figure 2 shows an example that consists of three if-then decision rules used for binary classification of tropical disturbances into DEV or non-DEV. This case selected only two variables of wind_cv_fix, and wind_ave. As mentioned, these are the most frequent variables in the DT training process and they were also used in Park et al. [15].

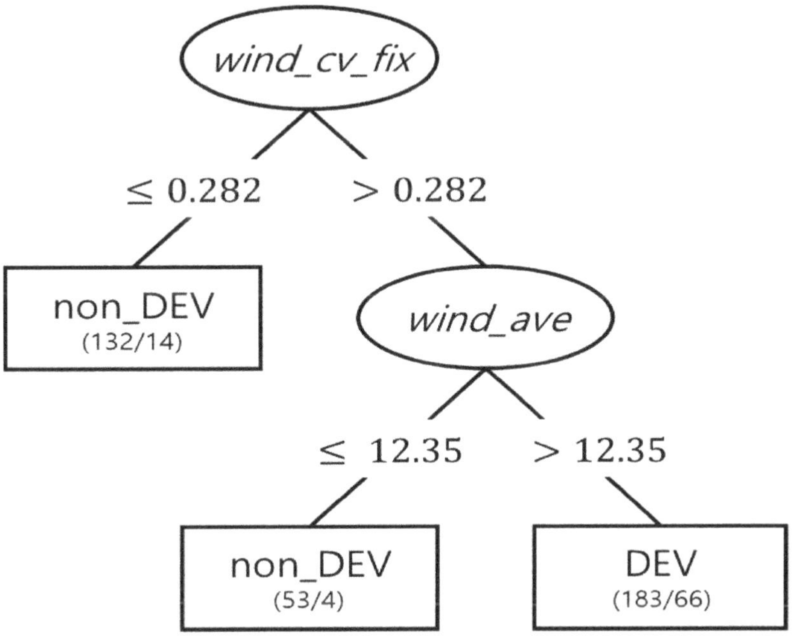

Figure 2. An example of the decision tree that classify developing (DEV) or non-DEV tropical disturbances, built by the decision trees (DT) model trained with a subset data for 2005, 2007, and 2009. The eclipses contained selected predictors and rectangles represented the number of corrected classified and misclassified sample. Two predictors—wind_cv_fix, and wind_ave—were selected from the training with the percentage value of importance 100%, and 68%, respectively.

In developing a RF model with eight predictors, Figure 3 showed the mean decrease accuracy for each predictor obtained from training with 1000 sample datasets. The predictors of wind_cv_fix and wind_ave were revealed to be most important with the mean decrease of the accuracy of 38.60 and 30.19, respectively. The accuracy of RF was 100% for all tested sample datasets. This is common in the calibration stage for RF as it consists of not-pruned decision trees [40].

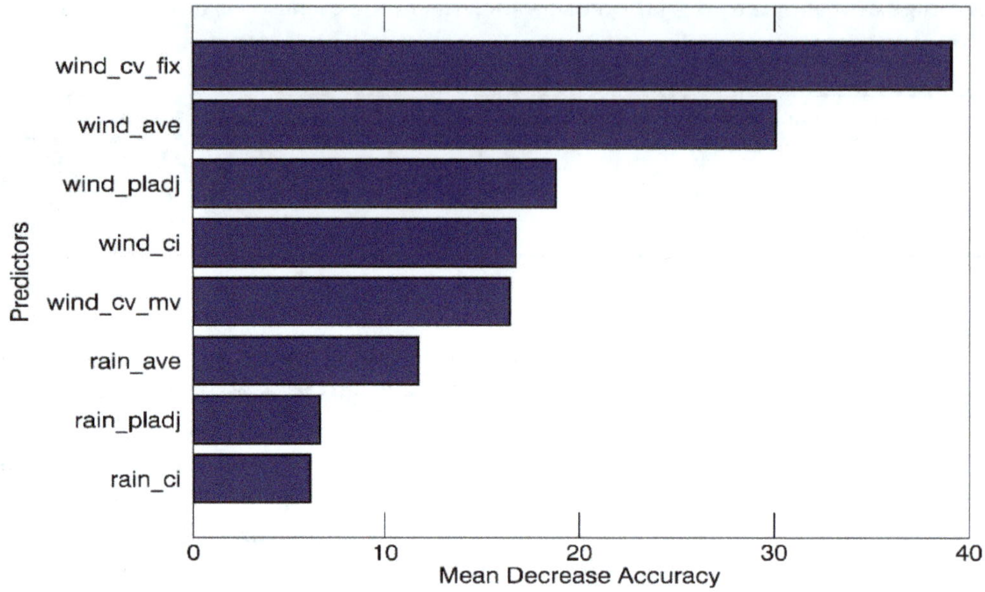

Figure 3. The averaged relative importance of predictors in random forest (RF) using 1000 different training datasets. The importance is measured as the mean decrease accuracy, which indicates the accuracy decrease when a specific predictor is deselected.

As mentioned in the model construction part, three different types of kernel functions were tested to figure out the best (Figure 4). In the case of HR, the radial basis function kernel showed much higher HR (91.6%) than the other two kernels, with the lowest FAR (22.0%). As a result, it showed the best performance based on PSS. Figure 5 shows F-score values from the SVM training results using the radial basis function. Each F-score value was measured using seven predictors without denoted target predictor so that a lower value of F-score indicates lower classification ability when the specific predictor is excluded. As a result, wind_cv_fix and wind_ave were identified as the most important predictors in detecting tropical cyclone formation which was consistent with the cases of other machine learning models.

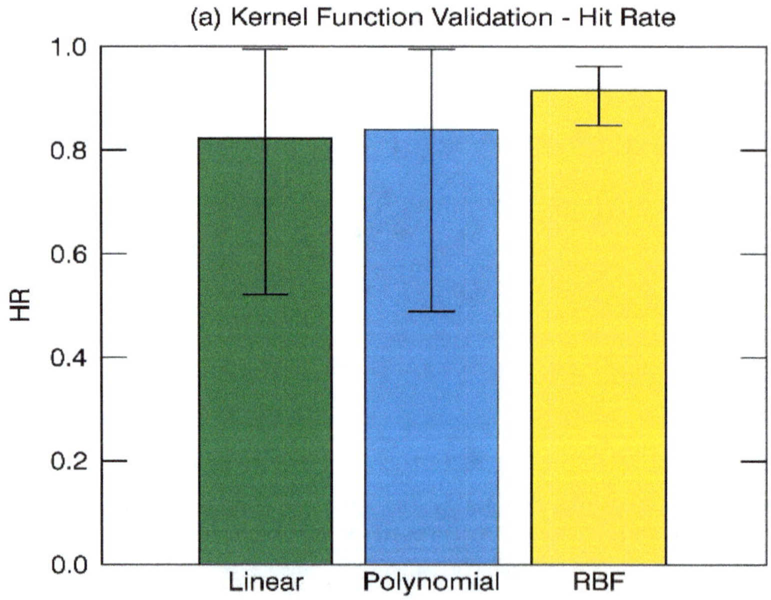

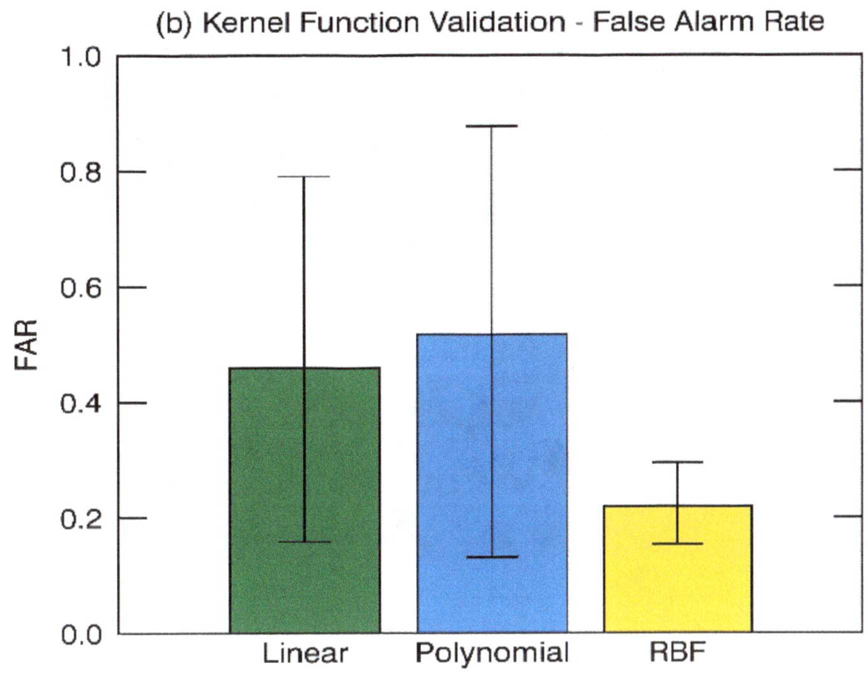

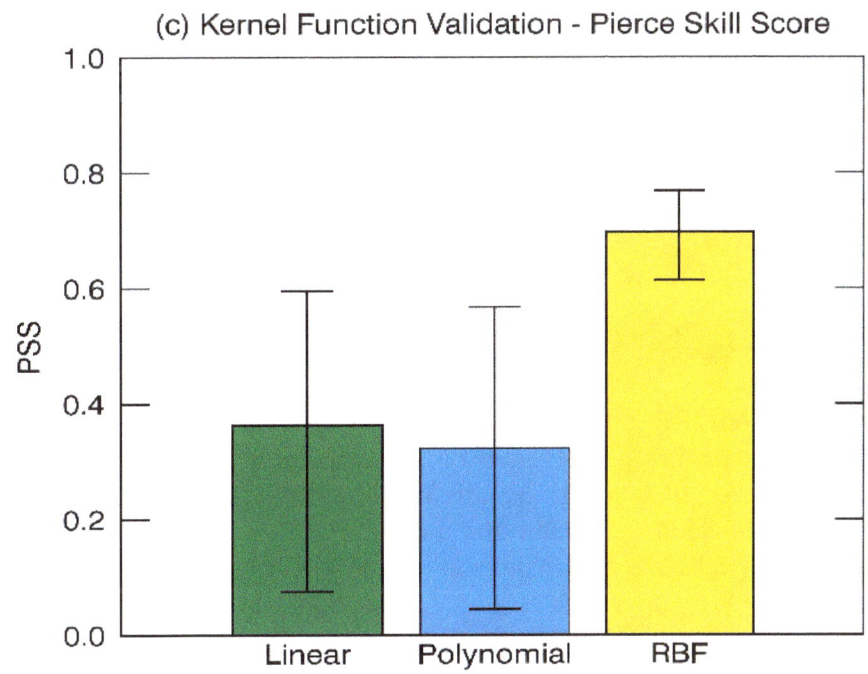

Figure 4. The comparison of the kernel function selected as linear (green), polynomial (blue), and radial basis function (yellow) in the support vector machines (SVM) model. Filled bars in (**a**) HR, (**b**) false alarm rate (FAR), and (**c**) Peirce Skill Score (PSS) represents the median values from the calibration tests for 1000 times. The 90 and 10 percentile values are also shown in each bar graph.

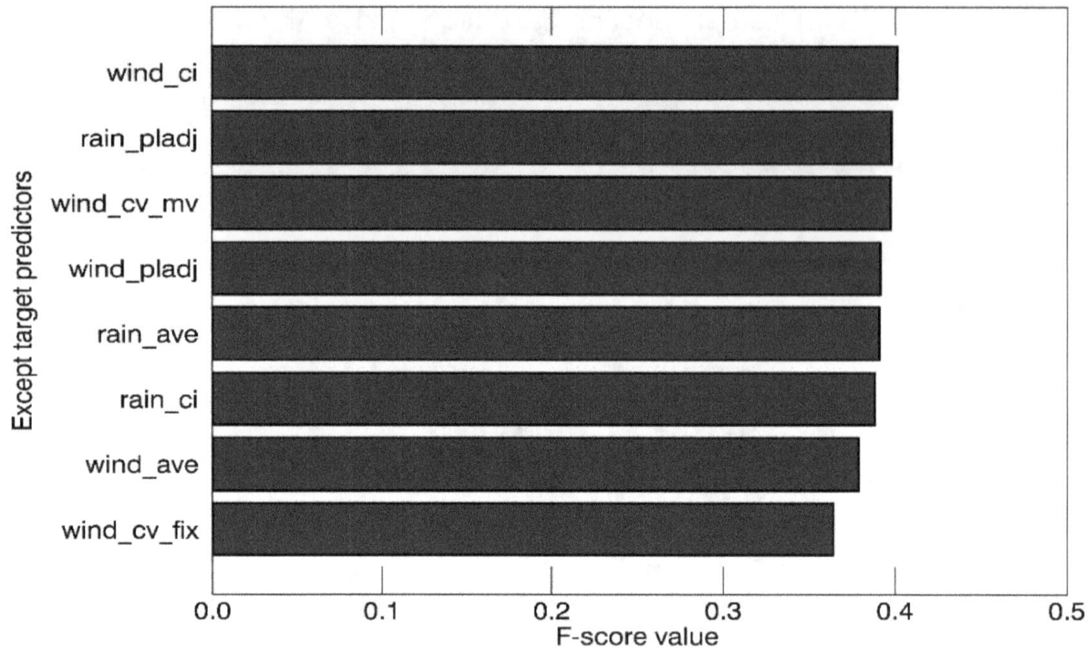

Figure 5. The average of F-score values in the training of the SVM model with the radial basis function. Each F-score value was obtained from the calibration using seven predictors without denoted target predictor.

3.2. Verification

Figure 6 compared the performance skill for LDA and the three ML approaches from all cases of validation (i.e., 1000 cases), in terms of HR, FAR, and PSS. Figure 6a is the comparison of HR for the four different models, in which the total averaged HR of three ML approaches (94%) is consistently higher HR than that from LDA (77%). The ML approaches produced relatively smaller HR variation due to the different sample years compared with that in the LDA (16%), suggesting that LDA had larger discrepancies along with the validation period. A few cases in ML approaches could hit TC genesis perfectly (e.g., 0509 in DT and RF, 0507 in RF, and 0508 in SVM).

As shown in Figure 6b, LDA (13%) showed lower FAR than ML approaches, although a few cases showed comparable values (e.g., 0506 in Figure 6b). The FAR in LDA changes relatively little with 1–11% variation using different validating data samples, while that in ML approaches vary more with 2–23% variation. FAR changes more at the change of validation samples, about 10% in LDA and 10–17% in ML approaches, respectively. This result indicated that LDA could classify DEV and non-DEV with small variances and little affected by the validating samples. There was a small difference among ML approaches in the case of HR, but in contrast, FAR showed a large variation among ML approaches. DT (28%) had the largest FAR compared with RF (23%) and SVM (21%).

In the case of HR, ML approaches had relatively higher performance than LDA, on the other hand, LDA showed much lower FAR than ML approaches. To consider both HR and FAR simultaneously, PSS was introduced in Figure 6c. The PSS, as mentioned above, estimated the goodness of a classifier in the binary classification. The overall results showed that ML approaches, in general, had higher PSS than LDA. Differences depending on the validation samples showed that ML approaches (10–17%) produced less variation than LDA (20%), which is consistent with the case of model calibration results (Section 3.1) exhibiting 7–17% variation for ML approaches and 18–29% for LDA. In the case of RF and SVM, they gave a higher classifying ability between DEV and non-DEV. However, DT had comparable skills with LDA.

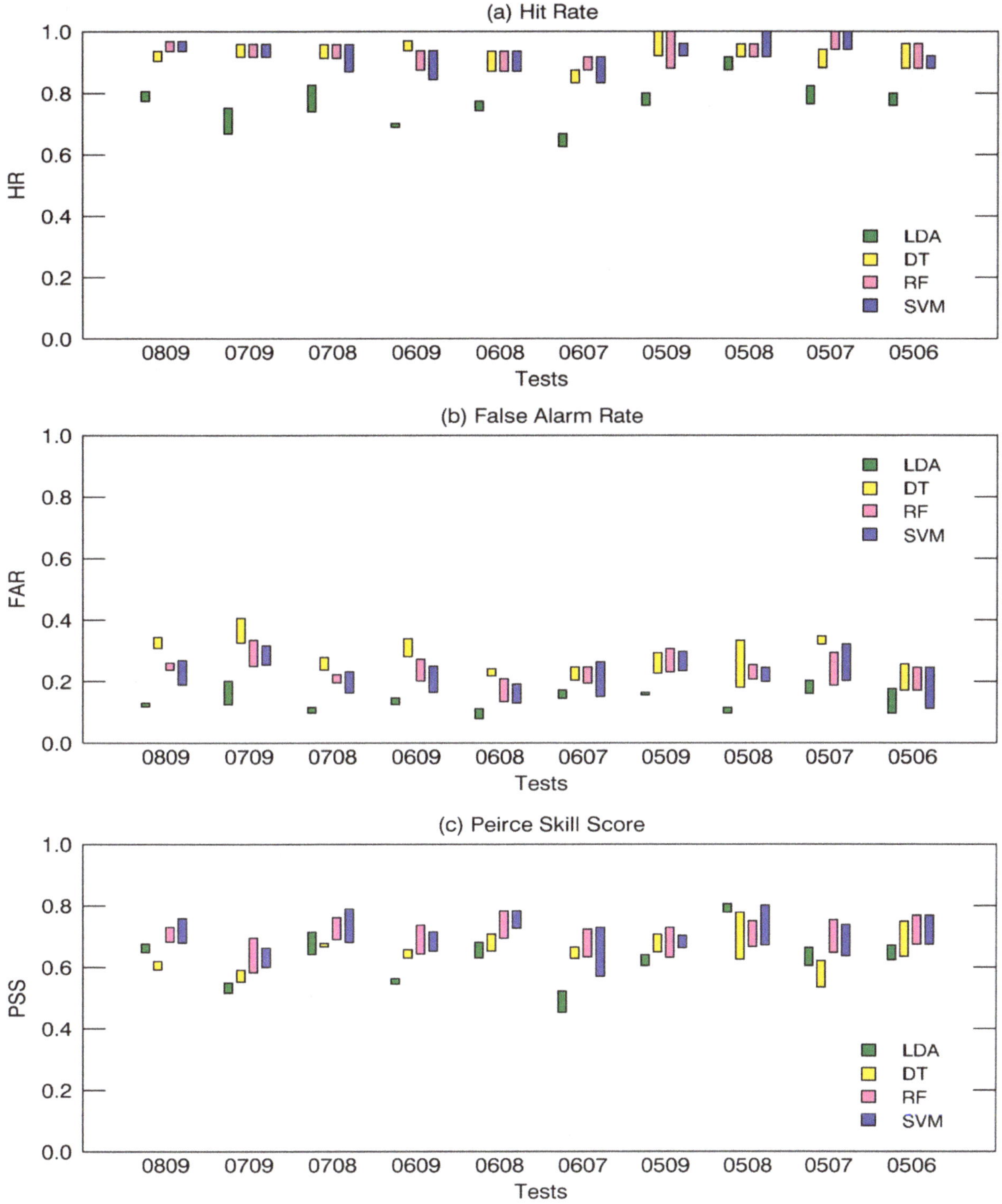

Figure 6. Forecast verification of (**a**) hit rate, (**b**) false alarm rate, and (**c**) Peirce skill score for LDA (green), DT (yellow), RF (pink), and SVM (blue) from validation sets denoted on the x-axis. The upper and lower bounds of each bar represent 90 and 10 percentile values, respectively.

To assess a prediction ability of each model, Figure 7 shows histograms of the lead time classified as TC by LDA (green), DT (yellow), RF (pink), and SVM (blue), and of the first time the WindSat is available (gray). In this figure, zero lead time represents the first time to reach an intensity of TD according to the JTWC best track, and early detection of TC has negative (positive) lead time which located left (right) relative to zero lead time. The frequency of the first observation from WindSat showed the maximum between 0 and −24 h, and radically decreased in the early time before −72 h. A smaller observation number from WindSat in earlier developing stage was in part due to the fact that the disturbance tracking backward in the time needed to be stopped in the premature or less-organized stage of disturbances. The frequency of the first detected time by the WindSat observation (gray) and that by the LDA and ML models showed a similar distribution. Although all models were able to predict TC formation when the WindSat observation was available in advance, the result demonstrated that ML approaches were able to detect TC formation earlier than LDA (Figure 7).

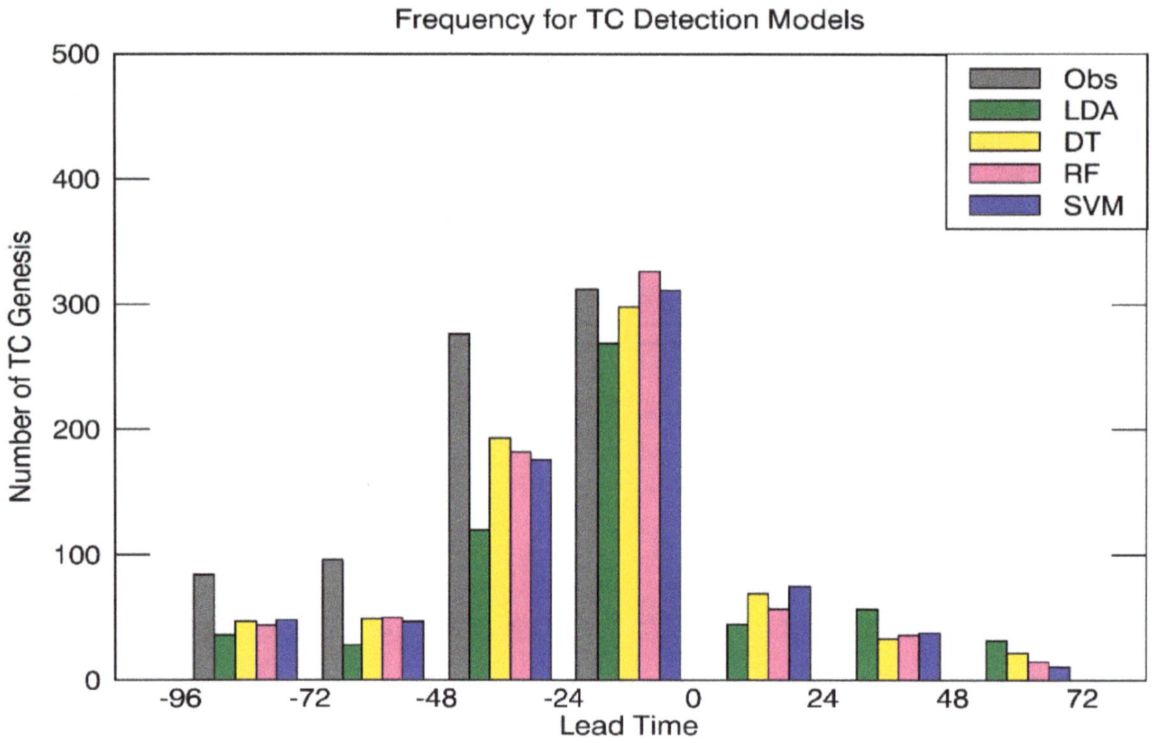

Figure 7. Frequency histogram of lead time for LDA (green), DT (yellow), RF (pink), and SVM (blue) from 1000 validation sets, also shown with the first observed time of WindSat (gray). The zero lead time is the actual time of TC formation that intensity reaches 13 m s^{-1} and negative (positive) values present early (late) detecting the time of TC formation.

Figure 8 summarized the results of four different TC formation detecting models in terms of HR, FAR, PSS, and lead time. Each box plot represented median values among 1000 tests and upper and lower boundaries represented 90 and 10 percentiles, respectively. Median HR showed that ML approaches (0.94–0.96) outperformed LDA (0.77). RF had the lowest (0.09) and LDA had the highest (0.25) error boundary. Figure 8b showed that LDA (0.13) had a better FAR than ML approaches (0.21–0.28) although they occasionally had the similar ability (e.g., 0506 in Figure 7b). Among the ML approaches, they showed different FAR and variances. In particular, DT tends to have more missed cases (28%) than RF (23%) and SVM (21%). The PSS results may be divided into two groups according to their skill score. The RF (72%) and SVM (73%) showed relatively higher score than DT (65%) and LDA (64%). The PSS is directly connected with HR and FAR in that is calculated by the difference between HR and FAR. Two high PSS models had a high HR with low FAR, however, the two lower models had different reasons that DT had the highest FAR although it had high HR and LDA had the lowest FAR with lower HR (Figure 8). Nevertheless, DT had a lower variance of validating results rather than LDA. In the case of lead time in Figure 8d, even though the same satellite observations are given, ML approaches (26–30 h earlier) could have earlier detection of TC genesis than LDA (21 h earlier). All models showed a large difference in lead time due to the infrequent WindSat sample. LDA sometimes showed after detecting cases (positive lead time).

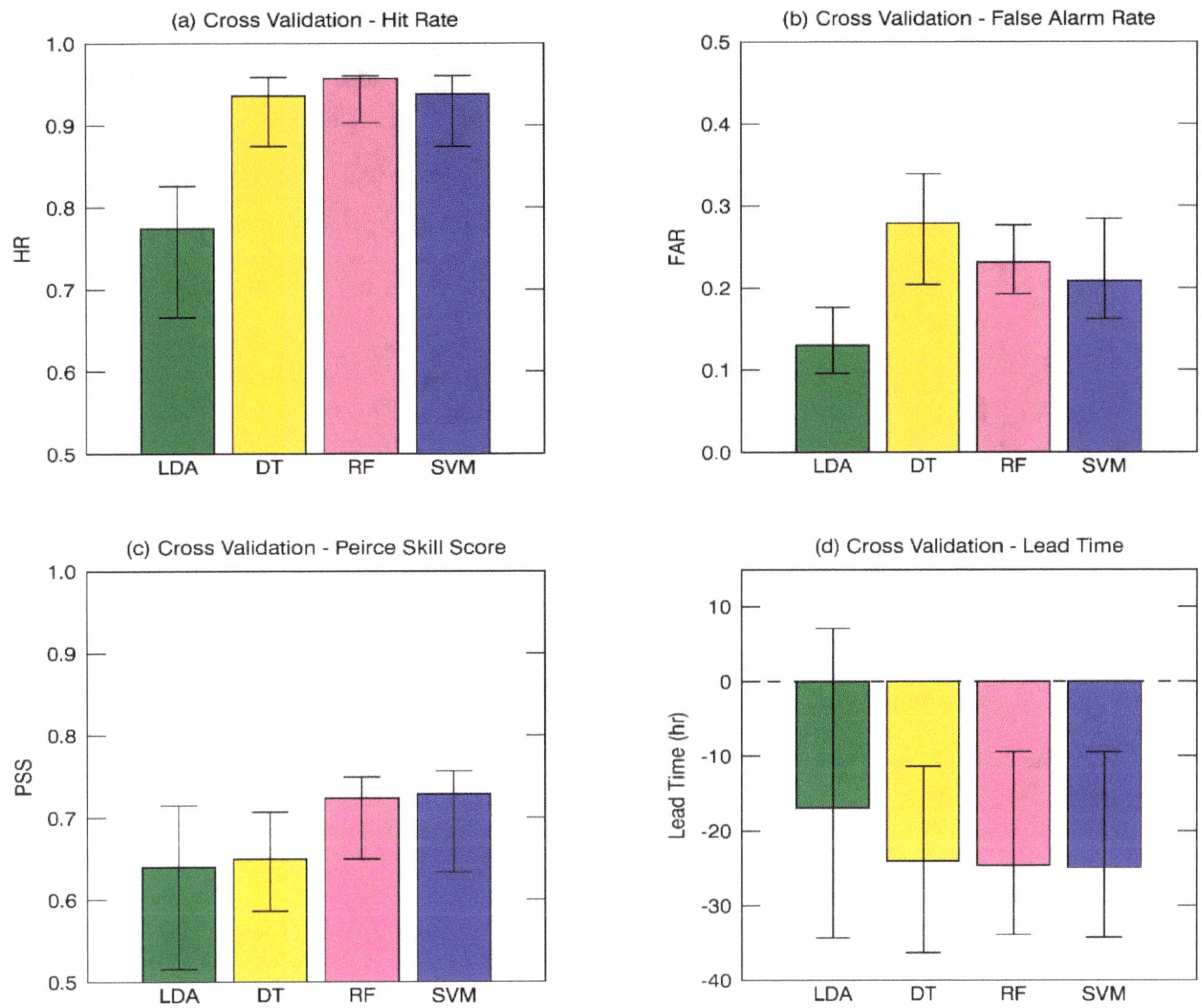

Figure 8. The comparisons of (**a**) HR, (**b**) FAR, (**c**) PSS, and (**d**) lead time among TC formation detecting models, LDA (green), DT (yellow), RF (pink), and SVM (blue). Median values are represented by box plot with 90 and 10 percentiles from the verification tests. The zero lead time is actual time of TC formation that intensity reached 13 m s^{-1} and negative (positive) values presented early (late) detecting the time of TC formation.

As alternative ways to compare the performance of four TC genesis detection models, two-dimensional histogram plot was made with FAR (x-axis) and HR (y-axis), which results were given in Figure 9. A skillful model that achieves much higher HR and less FAR, so that it tends to be located on the top left corner of the plot. A model which has no skill, on the other hand, lies along or under the diagonal (i.e. the model has no skill better than arbitrary choice). Each distribution from the three ML approaches (Figure 9b–d) was clearly separated from that of LDA (Figure 9a). The ML approaches had the advantage of better HR, whereas LDA tended to have smaller FAR. It was also noted that RF and SVM were located more to the left than DT, ensuring better performance in terms of FAR. Because RF was made for complementing the defect of DT, RF outperformed DT in general. It is possible that training procedures are affected by the lack of samples and quality even though applying random sampling methodology. There are many studies suggesting that SVM performs better with small samples [48,56–58], which is consistent with the results of this study.

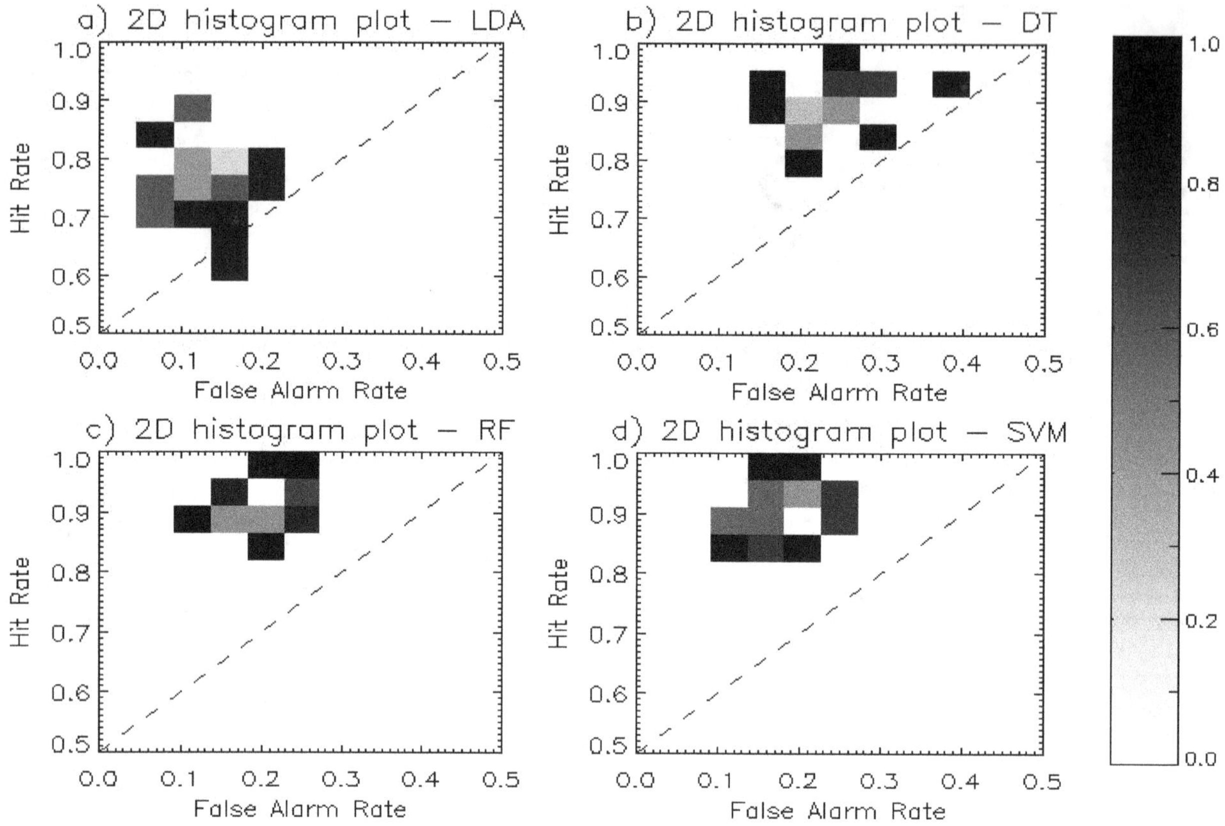

Figure 9. Two-dimensional normalized histogram plot for (**a**) LDA, (**b**) DT, (**c**) RF, and (**d**) SVM of different validation tests. The x-axis represents a false alarm rate versus y-axis represents a hit rate from 2 × 2 contingency table of observed and model-forecasted DEVs and non-DEVs.

4. Discussions

The formation of TC involves various dynamical and physical processes in multiple spatial and time scales [5,10]. Previous regression-based models [6] may not be appropriate to deal with complicated nonlinearities involved in the TC formation process. This study suggested an alternative approach for calibrating such an intricate formation process based on the ML approaches. The HR by ML approaches tested in this study was similar to values ranging from 94% to 96%, which were in general higher than that by LDA. Despite the ML approaches showed a higher FAR ranging 21–28% than that by LDA (~13%), they outperformed in the classification skill measured by PSS. The ML approaches also demonstrated a useful skill in the detection time, which was 26–30 h earlier than actual TC formation. The detection time by MLs was even earlier by 5–9 h compared with that by LDA.

The detection performance of the models developed from this study can be compared with those from existing studies, although an exact comparison may not be possible due to the differences in data and methods. The average hit rate of 95% by ML adopted in this study is obviously higher than that of numerical TC forecast models. Multi-model numerical weather forecasts were reported to show an average of 20% of the conditional probability of hit over the period of 2007–2011 [3], although they were tested with more strict criteria for the detection of tropical cyclones such as the more limited allowance of detection time and the genesis location. The results also exhibited a large dependence on the numerical models tested. By and large the statistical models such as ones used in this study show a more consistent and reliable performance than the numerical models. In our study, RF and SVM showed comparable skills, although SVM slightly underpredicted the tropical cyclone formation.

Our results can be compared with the skills from other ML approaches. The DT model in this study performed 93.5% of HR and 27.9% of FAR, while those from Park et al. [15] showed 95.3% and 28.5%, respectively. Although both studies used the identical DT algorithm, the previous study tested with fewer samples compared with the current study. The overall accuracy of classification by Zhang et al. [16] using DT was 84.6%, which was slightly lower than that from this study (~95%). The skill difference seemed to be partly related to the difference in the number of samples, where Zhang et al. [16] used approximately 2000 samples while this study used 1000 samples or more. The lead time for detecting TC formation by DT was 26 h in this study, which was comparable to that by Zhang et al. [16] with 24 h. Another reason for skill difference seemed to be related with the selection of predictors, where Zhang et al. [16] used large-scale environmental predictors derived from global atmospheric reanalysis and SST while this study used the direct information for disturbances in terms of their surface wind and rain rate.

Due to the complexity of RF- and SVM-based algorithms, they are difficult to visualize in terms of decision rules in detail. One can examine the relative importance of each input variable, instead, to identify the important dynamical or physical characteristics in the TC formation. DT and RF commonly revealed that the wind intensity average (wind_ave) and the degree of symmetry (wind_cv_fix) in the ocean-surface wind pattern near the disturbance center were the most critical predictors, even though their training procedures were independent. In the LDA model, the linear equation was composed with the degree of the strong wind organization (wind_pladj) in addition to the foregoing two variables.

The lead time how early the model can predict the TC formation depends strongly on the time of data availability. This study used the polar-orbiting satellite observations which were often missing or too late for detecting the formation potential. There are at least two ways for further advancing TC genesis prediction time such as 1) to advance the first observation time over a disturbance by utilizing other sources of ocean surface wind measurements such as the ASCAT [59] and 2) to advance the first time by developing an algorithm using a geostationary satellite for continuous monitoring of tropical disturbances.

The comparison among the three ML approaches revealed a slight difference in the performance. SVM exhibited less FAR compared with the other ML approaches, although all showed comparable skill in terms of HR. The results are consistent with previous studies [19,36] in different application also showed that RF performed better than DT. However, it is admitted that our result is still preliminary in terms of limited sample datasets from satellite. More extensive studies are required for the performance comparison between ML approaches, once more data are available in the near future.

5. Conclusions

This study constructed the TC formation detection models independently using the identical dataset but differing the algorithm based on LDA and the three ML approaches—DT, RF, and SVM. The primary purposes of this study were to evaluate the overall performance of each model depending on the sample datasets, and to examine if ML approaches would outperform the conventional LDA-based statistical method. Eight predictors quantifying the potential of TC formation were derived from the WindSat surface wind and rain rate observations over 1325 tropical disturbances in the western North Pacific during 2005–2009. The sample dependence was extensively tested by cross-validation with multiple sets of calibration and validation datasets for five-year WindSat observations. The performance skill was measured using HR, FAR, and PSS based on a 2×2 contingency table, and the detection lead time.

Overall, the ML algorithms showed better performance with significantly higher hit rates (~95%) compared with that from the LDA-based model (~77%), although false alarm rates were slightly higher in MLs (21–28%) than that by LDA (~13%). Combining HR and FAR, the ML approaches showed higher PSS than LDA. In addition, the detection time of TC formation by the ML algorithms was as early as 26–30 h before the actual time of TC formation, and it was also 5 to 9 h earlier than that by LDA. The models showed large variation in detection skills and the lead time depending on the tested sub-samples, presumably due to the limitation in sampling from the orbiting satellite. Nevertheless, the detection skill was less dependent on the ML algorithms, showing relatively small skill differences across MLs. The result obtained from this study demonstrates well that the ML approach provides skillful detection of TC formation with high accuracy and ahead of time before the actual TC formation, and the approach could be practically useful for operational use, compared with conventional linear approaches or numerical forecast models.

In operation by official TC centers, designating tropical cyclones remains quite subjective depending on individual forecaster's decision. Although the subjectivity in declaration could be important and helps explain potentially complex variable interactions based on an expert's own intuition and long-term experiences, the subjective forecast is often incoherent and difficult to make systematic improvement in the forecast skill. Officially, the U.S. JTWC initiates the warning of TC formation when a tropical disturbance develops into the category of tropical depression, in which "the maximum sustained surface wind speed (MSW)" within "a closed tropical circulation" meets or exceeds 13 m s^{-1} in the North Pacific (http://www.usno.navy.mil). The ML methods approached in this study all make use of the wind intensity average and the degree of symmetry. From the perspective of using wind intensity, it is consistent with the conventional operation. What is different is that the current ML-based methodology based on the WindSat observations uses the area-averaged wind speed rather than MSW. The MSW is an important metric to identify TCs at many warning centers such as JTWC and US National Hurricane Center. Although it is useful to characterize the stage of development of tropical disturbances with no dependence on the storm size and shape by definition, this maximum wind speed at a local point may contain large errors and uncertainties in the satellite-based wind estimation. Alternatively, the area-averaged wind speed is used to minimize satellite retrieval errors as used in the current study, and it makes the forecast system be more objective in declaring TC formation.

Based on our cross-validation results, this study suggests that an application of ML approaches provides a detection model for TC formation with better accuracy and with more extended forecast lead time compared with the conventional LDA-based model, regardless of the ML algorithms. This indicates that MLs have an advantage for earlier detection of TC formation. However, further investigations are needed with various sources of remote sensing observations (e.g., cloud, atmospheric temperature sounding, and precipitation signals both from microwave and scatterometer) to explore the full advantages in the ML models, rather than a single satellite observation of low-level circulation and precipitation by WindSat.

Author Contributions: M.K. led manuscript writing and contributed to the data analysis and research design. M.-S.P. and J.I. contributed to data analysis and manuscript writing. S.P. contributed to data processing and the discussion of the results. M.-I.L. supervised this study, contributed to the research design and manuscript writing, and served as the corresponding author.

Funding: This work was supported by the Korea Meteorological Administration Research and Development Program under Grant KMI2017-02410, by Next-Generation Information Computing Development Program through the National Research Foundation of Korea (NRF) funded by the Ministry of Science, ICT(NRF-2016M3C4A7952637), by a grant (2019-MOIS32-015) of Disaster-Safety Industry Promotion Program funded by Ministry of Interior and Safety (MOIS), Korea, and by the Ministry of Science and ICT (MSIT), Korea, under the Information Technology Research Center (ITRC) support program (IITP-2019-2018-0-01424) supervised by the Institute for Information & communications Technology Promotion (IITP). This research was a part of the project titled "Study on Air-Sea Interaction and Process of Rapidly Intensifying Typhoon in the Northwestern Pacific" funded by the Ministry of Oceans and Fisheries, Korea.

Acknowledgments: WindSat data are produced by Remote Sensing Systems and sponsored by the NASA Earth Science MEaSUREs DISCOVER Project and the NASA Earth Science Physical Oceanography Program. RSS WindSat data are available at www.remss.com.

References

1. Pielke, R.A., Jr.; Gratz, J.; Landsea, C.W.; Collins, D.; Saunders, M.A.; Musulin, R. Normalized hurricane damage in the united states: 1900–2005. *Nat. Hazards Rev.* **2008**, *9*, 29–42. [CrossRef]
2. Zhang, Q.; Liu, Q.; Wu, L. Tropical Cyclone Damages in China 1983–2006. *Am. Meteorol. Soc.* **2009**, *90*, 489–496. [CrossRef]

3. Halperin, D.J.; Fuelberg, H.E.; Hart, R.E.; Cossuth, J.H.; Sura, P.; Pasch, R.J. An Evaluation of Tropical Cyclone Genesis Forecasts from Global Numerical Models. *Weather. Forecast.* **2013**, *28*, 1423–1445. [CrossRef]
4. Burton, D.; Bernardet, L.; Faure, G.; Herndon, D.; Knaff, J.; Li, Y.; Mayers, J.; Radjab, F.; Sampson, C.; Waqaicelua, A. Structure and intensity change: Operational guidance. In Proceedings of the 7th International Workshop on Tropical Cyclones, La Réunion, France, 15–20 November 2010.
5. Park, M.-S.; Elsberry, R.L. Latent Heating and Cooling Rates in Developing and Nondeveloping Tropical Disturbances during TCS-08: TRMM PR versus ELDORA Retrievals*. *J. Atmos. Sci.* **2013**, *70*, 15–35. [CrossRef]
6. Schumacher, A.B.; DeMaria, M.; Knaff, J.A. Objective Estimation of the 24-h Probability of Tropical Cyclone Formation. *Weather. Forecast.* **2009**, *24*, 456–471. [CrossRef]
7. Kalnay, E.; Kanamitsu, M.; Kistler, R.; Collins, W.; Deaven, D.; Gandin, L.; Iredell, M.; Saha, S.; White, G.; Woollen, J.; et al. The NCEP/NCAR 40-Year Reanalysis Project. *Am. Meteorol. Soc.* **1996**, *77*, 437–471. [CrossRef]
8. Hennon, C.C.; Hobgood, J.S. Forecasting Tropical Cyclogenesis over the Atlantic Basin Using Large-Scale Data. *Mon. Weather. Rev.* **2003**, *131*, 2927–2940. [CrossRef]
9. Fan, K. A Prediction Model for Atlantic Named Storm Frequency Using a Year-by-Year Increment Approach. *Weather. Forecast.* **2010**, *25*, 1842–1851. [CrossRef]
10. Houze, R.A., Jr. Clouds in tropical cyclones. *Mon. Weather Rev.* **2010**, *138*, 293–344. [CrossRef]
11. Hennon, C.C.; Marzban, C.; Hobgood, J.S. Improving Tropical Cyclogenesis Statistical Model Forecasts through the Application of a Neural Network Classifier. *Weather. Forecast.* **2005**, *20*, 1073–1083. [CrossRef]
12. Rhee, J.; Im, J.; Carbone, G.J.; Jensen, J.R. Delineation of climate regions using in-situ and remotely-sensed data for the Carolinas. *Remote. Sens. Environ.* **2008**, *112*, 3099–3111. [CrossRef]
13. DeFries, R. Multiple Criteria for Evaluating Machine Learning Algorithms for Land Cover Classification from Satellite Data. *Remote. Sens. Environ.* **2000**, *74*, 503–515. [CrossRef]
14. Mountrakis, G.; Im, J.; Ogole, C. Support vector machines in remote sensing: A review. *Isprs J. Photogramm. Sens.* **2011**, *66*, 247–259. [CrossRef]
15. Park, M.-S.; Kim, M.; Lee, M.-I.; Im, J.; Park, S. Detection of tropical cyclone genesis via quantitative satellite ocean surface wind pattern and intensity analyses using decision trees. *Remote. Sens. Environ.* **2016**, *183*, 205–214. [CrossRef]
16. Zhang, W.; Fu, B.; Peng, M.S.; Li, T. Discriminating Developing versus Nondeveloping Tropical Disturbances in the Western North Pacific through Decision Tree Analysis. *Weather. Forecast.* **2015**, *30*, 446–454. [CrossRef]
17. Bayler, G.; Lewit, H. The Navy Operational Global and Regional Atmospheric Prediction Systems at the Fleet Numerical Oceanography Center. *Weather. Forecast.* **1992**, *7*, 273–279. [CrossRef]
18. Gaiser, P.; Germain, K.S.; Twarog, E.; Poe, G.; Purdy, W.; Richardson, D.; Grossman, W.; Jones, W.L.; Spencer, D.; Golba, G.; et al. The WindSat spaceborne polarimetric microwave radiometer: sensor description and early orbit performance. *IEEE Trans. Geosci. Sens.* **2004**, *42*, 2347–2361. [CrossRef]
19. Han, H.; Lee, S.; Im, J.; Kim, M.; Lee, M.-I.; Ahn, M.H.; Chung, S.-R. Detection of Convective Initiation Using Meteorological Imager Onboard Communication, Ocean, and Meteorological Satellite Based on Machine Learning Approaches. *Remote. Sens.* **2015**, *7*, 9184–9204. [CrossRef]
20. Kim, D.H.; Ahn, M.H. Introduction of the in-orbit test and its performance for the first meteorological imager of the Communication, Ocean, and Meteorological Satellite. *Atmos. Meas. Tech.* **2014**, *7*, 2471–2485. [CrossRef]
21. Sesnie, S.E.; Finegan, B.; Gessler, P.E.; Thessler, S.; Bendana, Z.R.; Smith, A.M.S. The multispectral separability of Costa Rican rainforest types with support vector machines and Random Forest decision trees. *Int. J. Sens.* **2010**, *31*, 2885–2909. [CrossRef]
22. Ritchie, E.A.; Holland, G.J. Scale Interactions during the Formation of Typhoon Irving. *Mon. Weather. Rev.* **1997**, *125*, 1377–1396. [CrossRef]
23. Usama, F. Data mining and knowledge discovery in databases: Implications for scientific databases. In Proceedings of the 9th International Conference on Scientific and Statistical Database Management (SSDBM'97), Olympia, WA, USA, 11–13 August 1997.
24. Quinlan, J.R. Simplifying decision trees. *Int. J. Man-Mach. Stud.* **1987**, *27*, 221–234. [CrossRef]

25. Helms, C.N.; Knapp, K.R.; Bowen, A.R.; Hennon, C.C. An Objective Algorithm for Detecting and Tracking Tropical Cloud Clusters: Implications for Tropical Cyclogenesis Prediction. *J. Atmos. Ocean. Technol.* **2011**, *28*, 1007–1018.
26. McGarigal, K.; Cushman, S.A.; Ene, E. Fragstats v4: Spatial Pattern Analysis Program for Categorical and Continuous Maps. Computer software program produced by the authors at the University of Massachusetts, Amherst. Available online: http://www.umass.edu/landeco/research/fragstats/fragstats.html (accessed on 18 May 2019).
27. Refaeilzadeh, P.; Tang, L.; Liu, H. Cross-validation. In *Encyclopedia of Database Systems*, 1st ed.; Liu, L., Özsu, M.T., Eds.; Springer US: New York, NY, USA, 2009; Volume 1, pp. 532–538.
28. Bradford, J.P.; Kunz, C.; Kohavi, R.; Brunk, C.; Brodley, C.E. Pruning decision trees with misclassification costs. In Proceedings of the European Conference on Machine Learning, Chemnitz, Germany, 21–23 April 1998.
29. Cao, J.; Liu, K.; Liu, L.; Zhu, Y.; Li, J.; He, Z. Identifying Mangrove Species Using Field Close-Range Snapshot Hyperspectral Imaging and Machine-Learning Techniques. *Remote. Sens.* **2018**, *10*, 2047. [CrossRef]
30. Switzer, P. Extensions of linear discriminant analysis for statistical classification of remotely sensed satellite imagery. *Math. Geosci.* **1980**, *12*, 367–376. [CrossRef]
31. Roth, V.; Steinhage, V. Nonlinear discriminant analysis using kernel functions. In Proceedings of the Advances in Neural Information Processing Systems, Denver, CO, USA, 29 November 1999.
32. Peirce, C.S. The numerical measure of the success of predictions. *Science* **1884**, *4*, 453–454. [CrossRef] [PubMed]
33. Lee, S.; Han, H.; Im, J.; Jang, E.; Lee, M.-I. Detection of deterministic and probabilistic convection initiation using Himawari-8 Advanced Himawari Imager data. *Atmos. Meas. Tech.* **2017**, *10*, 1859–1874. [CrossRef]
34. Im, J.; Jensen, J.R.; Coleman, M.; Nelson, E. Hyperspectral remote sensing analysis of short rotation woody crops grown with controlled nutrient and irrigation treatments. *Geocarto Int.* **2009**, *24*, 293–312. [CrossRef]
35. Kim, Y.H.; Im, J.; Ha, H.K.; Choi, J.-K.; Ha, S. Machine learning approaches to coastal water quality monitoring using GOCI satellite data. *GISci. Remote Sens.* **2014**, *51*, 158–174. [CrossRef]
36. Rhee, J.; Im, J. Meteorological drought forecasting for ungauged areas based on machine learning: Using long-range climate forecast and remote sensing data. *Agric. Meteorol.* **2017**, *237*, 105–122. [CrossRef]
37. Lu, Z.; Im, J.; Quackenbush, L. A Volumetric Approach to Population Estimation Using Lidar Remote Sensing. *Photogramm. Eng. Sens.* **2011**, *77*, 1145–1156. [CrossRef]
38. Lee, S.; Im, J.; Kim, J.; Kim, M.; Shin, M.; Kim, H.-C.; Quackenbush, L.J. Arctic Sea Ice Thickness Estimation from CryoSat-2 Satellite Data Using Machine Learning-Based Lead Detection. *Remote. Sens.* **2016**, *8*, 698. [CrossRef]
39. Quinlan, J. C5. 0 Online Tutorial. Available online: www.rulequest.com (accessed on 18 May 2019).
40. Breiman, L. Random forests. *Mach. Learn.* **2001**, *45*, 5–32. [CrossRef]
41. Guo, Z.; Du, S. Mining parameter information for building extraction and change detection with very high-resolution imagery and gis data. *GISci. Remote Sens.* **2017**, *54*, 38–63. [CrossRef]
42. Kim, M.; Im, J.; Han, H.; Kim, J.; Lee, S.; Shin, M.; Kim, H.-C. Landfast sea ice monitoring using multisensor fusion in the Antarctic. *GISci. Remote Sens.* **2015**, *52*, 239–256. [CrossRef]
43. Lu, Z.; Im, J.; Rhee, J.; Hodgson, M. Building type classification using spatial and landscape attributes derived from LiDAR remote sensing data. *Landsc. Urban Plan.* **2014**, *130*, 134–148. [CrossRef]
44. Richardson, H.J.; Hill, D.J.; Denesiuk, D.R.; Fraser, L.H. A comparison of geographic datasets and field measurements to model soil carbon using random forests and stepwise regressions (British Columbia, Canada). *GISci. Remote Sens.* **2017**, *17*, 1–19. [CrossRef]
45. Park, S.; Im, J.; Jang, E.; Rhee, J. Drought assessment and monitoring through blending of multi-sensor indices using machine learning approaches for different climate regions. *Agric. Meteorol.* **2016**, *216*, 157–169. [CrossRef]
46. Liaw, A.; Wiener, M. Classification and regression by randomforest. *R News* **2002**, *2*, 18–22.
47. Chu, H.-J.; Wang, C.-K.; Kong, S.-J.; Chen, K.-C. Integration of Full-waveform LiDAR and Hyperspectral Data to Enhance Tea and Areca Classification. *GISci. Remote Sens.* **2016**, *53*, 542–559. [CrossRef]
48. Zhang, C.; Smith, M.; Fang, C. Evaluation of Goddard's LiDAR, hyperspectral, and thermal data products for mapping urban land-cover types. *GISci. Remote Sens.* **2018**, *55*, 90–109.

49. Tesfamichael, S.; Newete, S.; Adam, E.; Dubula, B. Field spectroradiometer and simulated multispectral bands for discriminating invasive species from morphologically similar cohabitant plants. *GIScі. Remote Sens.* **2018**, *55*, 417–436. [CrossRef]
50. Liu, T.; Abd-Elrahman, A.; Morton, J.; Wilhelm, V.L.; Jon, M. Comparing fully convolutional networks, random forest, support vector machine, and patch-based deep convolutional neural networks for object-based wetland mapping using images from small unmanned aircraft system. *GIScі. Remote Sens.* **2018**, *55*, 243–264. [CrossRef]
51. Wylie, B.; Pastick, N.; Picotte, J.; Deering, C. Geospatial data mining for digital raster mapping. *GIScі. Remote Sens.* **2019**, *56*, 406–429. [CrossRef]
52. Xun, L.; Wang, L. An object-based SVM method incorporating optimal segmentation scale estimation using Bhattacharyya Distance for mapping salt cedar (Tamarisk spp.) with QuickBird imagery. *GIScі. Remote Sens.* **2015**, *52*, 257–273. [CrossRef]
53. Zhu, X.; Li, N.; Pan, Y. Optimization Performance Comparison of Three Different Group Intelligence Algorithms on a SVM for Hyperspectral Imagery Classification. *Remote. Sens.* **2019**, *11*, 734. [CrossRef]
54. Powers, D.M. Evaluation: from precision, recall and F-measure to ROC, informedness, markedness and correlation. *J. Mach. Learn. Technol.* **2011**, *2*, 37–63.
55. Chang, C.-C. Libsvm: A Library for Support Vector Machines, Acm Transactions on Intelligent Systems and Technology, 2: 27: 1–27: 27, 2011. Available online: http://www.csie.ntu.edu.tw/~{}cjlin/libsvm (accessed on 18 May 2019).
56. Chen, Y.; Lin, Z.; Zhao, X. Optimizing Subspace SVM Ensemble for Hyperspectral Imagery Classification. *IEEE J. Sel. Top. Appl. Earth Obs. Sens.* **2014**, *7*, 1295–1305. [CrossRef]
57. Sonobe, R.; Yamaya, Y.; Tani, H.; Wang, X.; Kobayashi, N.; Mochizuki, K.-I. Assessing the suitability of data from Sentinel-1A and 2A for crop classification. *GIScі. Remote Sens.* **2017**, *54*, 1–21. [CrossRef]
58. Georanos, S.; Grippa, T.; Vanhuysse, S.; Lennert, M.; Shimoni, M.; Kalogirou, S.; Wolff, E. Less is more: optimizing classification performance through feature selection in a very-high-resolution remote sensing object-based urban application. *GIScі. Remote Sens.* **2018**, *55*, 221–242.
59. Figa-Saldaña, J.; Wilson, J.J.; Attema, E.; Gelsthorpe, R.; Drinkwater, M.; Stoffelen, A. The advanced scatterometer (ASCAT) on the meteorological operational (MetOp) platform: A follow on for European wind scatterometers. *Can. J. Sens.* **2002**, *28*, 404–412. [CrossRef]

6

Estimating Live Fuel Moisture from MODIS Satellite Data for Wildfire Danger Assessment in Southern California USA

Boksoon Myoung [1], Seung Hee Kim [2,*], Son V. Nghiem [3], Shenyue Jia [2], Kristen Whitney [2] and Menas C. Kafatos [2]

1 APEC Climate Center, 12 Centum 7-ro, Haeundae-gu, Busan 48058, Korea; bmyoung@apcc21.org
2 Center of Excellence in Earth Systems Modeling and Observations, Chapman University, Orange, CA 92866, USA; sjia@chapman.edu (S.J.); whitn111@mail.chapman.edu (K.W.); kafatos@chapman.edu (M.C.K.)
3 Jet Propulsion Laboratory, California Institute of Technology, Pasadena, CA 91109, USA; Son.V.Nghiem@jpl.nasa.gov
* Correspondence: sekim@chapman.edu.

Abstract: The goal of the research reported here is to assess the capability of satellite vegetation indices from the Moderate Resolution Imaging Spectroradiometer onboard both Terra and Aqua satellites, in order to replicate live fuel moisture content of Southern California chaparral ecosystems. We compared seasonal and interannual characteristics of in-situ live fuel moisture with satellite vegetation indices that were averaged over different radial extents around each live fuel moisture observation site. The highest correlations are found using the Aqua Enhanced Vegetation Index for a radius of 10 km, independently verifying the validity of in-situ live fuel moisture measurements over a large extent around each in-situ site. With this optimally averaged Enhanced Vegetation Index, we developed an empirical model function of live fuel moisture. Trends in the wet-to-dry phase of vegetation are well captured by the empirical model function on interannual time-scales, indicating a promising method to monitor fire danger levels by combining satellite, in-situ, and model results during the transition before active fire seasons. An example map of Enhanced Vegetation Index-derived live fuel moisture for the Colby Fire shows a complex spatial pattern of significant live fuel moisture reduction along an extensive wildland-urban interface, and illustrates a key advantage in using satellites across the large extent of wildland areas in Southern California.

Keywords: wildfire; satellite vegetation indices; live fuel moisture; empirical model function; Southern California; chaparral ecosystem

---

1. Introduction

Wildfires in Southern California (SoCal) are part of the natural cycle under Mediterranean climatic conditions. However, excessive urban growth in SoCal significantly increases the wildland-urban interface, and thus seriously compounds wildfire hazards, resulting in loss of human life and property [1,2]. Thus, improving fire danger assessment systems with a high spatial resolution and a wide coverage across the vast wildland is essential for decision makers and fire agencies to develop and implement pro-active policies. To assess wildfire danger, the United States Forest Service (USFS) has developed and utilized the National Fire Danger Rating System (NFDRS) [3], for which vegetation moisture is a key input.

While the moisture content of dead vegetation in NFDRS can be rather easily obtained from weather-dependent models since dead fuels are dependent on atmospheric variability [4], estimating the moisture content of live vegetation is more complicated because it depends on physiological properties that may significantly vary among different plant species [5]. To quantify moisture content of live vegetation, live fuel moisture (LFM) is defined as the percentage ratio of the difference between wet and dry weight to the dry weight of a vegetation sample [6].

In general, LFM is closely related to fire ignition, propagation, and intensity [7–9]. LFM has been incorporated into many fire behavior models (e.g., Fire Area Simulator or FARSITE model). Per Weise et al. [5], wildfire danger can be categorized with LFM levels (e.g., low: greater than 120%, moderate: between 80% and 120%, high: between 60% and 80%, and critical; less than 60%). Dennison et al. [9] have suggested that LFM lower than 77% appears to be historically associated with large fires in the Santa Monica Mountains of Los Angeles County, CA. Understanding seasonal trends of LFM can improve seasonal outlooks of LFM change and help to improve effective wildfire management as fire agencies operationally rely on field observations of LFM [6].

Currently, spatial coverage and temporal sampling of LFM data are severely limited as fieldwork for LFM measurements is labor intensive. LFM is manually measured weekly, biweekly, or monthly at a limited number of sampling sites across SoCal. For example, the Los Angeles County Fire Department typically samples LFM only at 11 disparate sites in its jurisdiction once every two weeks, leaving large data voids in areas where weather and geophysical variations can substantially affect LFM. In this regard, the capability of satellite data to observe LFM in each area around a given LFM site on a nearly daily basis, as compared to the weekly-monthly data from the manual method, can be a major advantage that is beneficial to fire agencies.

A potential approach to overcome the spatial and temporal limitations of manual measurements of LFM is to use vegetation indices (VIs) derived from satellite data. Satellite VI-based LFM estimations that have been attempted in the past were mostly for chamise ecosystems in California [10–13] and in Spain [8]. However, many studies were based on short-term records and statistical relationships without investigating seasonal and interannual characteristics of LFM and VIs based on obsolete satellite data collections with an inaccurate calibration.

Physically, LFM is dependent on precipitation, soil moisture, evapotranspiration, and the physiology of plants [6,14]. VIs retrieved from satellite remote sensing measurements are related to surface greenness and biomass of vegetation represented by the green leaf area index [15], which are impacted by and thereby correlated with LFM. Thus, VI and LFM are interdependent variables with similar seasonal and interannual trends, which suggest a possible estimation of LFM from remotely sensed VI. However, dissimilarities between them also exist. For example, plant growth requires not only moisture, but also optimal temperature and solar radiation, and vegetation moisture can also vary during the complex photosynthetic and xylem embolism processes in different plant species [16,17]. Therefore, to retrieve LFM from satellite VI data, it is necessary to conduct a careful investigation of LFM and VI characteristics by utilizing decadal datasets of in-situ and satellite measurements.

Previous studies have attempted to make a point-to-point comparison between LFM and VI or a combination of other VIs (e.g., [10–13]). Here, we examine the validity of multiple remotely sensed products used to estimate LFM, and thus we will investigate confounding factors and additional physical parameters necessary in the development of LFM model functions. Moreover, a review of past analyses raised concerns in remotely sensed LFM products [18]; however, many past results based on Moderate Resolution Imaging Spectroradiometer (MODIS) Collection 5 or earlier versions suffered from serious calibration problems [19,20], which caused significant errors in the remotely sensed VI products as recently published by Zhang et al. [21]. Such calibration problems necessitate a re-evaluation of the use of remotely sensed products to estimate LFM. Our novel approach is to test the LFM relationship with enhanced vegetation index (EVI) that is averaged over various spatial extents centered at each in-situ LFM sampling location. The objectives of this study are to: (1) Compare seasonal and interannual characteristics of LFM with those of VIs calculated from satellite data in SoCal; (2) develop an empirical model function of LFM based on an optimal vegetation index together with temperature data; and, (3) evaluate the feasibility, as well as limitations of the empirical model for wildfire danger assessments.

2. Methods and Materials

2.1. Live Fuel Moisture

Moisture content in live biomass is quantitatively characterized by LFM. LFM is defined as the percentage difference between wet and dry vegetation material over the dry mass of vegetation. Explicitly,

$$\text{LFM}(\%) = \frac{m_w - m_d}{m_d} \times 100, \tag{1}$$

where m_w is weight of the sampled vegetation, and m_d is the dry weight of the same sample. Our analysis was carried out primarily on chamise chaparral (*Adenostoma fasciculatum*), the most common shrub in the chaparral and regarded as an important fuel component in SoCal. The in-situ LFM dataset was obtained from the national live fuel moisture database (http://www.wfas.net/index.php/national-fuel-moisture-database-moisture-drought-103). LFM data are collected regularly every one or two weeks; however, the intervals can be longer during wet seasons when leaves and twigs remain wet after rainfall events. In these cases, fire agencies postpone their LFM sampling by a few days to avoid errors in LFM caused by excessive rainwater onto vegetation. To be compared to VIs, the LFM dataset was linearly interpolated at a daily time scale.

Among the 24 LFM sampling sites in Los Angeles, Ventura, and Orange County, 16 sites had data coverage for more than three years (Table 1). The data from these 16 sites were selected for the regression analysis between LFM and VIs. For the longer-term analysis, data were selected from seven sites having more than 10 years of record from 2002 (bold characters in Table 1 and Figure 1). Four sites (Bitter, Placerita, La Tuna, and Laurel) were in inland areas (inland sites, hereafter), whereas the other three sites (Trippet, Schueren, and Clark) were in coastal areas (coastal sites, hereafter). Bitter and Schueren sites had corresponding meteorological stations, called Remote Automatic Weather Stations (RAWS); therefore, LFM and EVI comparisons with their corresponding atmospheric conditions were also investigated at these two sites. While utilizing LFM data at all of the sites to investigate and determine an overall universal LFM-EVI function may be desirable in principle, it is cautious that long-term data records at all the sites are required to ensure sufficient statistical sampling over the vast wildland in SoCal.

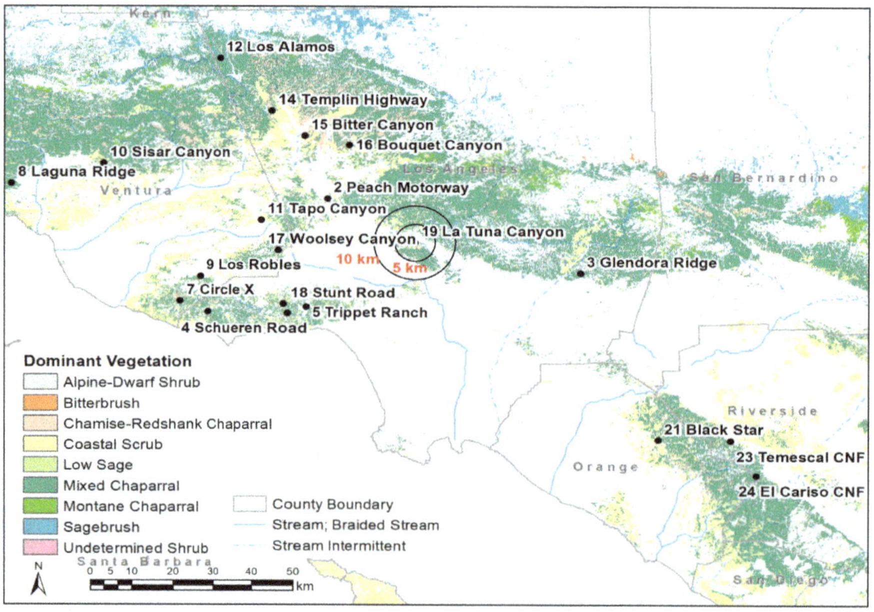

Figure 1. 16 live fuel moisture sampling sites overlaid on a Fire and Resource Assessment Program vegetation map. The colors indicate dominant vegetation species; only chamise-dominant areas (e.g., shrubland and scrubland) are shown. Two circles at La Tuna Canyon with a radius of 5 km and 10 km are also shown.

Table 1. Live fuel moisture (LFM) stations used in the study. Underlines indicate short names of the seven main research sites.

Name	Site Number	Latitude	Longitude	Fire Agency
Bitter Canyon	15	34.510000	−118.594444	LA County
Placerita Canyon	1	34.375278	−118.438889	LA County
La Tuna Canyon	19	34.246667	−118.302778	LA County
Laurel Canyon	20	34.124722	−118.368889	LA County
Trippet Ranch	5	34.093333	−118.597778	LA County
Schueren Road	4	34.078889	−118.644722	LA County
Clark Motorway	6	34.084444	−118.862500	LA County
Peach Motorway	2	34.355556	−118.534722	LA County
Bouquet Canyon	16	34.486111	−118.472778	LA County
Glendora Ridge	3	34.165278	−117.865000	LA County
CircleX	7	34.110833	−118.937222	Ventura County FD
Laguna Ridge	8	34.400000	−119.378889	Ventura County FD
Los Robles	9	34.171667	−118.882222	Ventura County FD
Tapo Canyon	11	34.306389	−118.710278	Ventura County FD
Sisar Canyon	10	34.447500	−119.135278	Ventura County FD
Black Star	21	33.754722	−117.670833	Orange County FD

2.2. Remote Sensing Data

The present study focuses on the two most relevant VIs among an array of many VIs defined and used for different purposes: the normalized difference vegetation index (NDVI) and the EVI. NDVI and EVI were derived from the MODIS' Vegetation Indices 16-Day L3 Global 250 m (MOD13Q1 and MYD13Q1)' products from both the Terra and Aqua satellites [22]. The datasets were provided by the NASA EOSDIS Land Processes Distributed Active Archive Center (LP DAAC) at the USGS/Earth Resources Observation and Science (EROS) Center. The analysis period covered 10 years between October 2002 and September 2012, as VIs from MODIS have been available since 2001 for Terra and since 2002 for Aqua. We also investigated other VIs derived from MODIS land surface reflectance products (MOD09A1) for the same sites, including normalized difference water index (NDWI), normalized difference infrared index (NDII), and visible atmospherically resistant index (VARI) [23] (e.g., Table S1). These three VIs were recognized as effective indicators of vegetation water content and soil moisture [24,25].

To test the sensitivity of the LFM relationship with different areal averages of VIs, the values of the VIs were averaged over circular areal extents with various radii, ranging from 0.5 to 25 km. Then, the averaged VIs were used to correlate with LFM. This method allows for an independent assessment of the spatial extent where the in-situ LFM measurements are valid beyond the central sampling point. This is important because fire agencies intentionally select their LFM sampling locations to be representative of the surrounding vegetation conditions as far as possible so that measured LFM values are representative over an extensive area instead of being valid only at each sampling site. The sensitivity test results showed that a slightly higher correlation is observed at the 10-km radius (correlation coefficient of about 0.79) than that at 0.5-km radius (about 0.72 correlation coefficient). This suggests that a spatial average of VIs over a larger extent (~10-km radius) around each LFM location includes a larger ensemble of VI data, which statistically reduces satellite measurement noises as well as the effects of heterogeneous mixtures of different plant species within each sampling area. Thus, in this study, results for the areal extent of a 10-km radius, having the highest correlations, were selected to carry out the analysis.

2.3. Empirical Model

First, the Pearson correlation analysis is carried out to investigate the relationship between LFM and multiple VIs at 16 LFM sites to find the VI with the highest correlation against LFM. This VI is later employed as the major MODIS-derived indicator of vegetation water content for further analysis.

LFM data available at the seven LFM sites in a 10-year period were separated into two different groups, representing the inland region and coastal region. Regional characteristics of LFM and EVI between inland and coastal regions were investigated together with their interannual variations. We then examined possible reasons for the different regional characteristics.

Next, linear regression models of VIs for LFM at the seven sites with decadal records were developed and evaluated across the 10-year data period with respect to the averages and inter-annual variability of maxima, minima, and transitional levels of LFM. We also tested non-linear models with a quadratic term or log transformation of the predictor, but a substantial improvement was not found. Therefore, in this study, two linear models are developed and tested. The first model uses each VI as a sole predictor (Equation (2)), while the second model includes a composite of collocated and contemporaneous VI and meteorological variables as predictors to account for the environmental dependence of LFM (Equation (3)), as follows:

$$\mathrm{LFM_i} = \beta_0 + \beta_1\,\mathrm{VI_i} + \varepsilon_i, \tag{2}$$

$$\mathrm{LFM_i} = \beta_0 + \beta_1\,\mathrm{VI_i} + \beta_2\,\mathrm{MI_i} + \varepsilon_i, \tag{3}$$

where MI is a meteorological factor with index i refers to various observations ($i = 1, \ldots, N$), and ε_i is a residual error term.

VIs alone may not be sufficient to fully replicate LFM since using them is an indirect approach to infer the vegetation moisture. The other factors that were related to the dryness of vegetation conditions were selected for a test as an independent variable in addition to VIs. In this study, meteorological observations such as daily temperature (minimum, maximum and mean), relative humidity, and precipitation are chosen as additional variables in the composite estimation model. Due to large fluctuations in daily data, we used a 15-day running mean on LFM, EVI, and meteorological data in our analyses.

Finally, the capability of our satellite derived LFM model is tested in the case of the 2014 Colby Fire. The Colby Fire was ignited by an illegal campfire along the Colby Truck Trail in the San Gabriel Mountains of the Angeles National Forest on 16 January 2014 [26]. Fanned by dry and powerful Santa Ana winds, it burned over 1962 acres by 25 January at 98% containment. The fire destroyed five homes, damaged 17 other structures, injured one person, and forced an evacuation of 3600 people in the cities of Glendora and Azusa, California.

3. Results

3.1. Comparison of LFM and VIs

In-situ LFM and VIs showed similar interannual patterns with different amplitudes (Figure S1). Among the VIs, EVI showed the highest correlation (Figure S2). In fact, MODIS EVI was developed to enhance the sensitivity to a wider range of vegetation conditions and to improve vegetation monitoring through a decoupling of the canopy background signal and a reduction in atmospheric influences.

Regarding the difference between Aqua and Terra, the highest correlation to in-situ LFM was EVI derived from Aqua data when compared to the Terra data, and the lowest correlation with NDVI resulted from Terra data. In-situ LFM measurements were collected between 12 p.m.–4 p.m., spanning the local overpass time of Aqua (around 1:30 p.m.), while the data acquisition local time of Terra (around 10:30 a.m.) caused a mismatch with the timing of LFM sampling. Another issue was the gain drift problem in Terra data [19], which was corrected in MODIS Collection 6; however, the residual error remained larger in Terra products. Thus, we primarily used EVI from Aqua in the subsequent analyses in this paper.

Presented in Figure 2 are 10-year records of daily averaged LFM and the corresponding EVI at the coastal and inland sites. The results show distinctive differences between the two regional sites. For example, the moisture level of chamise at the coastal sites was higher than that at the inland sites, particularly in the moist-up phase (November-April) when LFM increased over the growing season (Figure 2a). Both coastal and inland LFMs attained their minima at the same time around the day-of-year (hereafter DOY) of 275. However, the coastal LFM reached its maximum earlier than that of the inland LFM, at DOY 87 and DOY 133, respectively.

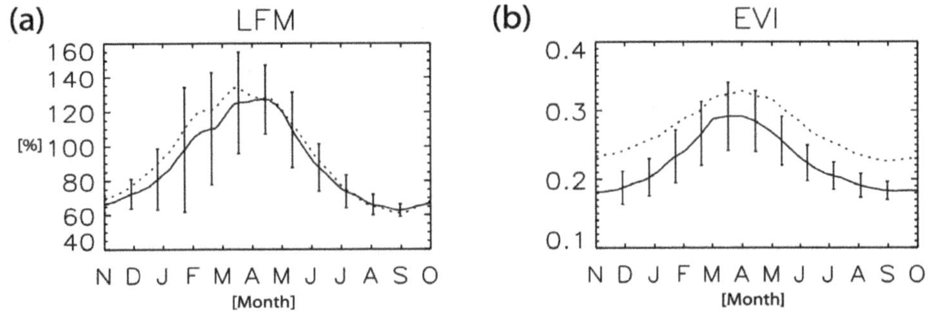

Figure 2. A 10-year daily mean of (**a**) live fuel moisture and (**b**) enhanced vegetation index for the inland sites (solid line) and the coastal sites (dashed line). The x-axis represents months from November to October. Error bars are indicated by vertical bars for the inland sites only.

Regarding EVI, the coastal EVI was consistently higher than the inland EVI, which was attributed to higher vegetation fractions at the coastal sites than those at the inland sites. The maxima occur almost simultaneously around DOY 105 in both regions, but date of minimum EVI at the costal sites was DOY 260, about 40 days earlier than that at the inland sites. While LFM exhibited strong variations in the moist-up phase, EVI had a more definitive peak toward the end of the growth period (Figure 2). These results reflected intrinsic differences between LFM and EVI characteristics.

The time series of LFM and EVI show significant interannual variations at both inland and coastal sites (Figure 3). Limited fuel moisture in chamise was most prevalent in 2007, and the vegetation at the inland sites experienced a greater moisture deficit than that at the coastal sites. In contrast, there was relatively higher fuel moisture in 2003 and 2005. Similar features were consistently found in EVI in both wet and dry years. In addition to the overall pattern, EVI also replicated the differences of LFM between the inland and coastal sites; e.g., higher LFM values at the coastal sites as compared to those at the inland sites during the 2011–2012 winter.

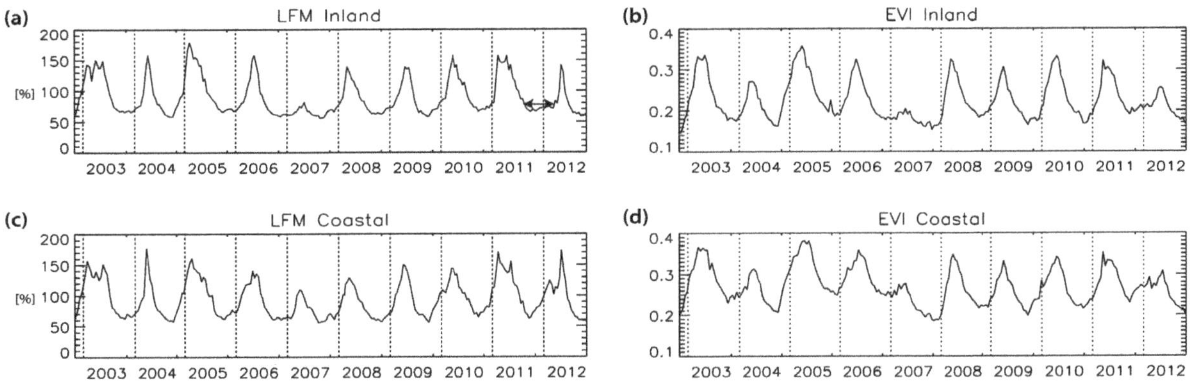

Figure 3. 10-year time series (2003–2012) of live fuel moisture (**a**,**c**) and enhanced vegetation index (**b**,**d**) at the inland sites (**a**,**b**) and the coastal sites (**c**,**d**); The arrow in (**a**) indicates the period of live fuel moisture that was less than 75% in the 2011–2012 fire season.

The time series also reveal different interannual characteristics of LFM and EVI (Figure 3). First, LFM showed higher values of maxima for most years, even during dry years, except for 2007.

For example, during the two dry years, 2004 and 2012 (77% and 27% winter precipitation received with respect to the 1950–2000 mean, respectively), maximum LFM values were close to those in wetter years. In these two years, the start of LFM growth period was delayed and began decreasing earlier, and these changes induced an enhanced kurtosis shape of the LFM time series, and vice versa in the wetter year. The variations in the LFM kurtosis were more pronounced at the inland sites. When compared to LFM, the kurtosis of EVI did not vary substantially, but maximum values fluctuated more interannually, while the rate of EVI seasonal change did not have a strong variation from year to year.

Minimum values of LFM stayed in the 50% range even in dry years (e.g., 2004, 2007, and 2012), albeit that EVI value dropped below normal in those years. Near minimum LFM values at the inland sites continued for seven months during the 2011–2012 fire season (e.g., the arrow in Figure 3a). The persistently low LFM values occurred as the plants sustained a minimal level of moisture for survival by tightly closing their stomata during dry and hot summers to minimize water loss through transpiration [8]. While there was an overall similarity in seasonal and interanual behavior of LFM and EVI, detailed differences in LFM and EVI characteristics would contribute to the uncertainty when estimating LFM from EVI to be discussed in the next section.

To investigate the responses of LFM and EVI to precipitation, we overlaid a time series of precipitation on the LFM and EVI records at Bitter (one of the inland sites) and Schueren (one of the coastal sites) as presented in Figure 4. The intensity and timing of rainfall were closely related with LFM behaviors. LFM started to increase after rainfall and peaked in early summer. LFM showed minimum in fall until rain would start in the next winter. When compared to the Bitter site, the precipitation rate at Schueren was higher and attributable to the higher levels of LFM in fall and winter. Minute amounts of precipitation in 2007 and 2012 caused abnormally low LFM. These results suggested that LFM would directly respond to the water availability from rainfall.

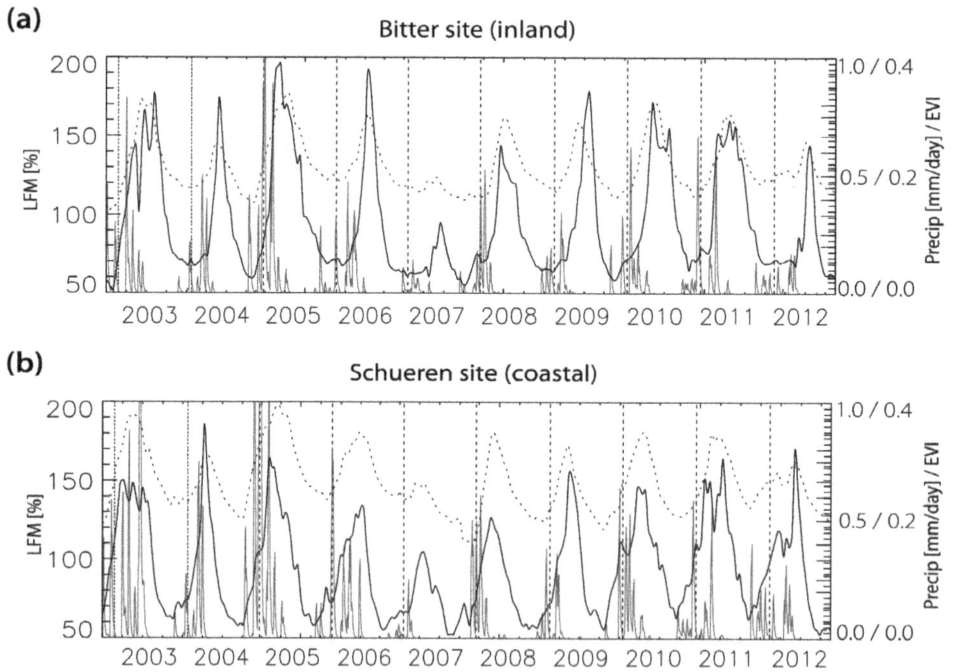

Figure 4. 10-year time series (November 2002–October 2012) of live fuel moisture (solid black) and enhanced vegetation index (dashed) with precipitation (solid gray) at (**a**) Bitter and (**b**) Schueren. The 15-day running mean is applied to each variable. Vertical dashed lines indicate beginning of each year.

EVI exhibits similar characteristics as LFM, except for a delayed response to initial precipitation events in some years. For example, there was a tendency for an increase of LFM prior to an increase of EVI in the early wet season, supported by consistent negative values of the difference between minimum LFM dates and minimum EVI dates, as shown in Figure 5a, for the difference of minimum dates (LFM-EVI). Ranging from −4 to −123 days, this pattern was observed in 68% of the cases at the Bitter and Schueren sites and more obvious at Schueren. When precipitation in a rainy season is significantly reduced, and thus seasonality of LFM becomes vague, the minimum date difference tends to be large (e.g., 2007/08 and 2012/13 cases in Schueren). Regarding the maxima, the timings of the EVI maxima were slightly earlier or later without any consistent bias of sign, compared to those of the LFM ranging from −39 to +68 days (Figure 5b). These results indicate that EVI captures the general seasonal trend of LFM. However, the discrepancy in the timing of minimum and maximum of EVI and LFM suggested that additional factors would be necessary in combination with EVI to accurately capture the seasonal behavior of LFM.

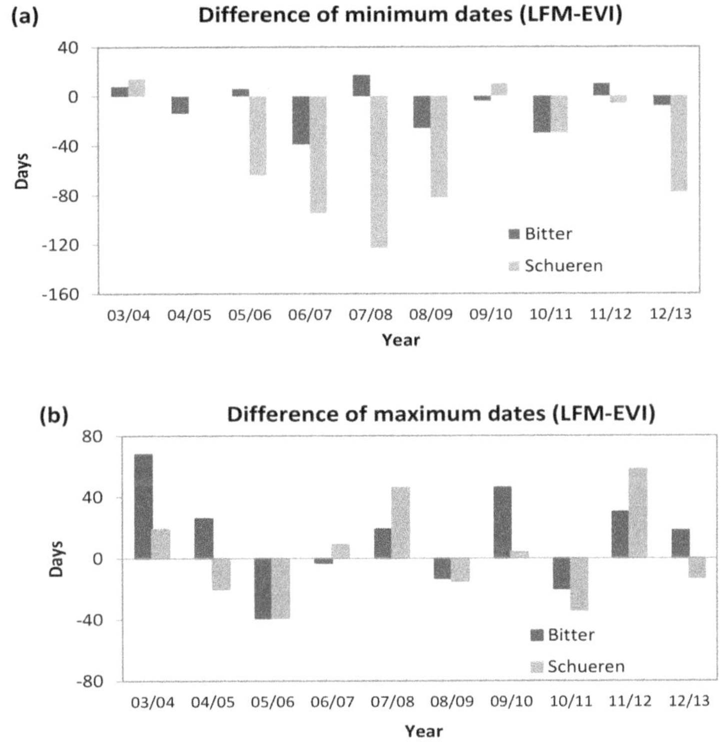

Figure 5. (**a**) Differences between minimum live fuel moisture and enhanced vegetation index dates for each year of the 10 years at Bitter (dark gray bar) and Schueren (light gray bar); (**b**) Same as (**a**) except differences between maximum dates.

3.2. Empirical Model for LFM Estimation

A linear regression model relating LFM to EVI, termed an empirical model function (EMF), was developed at each site, and the results are shown in Table 2. With EVI as a single predictor, constants and coefficients of the EMF were similar among several different sites. Figure 6a,c present time series of in-situ LFM and EVI-based estimates at Schueren and Bitter, respectively. These results highlighted the overall consistency across the 10-year period between in-situ LFM and EVI-estimated LFM. However, because of the significant discrepancies in the maxima and minima, it was necessary to quantitatively evaluate the performance of the EMF in characterizing 10-year averages and interannual variability of the timing, as well as in obtaining magnitudes of maxima and minima. In addition, the date when LFM reached 90% level was also examined. The 90% LFM value represents a transitional stage that approaches high fire danger and an active fire season [5].

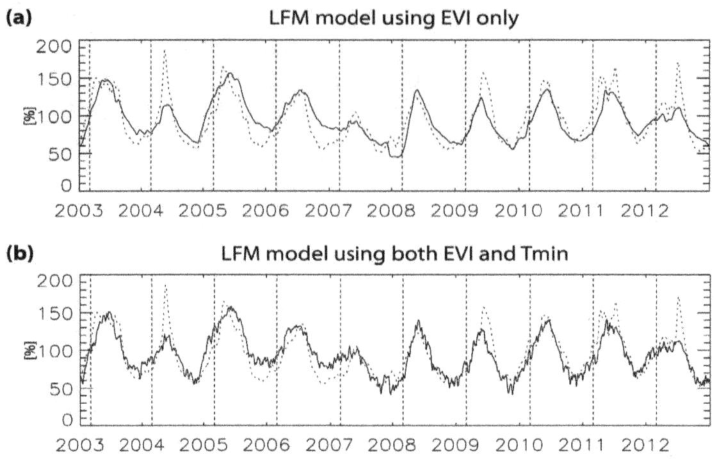

Figure 6. 10-year time series of (**a**) enhanced vegetation index (EVI)-estimated live fuel moisture (solid line) and in-situ live fuel moisture (dashed line) at Schueren; (**b**) is same as (**a**), except EVI and T_{min}-estimated live fuel moisture (solid line).

Table 2. Results of the linear regressions between LFM and EVI.

Site Name	Coefficient (β_1)	Constant (β_0)	R^2	Significance
Bitter	477.93	−3.98	0.73	<0.001
Placerita	669.43	−49.67	0.76	<0.001
La Tuna	538.50	−42.72	0.79	<0.001
Laurel	501.36	−38.76	0.70	<0.001
Trippet	468.59	−27.74	0.65	<0.001
Schueren	479.77	−33.36	0.67	<0.001
Clark	475.74	−27.73	0.74	<0.001

Table 3 represents the differences (estimated LFM minus in-situ LFM) of the 10-year mean values at each site. The results showed that the EVI-based model underestimated maximum LFM values by 10–20% when compared to in-situ LFM. The differences of the date of maximum LFM were usually less than 16 days without any systematic bias in signs, which were consistent with the lack of any systematic bias in the maximum dates of LFM and EVI (Figures 4 and 5). In contrast, EVI slightly overestimated minimum values by 0–7% of the in-situ LFM. Nevertheless, the differences in the minimum dates were significant (2~43 days), corresponding to the temporal lag or delay in EVI minima as compared to that of in-situ LFM minima.

Table 3. 10-year mean of differences between estimation and in-situ LFM for value and date of maximum and minimum LFM, and date of 90% LFM value at each site.

Site Name	Maximum LFM		Minimum LFM		90% LFM
	Value (%)	Date (Day)	Value (%)	Date (Day)	Date (Day)
Bitter	−16.7	−0.7	1.6	17.4	1.2
Placerita	−20.5	−16.2	1.8	10.3	−6.8
La Tuna	−10.7	1.1	1.0	28.8	−3.1
Laurel	−13.5	−3.0	0.5	26.9	−5.5
Trippet	−15.3	5.2	4.2	22.8	9.2
Schueren	−16.7	−6.7	7.0	42.7	1.5
Clark	−15.4	12.0	5.6	2.4	−4.9

With regards to the 10-year correlation (Table 4), interannual maximum values of in-situ LFM were significantly correlated with those of the estimates only in the inland regions (Bitter, Placerita, La Tuna, and Laurel) at the 95% confidence level. For the minimum values, only three sites (Placerita, Schueren, and Clark) showed statistically significant correlations. With respect to dates, the maximum dates of in-situ LFM matched well with those of the EVI-estimates for most of the sites, but minimum dates did not, as indicated by the low or negative correlation coefficients at most of the sites, except Placerita and Clark. These results reflected limitations of the EMF using EVI alone to replicate extrema (especially minima) in LFM values and dates.

Table 4. Interannual correlations between the estimated and in-situ LFM with respect to values and dates of maximum and minimum of LFM, and 90% LFM. Asterisks (*) indicate statistically significant correlations at the 95% confidence level.

Site Name	Maximum LFM		Minimum LFM		90% LFM
	Value	Date	Value	Date	Date
Bitter	0.84 *	0.45	0.32	0.57	0.85 *
Placerita	0.79 *	0.19	0.72 *	0.69 *	0.72 *
La Tuna	0.66 *	0.63 *	0.48	0.34	0.61 *
Laurel	0.70 *	0.82 *	0.42	0.12	0.78 *
Trippet	0.44	0.82 *	0.53	0.42	0.69 *
Schueren	0.27	0.71 *	0.63 *	−0.20	0.80 *
Clark	0.35	0.69 *	0.78 *	0.69 *	0.87 *

Regarding the issue of the underestimation of LFM maxima, a contributing factor was the small interannual variation observed in the maximum values of in-situ LFM (i.e., the large variations of LFM kurtosis in Figure 3) when compared to those of the EVI. The small variation in in-situ LFM maxima was likely a consequence of the LFM sampling practice of fire agencies during wet seasons; that is, when it rains during the period, the two-week interval of LFM measurements was often delayed by a few days to avoid errors that are caused by rainwater-residue on plants during or after rainfall events. Therefore, when fire agencies measure LFM after rainfall, LFM values were generally larger as the time delay allowed more absorption of ample moisture from precedent rainfall. As a result, high values of the LFM maxima were consistently observed for most of the years except for excessively dry years, such as 2007. In contrast, EVI is responsive to canopy physiological variation rather than just vegetation moisture. Furthermore, EVI values are less sensitive to precipitation owing to the spatial averages across large areas where precipitation may occur in different subsectors at different times. Because of the combined effects of these two factors, it is likely that LFM more directly and rapidly reacts to precipitation compared to the EVI response.

Our results also indicate that the estimation error of the EMF for minima is not negligible. Two reasons likely responsible for these limitations were: (1) The presence of the threshold of minimum LFM value (e.g., 50% range), unlike the behavior of the EVI; and, (2) the delayed increases of EVI when compared to LFM in early transition into the growing season, as presented in the previous section. In contrast, the 90% dates were well identified by EVI at all of the sites with respect to both the 10-year mean and variability. The 10-year average differences of the 90% dates were less than 10 days without any systematic bias (Table 3). As a result, the interannual correlations between the 90% LFM date and the corresponding estimates from the EMF were consistently and significantly high.

We also examined correlations between the modeled LFM and in-situ LFM where changing periods of the in-situ LFM reached at 100 to 70% (Table 5). The highest correlations at each site ranged between 0.78 and 0.92 and occurred at either a 90% or 80% wet-to-dry transitional threshold except at Placerita. The 90% and 80% LFM are related to the moderate or high fire danger levels in the 10-year analysis period, the dates of these thresholds varied significantly year to year. Therefore, the high correlation results validated a considerable capability of the LFM EMF to capture the dry vegetation transition into the high fire danger range. A previous study had pointed out that 77% of LFM is a threshold for large historical wildfires in SoCal [27]. This would indeed support that EVI-based estimations in this study could provide valuable information for determining the actual start dates of a fire season and potential dangers of large wildfires.

Table 5. Interannual correlations between the estimated and in-situ LFM with respect to of the dates of various LFM thresholds. Asterisks (*) indicate statistically significant correlations at the 95% confidence level. The highest correlation among the four LFM thresholds at each site is indicated in bold.

Site Name	Dates of In-Situ LFM Thresholds			
	100%	90%	80%	70%
Bitter	0.78 *	**0.85** *	0.81 *	−0.03
Placerita	0.64 *	0.72 *	0.68 *	**0.90** *
La Tuna	0.58 *	0.61 *	**0.78** *	0.60 *
Laurel	0.62 *	**0.78** *	0.68 *	0.40
Trippet	0.68 *	0.69 *	**0.80** *	0.69 *
Schueren	0.67 *	**0.80** *	0.59 *	−0.29
Clark	0.91 *	0.87 *	**0.92** *	0.64 *

3.3. Modified Empirical Model Using Temperature

As indicated in the previous section, EVI alone might not be sufficient to fully replicate LFM due to other factors that are involved in causing dryness of vegetation conditions. To improve the LFM EMF, we tested the EMF model using several meteorological variables from RAWS station as an independent variable in addition to EVI. The results indicate that among the daily temperatures, humidity, and precipitation, the model improvement was highest with the use of daily minimum temperature (T_{min}) (Table S2). It is notable that daily maximum and average temperature and humidity also result in the model improvement, while daily precipitation does not. The EMF using both EVI and T_{min} at Schueren is described in Table 6. The inclusion of T_{min} together with EVI substantially improved the results, especially for lower values (Figures 6 and 7), although the improvement in term of R^2 was not large.

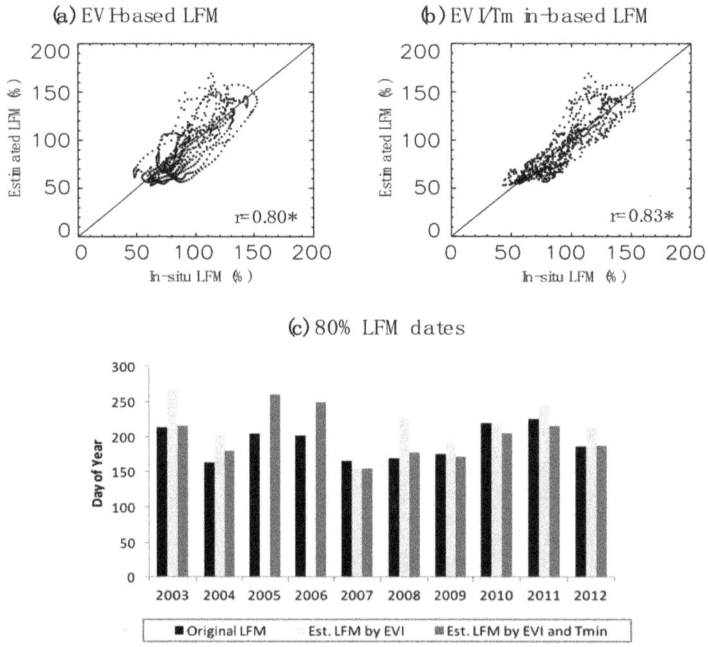

Figure 7. Daily time-scale scatter plots of the in-situ LFM in *x*-axis with (**a**) EVI-estimated LFM and (**b**) EVI and Tmin-estimated LFM in *y*-axis at Schueren; (**c**) Comparison of 80% LFM dates of in-situ LFM (black), EVI-estimated LFM (light gray), and EVI and T_{min}-estimated LFM (dark gray) at Schueren. There were no values for the EVI-estimated LFM in 2006 and 2007 since EVI-alone estimated LFM values were higher than 80% the entire wet-dry season.

Table 6. Results of the linear regressions of LFM at Schueren.

Independent Variable	Coefficient (β_1, β_2)	Constant (β_0)	Significance	R^2
EVI	478.801	−32.9543	<0.001	0.68
EVI, T_{min}	429.641, −1.100	40.5482	<0.001	0.73

In our modified EVI-T_{min} model, the coefficient of T_{min} was negative (−1.1), implying that a higher temperature was associated with lower LFM value for the same values of EVI. This suggested two possible processes that could be responsible: (1) Leaves closed their stomata under extremely hot conditions and surface temperatures consequently increased due to a lack of transpiration from the leaves [8]; and/or, (2) plants lost their moisture under hot and dry weather conditions through evaporation at their stomata or on the surface of their leaves. Anyway, the additional information carried in the minimum air temperature variable certainly enhanced the model performance, especially in the dry season. This is also supported by the closer association of the EVI-T_{min} derived values with the one-to-one line than that of the EVI-alone derived values (Figure 7a,b), especially for drier conditions, represented by LFM values lower than 110%.

Furthermore, wet-to-dry transition timing is better related to the EVI-T_{min} estimation. For example, 80% dates from the EVI-T_{min} estimation were better correlated with those from in-situ sampling than those from the EVI-alone estimation (Figure 7c). In particular, the EVI-alone model failed to detect the 80% date in 2006 and 2007, as the EVI-alone estimated LFM values that were were higher than 80% during the entire wet-dry season. In contrast, the EVI-T_{min} model significantly reduced errors in identifying the 80% date in 2006 and 2007 when compared to the results from the EVI-alone model. EVI-T_{min} model is also better suited for monitoring the transitional levels of LFM.

The improvement attained by adding the temperature variable was not uniform. For example, adding temperature as an additional independent variable at the Bitter site only slightly improved the performance of the EMF. Nevertheless, as low values of LFM signify higher fire danger levels, any improvement in the estimation of LFM, especially during dry seasons, can be valuable in enhancing the capability for fire danger assessment.

3.4. Applying LFM Model to a Real-Life Wildfire Case

The capability of satellite EVI to replicate LFM presented a potentially powerful methodology that could enable daily LFM observations over vast wildland areas in SoCal as well as similar climatic regions that are prone to wildfires around the world. Such capability would set forth a new era for fire danger assessment, one that uses satellite-estimated LFM validated by in-situ LFM. These data products enabled an improvement of more than one order of magnitude of temporal and spatial coverage as compared to the labor-intensive, manual methods currently conducted by fire agencies in their standard bi-weekly in-situ LFM sampling at sparsely selected local locations.

To demonstrate the satellite utility for LFM observations, the real-life case of the 2014 Colby Fire is highlighted as an example here. A quantitative measurement of LFM changes between 25 February 2013 and 8 January 2014 was illustrated over the regional topography in three dimensions, as shown Figure 8. This type of measurement was derived from Aqua MODIS EVI based on the EMF at Glendora Ridge, located within the burned area, i.e., LFM = (417.602 × EVI) + 6.78061. The EMF was developed in the same manner as EMFs were developed in Table 2, except using a nine-year dataset. This is due to the substantial amount of LFM data missing at Glendora Ridge in 2005. The red areas, pervasive mostly in the mountains, corresponded to a sharp decrease of over 80% in LFM on 8 January (Figure S3) from an LFM level of 140% on 25 February 2013. This happened during the 2013 fire season, which was anomalously prolonged into the first quarter of 2014 due to the severe winter drought of California in 2013 and 2014 [28]. Such a drastic plunge in LFM took the normally pre-fire season LFM levels in January in 2014 to below 60% by the week before the Colby Fire ignited. Note that 60%

LFM would be considered the critical fire danger level [5] as being used by fire agencies to implement pro-active fire preparedness measures.

The ignition point (flame symbol, Figure 8) of the Colby Fire was located on the east side of the fire perimeter (yellow contour, Figure 8) within vegetation with critically low LFM values, as observed by MODIS. Fortunately, the Colby Fire did not spread to many critical danger areas around the immediate vicinity due to the aggressive firefighting efforts of the Los Angeles County Fire Department (LACoFD). The LACoFD proactively decided to extend its normal fire season, typically ending around December, due to the extremely low LFM values that were measured throughout the wildlands in the county. The road network overlaid on the satellite LFM map (Figure 8) highlights the encroaching urban growth from several cities (Irwindale, Citrus, Azusa, Glendora, and San Dimas) sprawling into the Los Angeles County wildland. Note that Interstate 210, Route 66, and Route 57 clearly intersect the rough terrain where satellite LFM indicated critical red fire danger levels. This kind of map could be very useful for the National Weather Service (NWS), as well as for local fire agencies in SoCal for fire danger assessment in real-life operational environments. It could also be used by the commercial and private sectors, such as electric utility companies who might have power lines running over critical wildlands, and everyday homeowners who could use such a map to readily check their addresses for potential wildland fire danger in their local community [29].

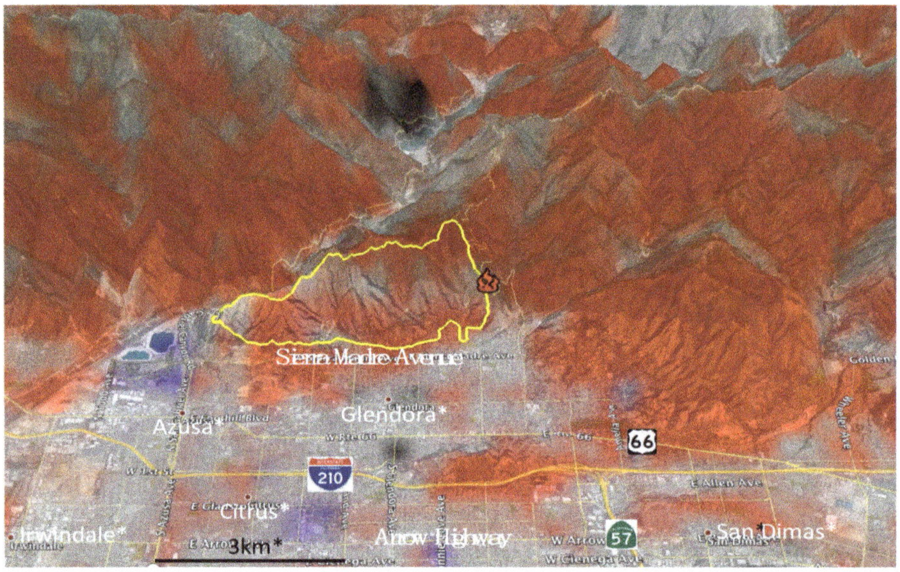

Figure 8. The case of the Colby Fire in January 2014: Fuel dry-up map derived from MODIS data acquired on 8 January 2014 from the satellite Aqua over the San Gabriel Mountains, CA, USA. It clearly shows adjacent cities encroaching into surrounding wildland at multiple wildland-urban interfaces because of urbanization. Red represents severe dry-up due to a ~80% decrease in LFM from a level of 140% on 25 February 2013 resulting in an LFM below 60%, the critical fire danger threshold. The ignition location is marked with the flame symbol on the east side of the fire perimeter denoted by the yellow contours. Since the LFM color map is made translucent to see the landscape features, the accurate full color bar and true LFM map are shown in Figure S3 in the supporting information.

4. Discussion

In this study, we have analyzed climatological, seasonal, and interannual characteristics of LFM and satellite VIs in SoCal in order to develop empirical model functions of LFM based on VIs together with air temperature data based on statistical analyses. Correlation results between LFM and various VIs indicated that LFM was most strongly correlated with EVI from Aqua. Unlike previous studies attempting a point-to-point comparison, we tested the LFM relationship with EVI averaged over different areal coverages in chamise-dominant grids (i.e., 0.5 km to 25 km radius circles), and found that LFM was well correlated with EVI averaged over large areas. It was most strongly correlated with the area of a 10-km radius centered around each in-situ LFM site. As LFM measurements represented information over a large spatial extent, LFM could have high correlations between the time-series data records at different locations as indicated in the high cross correlations. In addition, we found that the higher the cross correlation, the longer the distance between the LFM sampling sites (Table S3). This was an independent verification that measured LFM values at in-situ sites, strategically selected by fire agencies, are indeed a good representation of moisture levels over the extensive neighboring area.

A possible explanation of the better correlation between EVI and LFM is the co-varying leaf pigment concentrations with the change of vegetation water content in Southern California [18]. When plants are under water stress, depletion of chlorophyll may produce a decrease in reflectance in visible and NIR bands. Such change can be prominent in Mediterranean plants as they have a quick response under dehydration. This change may produce a stronger signal than the response in SWIR bands due to the change of vegetation water content.

We have developed the EMF models based on the actual relationship with in-situ LFM, and evaluated the model performance with respect to the 10-year averages and interannual variability. The EVI-alone model showed a limited ability in estimating the timings and magnitudes of annual maxima and minima of LFM primarily because of consistently high in-situ LFM values, even in dry years for the maxima and the delayed response of EVI to precipitation when compared to LFM for the minima. However, seasonal variations, especially wet-to-dry trends of LFM (e.g., 90% LFM date in spring and early summer), were well estimated even on an interannual time scale. This result was consistent with the fact that vegetation greenness represented by EVI was also sensitive to environmental dryness [30]. The study here also found that the EMF model performance for low LFM values during summer and fall could be improved by including an air temperature variable as an additional predictor. This implies that excessive loss of moisture in vegetation on extremely hot days is better captured with the temperature parameter in addition to EVI.

While the 15-day running averaged data were utilized during our model development, a partial autocorrelation error might become non-negligible. We have investigated a transformation method using the Cochrane-Orcutt procedure to adjust the excessive correlation introduced by the temporal autocorrelation at lag 1 [31]. The transformed model showed some reduction in adjusted values of R^2. However, the outcome indicated a similar pattern as results of non-transformed model, thus the temporal autocorrelation will not affect the overall conclusion of this study.

The high correlation results of time-series data at different locations supported the significance of high-resolution satellite data in advancing the capability for fire danger assessment. This is because high-resolution satellite data would enable: (1) A selection of the appropriate vegetation type while eliminating irrelevant land-use classes (e.g., lakes, bare soil, urban areas, etc.); and, (2) an estimation of vegetation moisture condition over the vast extent where in-situ LFM measurements would not be extensively and frequently possible. Moreover, the correlations of time-series data between different locations, while characterizing the seasonal behavior consistently pertaining to the chaparral ecosystem, would not necessarily imply a homogenous spatial pattern of the vegetation conditions. In fact, as shown in Figure 8, the spatial distribution of LFM, as enabled by satellite observations, could be quite variable across the vast wildland. Such observations clearly and independently justified efforts by fire agencies to make LFM measurements at multiple sites critical to fire danger assessment. This result indicates that satellite-derived vegetation data could provide useful information in estimating vegetation moisture levels in SoCal after reducing multiple errors by spatial averaging, and also that in-situ LFM measurements were valid over a large extent beyond the intermediate vicinity of individual sampling sites.

5. Conclusions

The example of LFM map derived by the EMF shown in Figure 8 demonstrates the utility of satellite-based vegetation information for fire danger assessment with a high spatial resolution in SoCal.

Such capability would improve more than one order of magnitude the current temporal and spatial coverage of the in-situ LFM measurement method conducted by fire agencies. The quality of such a kind of map could be further enhanced by LFM observations over the north-facing slopes and their modeling. Note that current LFM sampling sites were typically located in south-facing slopes, and thus the EMF model based on these data may be skewed towards the warmer and generally drier conditions that may result in an overestimation of fire risks. Therefore, LFM observations over the north-facing slopes and their modeling would be necessary for a more complete representation of LFM over complex terrain. Additional effective predictor(s), such as rainfall and soil moisture [32], should also be considered and tested for further improvement of the EMF model. Remotely sensed high-resolution soil moisture data, such as data from the Soil Moisture Active Passive (SMAP) mission [33], might provide additional information about regional soil moisture for synergistic enhancement of LFM estimation models.

There are uncertainties in both in-situ LFM and satellite VI data. First, a small sample size of the in-situ LFM may be insufficient for statistical analysis, resulting in large uncertainty. Weise et al. [5] reported that the uncertainty of LFM measurements varies significantly from ± 20 to $\pm 100\%$, depending on particular sites. Second, because of more intense insolation on the south sides of mountains, vegetation samples were only collected at south-facing mountain slopes; however, actual sample locations could change in different sampling excursions within an approximate three-acre lot selected by fire agencies where the topographic complexity might introduce more uncertainty. Regarding satellite data, VIs might have residual errors due to contaminations from clouds and aerosols that are not completely removed by the processing algorithms. Moreover, minor vegetation species might coexist within the chamise-dominant grid cells. There could be also uncertainties in the spatial coverage mismatch between LFM and VI data.

This study is only focused on the chaparral ecosystem in SoCal. However, the results can be applied to the Mediterranean region in Europe and elsewhere having similar climatic conditions via cross-validation process. In addition, our research framework can be adapted for applications to other wildfire-prone areas in the world with different climate conditions. Moreover, a universal model function approach should be considered in a future research, and such effort can be a key objective when sufficient statistical sampling across the extensive wildland and in-situ data records become sufficiently lengthened.

Supplementary Materials: Figure S1: Time series of vegetation indices and live fuel moisture measurements at Bitter Canyon station, Figure S2: Correlation between LFM and VIs at Bitter Canyon station, Figure S3: same as Figure 8 but with the full-scale color code for the EVI and LFM differences between 25 February 2013 and 8 January 2014, a week before the Colby Fire occurred, Table S1: Equations applied to calculate vegetation indices from MODIS MOD09A1 land surface reflectance product, Table S2: Results of the linear regressions of LFM using EVI and meteorological variables at Schueren, Table S3: LFM cross correlation coefficients among the 7 live fuel moisture sites.

Acknowledgments: The research carried out at the Jet Propulsion Laboratory, California Institute of Technology, was supported by the National Aeronautics and Space Administration (NASA). The support by the NASA Land-Cover and Land-Use Change (LCLUC) Program for the InterDisciplinary Science (IDS) research at JPL on urbanization and impacts, including urban-wildland interface in fire-prone regions, is acknowledged. We would like to thank Nikolas Hatzopoulos for his assistance in obtaining MODIS data sets. Especially, we very much appreciate the collaboration and participation in this research from the National Oceanic and Atmospheric Administration National Weather Service at the Los Angeles/Oxnard Office, the US Forest Service (USFS) Pacific Southwest Research Station in Riverside, the Los Angeles County Fire Department, the Orange County Fire Authority, the Ventura County Fire Department, the Jet Propulsion Laboratory Fire Department, the USFS Predictive Service Southern California Geographic Area Coordination Center, the USFS Angeles National Forest Headquarters, and the Sempra Energy Utility San Diego Gas and Electric. We also appreciate comments from homeowner associations in Los Angeles County.

Author Contributions: B.M., K.W. and S.V.N. performed experiment. B.M., S.V.N., S.J. and S.H.K. wrote this paper. S.V.N. and M.C.K. gave advice and overall comments.

References

1. Syphard, A.D.; Radeloff, V.C.; Keeley, J.E.; Hawbaker, T.J.; Clayton, M.K.; Stewart, S.I.; Hammer, R.B. Human Influence on California Fire Regimes. *Ecol. Appl.* **2017**, *17*, 1388–1402. [CrossRef]
2. Mell, W.E.; Manzello, S.L.; Maranghides, A.; Butry, D.; Rehm, R.G. The wildland-urban interface fire problem—Current approaches and research needs. *Int. J. Wildland Fire* **2010**, *19*, 238–251. [CrossRef]
3. Deeming, J.E.; Burgan, R.E.; Cohen, J.D. *The National Fire-Danger Rating System—1978*; General Technical Report INT-169; U.S. Department of Agriculture, Forest Service, Intermountain Forest and Range Experiment Station: Ogden, UT, USA, 1977; p. 63.
4. Cohen, J.D.; Deeming, J.E. *The National Fire-Danger Rating System: Basic Equations*; General Technical Report GTR-PSW-82; USDA Forest Service, Pacific Southwest Forest and Range Experiment Station: Berkeley, CA, USA, 1985.
5. Weise, D.R.; Hartford, R.A.; Mahaffey, L. Assessing live fuel moisture for fire management applications. In Proceedings of the Fire in Ecosystem Management: Shifting the Paradigm from Suppression to Prescriptiontall Timbers Fire Ecology, Boise, ID, USA, 7–10 May 1998; Pruden, T.L., Brennan, L.A., Eds.; Tall Timbers Research Station: Tallahassee, FL, USA, 1988; Volume 20, pp. 49–55.
6. Countryman, C.M.; Dean, W.H. *Measuring Moisture Content in Living Chaparral: A Field User's Manual*; General Technical Report PSW-36; USDA, Forest Service Pacific Southwest Forest and Range Experiment Station: Berkeley, CA, USA, 1979.
7. Dimitrakopoulos, A.; Papaioannou, K.K. Flammability assessment of Mediterranean forest fuels. *Fire Technol.* **2001**, *37*, 143–152. [CrossRef]
8. Chuvieco, E.; Cocero, D.; Riaño, D.; Martin, P.; Martínez-Vega, J.; de la Riva, J.; Pérez, F. Combining NDVI and surface temperature for the estimation of live fuel moisture content in forest fire danger rating. *Remote Sens. Environ.* **2004**, *92*, 322–331. [CrossRef]
9. Dennison, P.E.; Moritz, M.A.; Taylor, R.S. Examining predictive models of chamise critical live fuel moisture in the Santa Monica Mountains, California. *Int. J. Wildland Fire* **2008**, *17*, 18–27. [CrossRef]
10. Roberts, D.A.; Dennison, P.E.; Peterson, S.; Sweeney, S.; Rechel, J. Evaluation of Airborne Visible/Infrared Imaging Spectrometer (AVIRIS) and Moderate Resolution Imaging Spectrometer (MODIS) measures of live fuel moisture and fuel condition in a shrubland ecosystem in southern California. *J. Geophys. Res.* **2006**, *111*. [CrossRef]
11. Stow, D.; Niphadkar, M.; Kaiser, J. Time series of chaparral live fuel moisture maps derived from MODIS satellite data. *Int. J. Wildland Fire* **2006**, *15*, 347–360. [CrossRef]
12. Peterson, S.H.; Roberts, D.A.; Dennison, P.E. Mapping live fuel moisture with MODIS data: A multiple regression approach. *Remote Sens. Environ.* **2008**, *112*, 4272–4284. [CrossRef]
13. Dennison, P.E.; Roberts, D.A.; Peterson, S.H; Rechel, J. Use of Normalized Difference Water Index for monitoring live fuel moisture. *Int. J. Remote Sens.* **2005**, *26*, 1035–1042. [CrossRef]
14. Bowyer, P.; Danson, F.M. Sensitivity of spectral reflectance to variation in live fuel moisture content at leaf and canopy level. *Remote Sens. Environ.* **2004**, *92*, 297–308. [CrossRef]
15. Moran, M.S.; Vidal, A.; Troufleau, D.; Qi, J.; Clarke, T.R.; Pinter, P.J., Jr.; Mitchell, T.A.; Inoue, Y.; Neale, C.M.U. Combining multifrequency microwave and optical data for crop management. *Remote Sens. Environ.* **1997**, *61*, 96–109. [CrossRef]
16. Foley, J.A.; Prentice, I.C.; Ramankutty, N.; Levis, S.; Pollard, D.; Sitch, S.; Haxeltine, A. An integrated biosphere model of land surface processes, terrestrial carbon balance, and vegetation dynamics. *Glob. Biogeochem. Cycles* **1996**, *10*, 603–628. [CrossRef]
17. Myoung, B.; Choi, Y.S.; Hong, S.; Park, S.K. Inter- and intra-annual variability of vegetation in the Northern Hemisphere and its association with precursory meteorological factors. *Glob. Biogeochem. Cycles* **2013**, *27*, 31–42. [CrossRef]
18. Yebra, M.; Dennison, P.E.; Chuvieco, E.; Riaño, D.; Zylstra, P.; Hunt, E.R.; Danson, F.M.; Qi, Y.; Jurdao, S. A global review of remote sensing of live fuel moisture content for fire danger assessment: Moving towards operational products. *Remote Sens. Environ.* **2013**, *136*, 455–468. [CrossRef]
19. Wang, D.; Morton, D.; Masek, J.; Wu, A.; Nagol, J.; Xiong, X.; Levy, R.; Vermote, E.; Wolfe, R. Impact of sensor degradation on the MODIS NDVI time series. *Remote Sens. Environ.* **2012**, *119*, 55–61. [CrossRef]

20. Lyapustin, A.; Wang, Y.; Xiong, X.; Meister, G.; Platnick, S.; Levy, R.; Franz, B.; Korkin, S.; Hilker, T.; Tucker, J.; et al. Scientific impact of MODIS C5 calibration degradation and C6+ improvements. *Atmos. Meas. Tech.* **2014**, *7*, 4353–4365. [CrossRef]
21. Zhang, Y.; Song, C.; Band, L.E; Sun, G. Reanalysis of global terrestrial vegetation trends from MODIS products: Browning or greening? *Remote Sens. Environ.* **2017**, *191*, 145–155. [CrossRef]
22. Solano, R.; Didan, K.; Jacobson, A.; Huete, A. *MODIS Vegetation Index User's Guide*; Vegetation Index and Phenology Lab, The University of Arizona: Tucson, AZ, USA, 2010.
23. Gitelson, A.A.; Kaufman, Y.J.; Stark, R.; Rundquist, D. Novel algorithms for remote estimation of vegetation fraction. *Remote Sens. Environ.* **2002**, *80*, 76–87. [CrossRef]
24. Serrano, L.; Ustin, S.L.; Roberts, D.A.; Gamon, J.A.; Penuelas, J. Deriving water content of chaparral vegetation from AVIRIS data. *Remote Sens. Environ.* **2000**, *74*, 570–581. [CrossRef]
25. Thenkabail, P.S.; Lyon, J.G.; Huete, A. Advances in Hyperspectral Remote Sensing of Vegetation and Agricultural Croplands. In *Hyperspectral Remote Sensing of Vegetation*, 1st ed.; Thenkabail, P.S., Lyon, J.G., Huete, A., Eds.; CRC Press: Boca Raton, FL, USA, 2011. ISBN 9781439845370.
26. InciWeb, Incidence Information System, Colby Fire. Available online: http://inciweb.nwcg.gov/incident/3766/ (accessed on 1 November 2017).
27. Dennison, P.E.; Moritz, M.A. Critical live fuel moisture in chaparral ecosystems: A threshold for fire activity and its relationship to antecedent precipitation. *Int. J. Wildland Fire* **2009**, *18*, 1021–1027. [CrossRef]
28. Wang, S.-Y.; Hipps, L.; Gillies, R.R.; Yoon, J.-H. Probable causes of the abnormal ridge accompanying the 2013–2014 California drought: ENSO precursor and anthropogenic warming footprint. *Geophys. Res. Lett.* **2014**, *41*, 3220–3226. [CrossRef]
29. Nghiem, S.V.; Neumann, G.; Kwan, J.; Chan, S.; Kafatos, M.; Myoung, B.; Hatzopoulos, N.; Kim, S.H.; Liu, X.; Calderon, S.; et al. *Enhancing Wildland Fire Decision Support and Warning Systems*; A NASA Project for Wildfire Applications Using Satellite Data, Final Report; Jet Propulsion Laboratory, California Institute of Technology: Pasadena, CA, USA, 2013.
30. Schnur, M.T.; Hongjie, X.; Wang, X. Estimating root zone soil moisture at distant sites using MODIS NDVI and EVI in a semi-arid region of southwestern USA. *Ecol. Inform.* **2010**, *5*, 400–409. [CrossRef]
31. Verbeek, M. *A Guide to Modern Econometrics*; John Wiley & Sons Ltd.: Hoboken, NJ, USA, 2004. ISBN 978-88-08-17054-5.
32. Qi, Y.; Dennison, P.E.; Spencer, J.; Riaño, D. Monitoring live fuel moisture using soil moisture and remote sensing proxies. *Fire Ecol.* **2012**, *8*, 71–87. [CrossRef]
33. Entekhabi, D.; Yueh, S.; O'Neill, P.E.; Kellogg, K.H.; Allen, A.; Bindlish, R.; Brown, M.; Chan, S.; Colliander, A.; Crow, W.T.; et al. *SMAP Handbook, Soil Moisture Active Passive, Mapping Soil Moisture and Freeze/Thaw from Space*; Jet Propulsion Laboratory, California Institute of Technology: Pasadena, CA, USA, 2014; 180p.

7

A New Remote Sensing Dryness Index Based on the Near-Infrared and Red Spectral Space

Jieyun Zhang [1,2], Qingling Zhang [1,3,*], Anming Bao [1] and Yujuan Wang [4]

1 State Key Laboratory of Desert and Oasis Ecology, Xinjiang Institute of Ecology and Geography, Chinese Academy of Sciences, Urumqi 830011, China; zhangjieyun16@mails.ucas.ac.cn (J.Z.); baoam@ms.xjb.ac.cn (A.B.)
2 University of Chinese Academy of Sciences, Beijing 100039, China
3 School of Aeronautics and Astronautics, Sun Yat-Sen University, Guangzhou 510275, China
4 China-Asean Environment Cooperation Center, Beijing 100875, China; wang.yujuan@chinaaseanenv.org
* Correspondence: zhangqling@mail.sysu.edu.cn.

Abstract: Soil moisture, as a crucial indicator of dryness, is an important research topic for dryness monitoring. In this study, we propose a new remote sensing dryness index for measuring soil moisture from spectral space. We first established a spectral space with remote sensing reflectance data at the near-infrared (NIR) and red (R) bands. Considering the distribution regularities of soil moisture in this space, we formulated the Ratio Dryness Monitoring Index (RDMI) as a new dryness monitoring indicator. We compared RDMI values with in situ soil moisture content data measured at 0–10 cm depth. Results showed that there was a strong negative correlation ($R = -0.89$) between the RDMI values and in situ soil moisture content. We further compared RDMI with existing remote sensing dryness indices, and the results demonstrated the advantages of the RDMI. We applied the RDMI to the Landsat-8 imagery to map dryness distribution around the Fukang area on the Northern slope of the Tianshan Mountains, and to the MODIS imagery to detect the spatial and temporal changes in dryness for the entire Xinjiang in 2013 and 2014. Overall, the RDMI index constructed, based on the NIR–Red spectral space, is simple to calculate, easy to understand, and can be applied to dryness monitoring at different scales.

Keywords: remote sensing; dryness monitoring; soil moisture; NIR–Red spectral space; Landsat-8; MODIS; Xinjiang province of China

---

1. Introduction

Drought is a frequent natural phenomenon that influences social and economic development, poses a series of environmental problems, and causes natural disasters, such as change of surface water circulation, destruction of agricultural productivity, and desertification [1,2]. Since the middle of the last century, there has been an extensive unanticipated land desertification in arid and semi-arid regions, worldwide, which has threatened the population of many countries [3]. Moreover, according to the Intergovernmental Panel on Climate Change (IPCC), the amount of severely dry land is predicted to increase in the near future. This will be more serious in continental regions, during the summer months [4]. In the current climate change scenario, the consequence of such dryness trends could be catastrophic. Therefore, detecting drought areas, monitoring drought grade, and evaluating the impacts of drought on agriculture, environment, and the economy are critical for regional drought risk control and sustainable development policies [5].

Due to the complexity of drought occurrence, during the last few decades, a great quantity of drought monitoring models and methods, based on ground observation sites, have been proposed [6,7], which mainly use years of ground measured precipitation and evaporation data [8]. These indices, such as the Palmer Drought Severity Index (PDSI) [8], the Rainfall Anomaly Index (RAI) [9], the Crop Moisture Index (CMI) [10], the Bhalme-Mooley Index (BMDI) [11], the Surface Water Supply Index (SWI) [12], the Standardized Anomaly Index (SAI) [13], and the Standardized Precipitation Index (SPI) [14], could accurately reflect the duration and intensity of drought. However, they are difficult to be applied on a large-scale, due to the dependence on site data. Additionally, from a global perspective, the number of observation sites and the duration of meteorological records in many areas are insufficient to detect spatiotemporal variations of drought-related variables [15]. With the development of meteorological grid data, these indices mentioned above have made a breakthrough for large-scale applications, but the coarse spatial resolution of the data is still hard to meet the needs of agricultural drought risk management.

The development of remote sensing technology has effectively compensated for the shortcomings of the traditional site-based monitoring methods. Remote sensing data has become the main data source for drought research because of its low acquisition cost, wide monitoring range, and strong data continuity [16,17]. Over the past decades, a large number of remote sensing data on precipitation, soil moisture, surface temperature, evapotranspiration, and water reserves, have played important roles in drought research [18,19].

Precipitation and soil moisture are the important factors affecting the severity and duration of drought. Currently, a variety of satellite precipitation data has been used for drought research [20–22]. The primary satellite precipitation products are shown in Table 1. Satellite soil moisture products are also widely used in drought research, especially in arid and semi-arid regions. Soil Moisture Ocean Salinity (SMOS) [23], Soil Moisture Active Passive (SMAP) [24], and Advanced Microwave Scanning Radiometer–Earth Observing System (AMSR-E) [25] are the main sources of soil moisture monitoring data based on satellite remote sensing. These data have been widely used in soil moisture inversion and drought risk assessment [26–29]. However, due to the coarse spatial resolutions of the above products, they are also hard to be applied on agricultural drought assessment.

Table 1. A summary of primary global satellite precipitation products.

Precipitation	Spatial Resolution	Temporal Resolution	Launch Time	References
TRMM-TMPA	0.25°	3 h	1998	[30,31]
CMORPH	0.25°	3 h	2002	[32]
GSMaP	0.10°	1 h	2005	[33]
PERSIANN	0.25°	3 h	2000	[34]
GPCP	0.25°	monthly	1979	[35]

More than 99% of the soil spectral change information can be described by the red, near-infrared and shortwave infrared bands [36], which provides an important theoretical support for the use of optical remote sensing data for drought assessment. Researchers have proposed a variety of soil moisture and drought assessment methods, based on hyper-spectral and multi-spectral data.

Studies have analyzed the relationship between hyper-spectral data and soil moisture, based on physical and statistical models [37,38], and have proposed several soil moisture inversion indices [39–41]. Avoiding the difference between complex conditions in the field and spectral data obtained under laboratory conditions, is the main problem for a quantitative monitoring of soil moisture in the future.

Multi-spectral data, which is abundant in source and has a high spatiotemporal resolution compared with hyper-spectral data, has been widely used in drought monitoring. Table 2 shows the major multi-spectral data sources for drought monitoring. However, due to limited number of bands, drought or dryness monitoring can be achieved through band combinations or statistics only, based on the sensitive bands of soil moisture or vegetation status. Gao established the Normalized Difference Water Index (NDWI), based on Landsat/TM Channel 4 and 5, showing a good performance on measuring liquid water molecules [42]. There are several indices based on Red(R) and Near Infrared

(NIR), which were widely used in drought/dryness monitoring with different satellite data. Ghulam proposed the Perpendicular Drought Index (PDI) [43] (which can reflect the soil moisture status) and had it verified at the Beijing Shunyi Remote Sensing Experimental Base. The results showed that there is a high correlation between PDI and in situ drought values calculated from 0–20 cm of soil moisture, with correlation coefficients of $R^2 = 0.57$ ($r = 0.75$). However, PDI might fail under conditions of medium and high vegetation cover, because it does not consider the impact of vegetation cover. Ghulam presented the Modified Perpendicular Drought Index (MPDI) through the elimination of the influence of the vegetation fraction, by introducing the fractional vegetation coverage variable (f_v) which represents the proportion of ground covered by canopies [44]. After correlation analysis with the drought values obtained from in situ soil moisture data, MPDI performed as well as PDI, in the early stages of vegetation growth, with a correlation coefficient of $R^2 = 0.8134$. MPDI showed a better performance under dense vegetation cover conditions, since it considered the vegetation condition in the modeling process. However, parameters such as the vegetation reflectances of red and near infrared band and the soil line, are generally fixed by experience [45].

Table 2. The major multispectral data sources for drought monitoring.

Platform/Sensor	Number of Bands	Spectral Range (μm)	Spatial Resolution	Temporal Resolution	Launch Time
Landsat/TM	7	0.4–2.35	30 m/120 m	16 d	1984
Landsat/ETM+	8	0.45–2.35	15 m/30 m/60 m	16 d	1999
Landsat/OLI	9	0.43–1.39	30 m	16 d	2013
Landsat/TIRS	2	10.6–12.5	100 m	16 d	
Terra/ASTER	14	0.52–11.65	90 m	16 d	1999
HJ-1B/CCD	4	0.43–0.90	30 m	4d	2008
HJ-1B/IRS	4	0.75–12.50	300 m	4 d	
Aqura/MODIS	36	0.4–14.40	250 m/1000 m	0.5 d	2002
Terra/MODIS				0.5 d	2000
NOAA/AVHRR	5	0.55–12.50	1100 m	0.5 d	1979

Since dryness is often related to temperature and vegetation conditions, researchers established several dryness indices based on one single factor or a combination of multiple factors, and the ability of these indices to reflect dryness spatial distribution have been tested [46,47]. Temperature-Vegetation Drought Index (TVDI) is a soil moisture monitoring index constructed on the basis of the Normalized Difference Vegetation Index—Land Surface Temperature (NDVI-LST) feature space, which considered the response of vegetation growth status and temperature, to the change of soil moisture. NDVI is vulnerable to soil background for sparse vegetation, while being insensitive to dense vegetation. As a result, TVDI might fail in areas with low or high NDVI values [48]. In addition, other researchers have combined existing dryness indices as indicators for assessing soil moisture, with vegetation indices or surface temperatures, to construct new indices. Amani developed a 3-Dimensional (3D) space of LST, Perpendicular Vegetation Index (PVI), and soil moisture (SM) to construct the Temperature–Vegetation–Soil Moisture Dryness Index (TVMDI) for dryness estimation and monitoring, and MPDI is the indicator of SM [49]. TVMDI has been evaluated in Yanco, Australia, and the result shows an acceptable correlation with in situ soil moisture data ($r = -0.65$). It was compared with other dryness indices based on satellite data, such as PDI, MPDI, and TVDI, and the result showed that TVMDI is the most accurate index. However, there are a few issues with this index. First, as an important factor in TVMDI, LST was estimated using the Single Channel method, which often reduces the accuracy of dryness assessment [50]. Second, PVI used in the TVMDI can be affected by soil background, reducing its ability to accurately characterize vegetation moisture status. Furthermore, TVMDI uses MPDI as the indicator of soil moisture. Similar to MPDI, there is the issue of using empirical values for the soil line parameters.

Through the analysis of the multi-indicator dryness indices and the indices based on the NIR–Red spectral space, we find that these indices mainly have the following disadvantages: (1) Error propagation during the inversion of each indicator in the multi-indicators index might reduce the accuracy of the dryness index. (2) The impact of vegetation cover on soil moisture estimation is not fully considered. (3) The effectiveness of these dryness indices in the arid and semi-arid regions have not been fully investigated. To overcome these shortcomings, the main objective of this paper is to propose a new operable dryness monitoring index—the Ratio Dryness Monitoring Index (RDMI)—and to evaluate its applicability in semi-arid regions. We constructed RDMI based on the NIR–Red spectral space, implemented it with Landsat-8 images, and validated it with field-measured soil moisture sampling data. Furthermore, we applied RDMI with Landsat-8 imagery in the region around Fukang of Xinjiang, China and with Moderate Resolution Imaging Spectroradiometer (MODIS) imagery in the entire Xinjiang province, in order to test its robustness at different scales.

2. Study Areas and Data

2.1. Study Area

The Xinjiang Uygur Autonomous Region, one of the driest regions on the earth, is located in Northwestern China and the hinterland of the Eurasian continent [51]. The surface landscape includes the Tianshan Mountains, Altai Mountains, and the Kunlun mountains, which are seperated by two vast desert basins, the Tarim Basin, and the Junggar Basin [52].

Xinjiang has a temperate continental climate. The annual average precipitation is 158 mm, and the annual mean temperature is 7.6 °C. In general, Xinjiang is divided into the Northern and Southern part, by the Tianshan Mountains. There are climate differences between Northern Xinjiang and Southern Xinjiang. Northern Xinjiang has a temperate, continental, arid, and semi-arid climate, affected by the Western wind belt. The annual precipitation is about 150–200 mm. Southern Xinjiang has a warm, temperate, continental, arid climate, and the warm and humid airflow is difficult to reach due to the limitations of the terrain. The annual precipitation is about 25–100 mm. The annual potential evaporation in Xinjiang is up to 1800 mm. The ecological environment in Xinjiang is fragile, and the climatic characteristics of low precipitation and high evaporation can often easily lead to drought events.

The second study area is a part of the main agricultural area on the Northern slope of the Tianshan Mountains. Cotton, winter wheat, and spring corn are the main crops in this area. Meanwhile, this area is connected to the Gurbantunggut Desert and is a typical oasis desert ecotone, where the main soil type is sandy soil. This area contains a variety of surface landscapes, such as farmland, wetlands, desert ephemerals, and deserts.

In order to verify the effectiveness of the proposed method, it is necessary to test its robustness for monitoring and assessing dryness, over time and at different scales. We chose the region around Fukang, located on the Northern slope of the Tianshan Mountains, to test the proposed method, at a medium scale, and the entire Xinjiang region, to test it at a large scale (Figure 1).

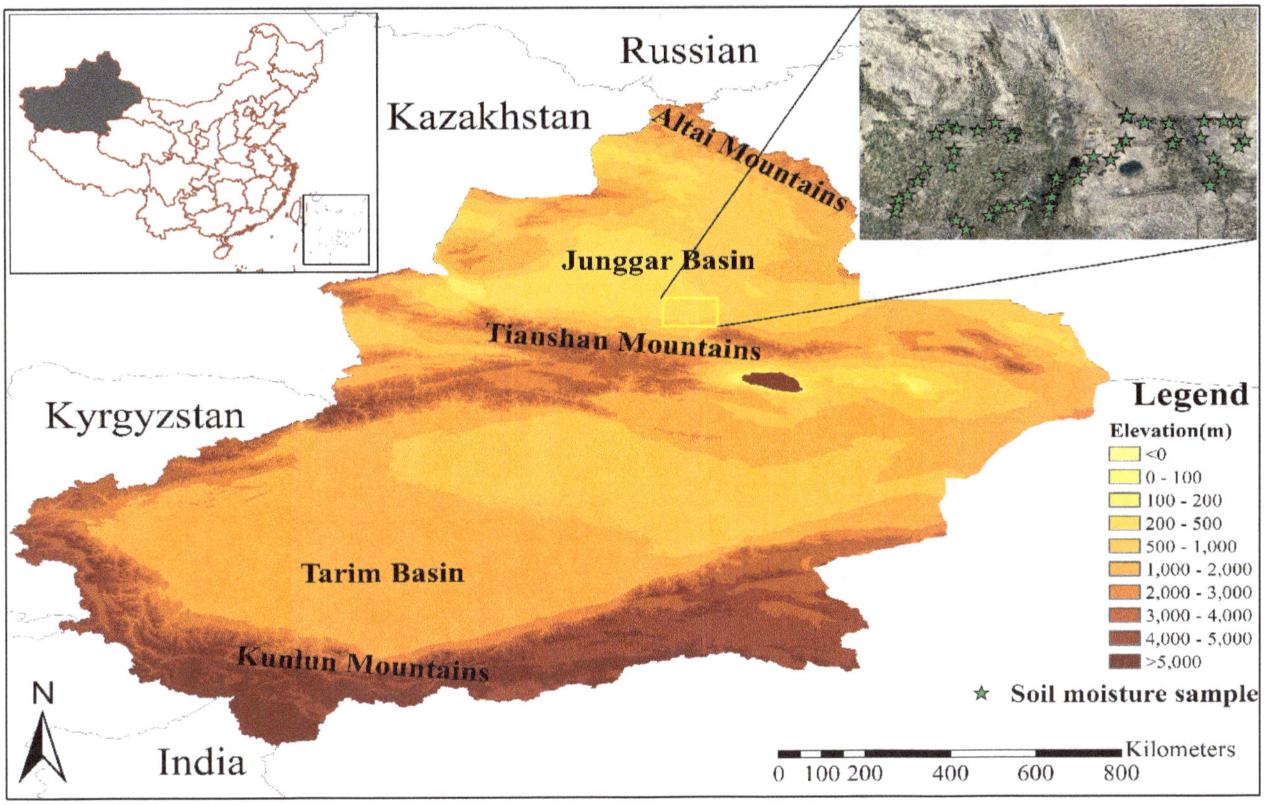

Figure 1. Study area (Xinjiang Uygur Autonomous Region, China). The inset image is the surrounding area of the Fukang City and locations of fifty-one soil samples were taken.

2.2. Field Measurement Data

In the ground soil moisture sampling, taking into account the impact of the cloud cover on the data quality, whenLandsat transits, we carried out multiple ground sampling experiments, according to the satellite transit time, to ensure the matching of the ground sampling time and the high-quality Landsat image. Therefore, from the end of April to mid-July 2014, soil moisture observation experiments were set up at the Fukang Experimental Stations of the Chinese Academy of Sciences, and the surrounding areas. The test sites were located in the surrounding areas of the Fukang City, on the Northern slope of the Tianshan Mountains, as shown in Figure 2. The main soil types in this area are tidal soil and lime desert soil.

In order to ensure the spatial consistency of the location of the sampling area and the Landsat pixel, we used a GPS device to select sampling areas, based on a comprehensive consideration of the sample land cover type and the Landsat pixel position. We selected a total of fifty-one sampling areas, with a minimum interval of one kilometer, between each sampling area. These sampling areas had different land cover types, including cotton (n = 19), winter wheat (n = 8), spring maize (n = 14), bare land (n = 8), and desert (n = 2). The size of each sampling area was 30 m × 30 m, which coincided with the Landsat pixel range, and five points were selected within each sampling area, including the center point. In each sampling area, five points of the GPS position information were first recorded, and then a soil drill was used to separately collect soil samples from 0–10 cm of the soil surface. Finally, soil samples from the five points were mixed and weighed, on site. The weights of the soils in each sampling area, were recorded. The soil samples were taken back into the laboratory, in sealed bags, and dried for 12 hours, in a 105 °C, constant, temperature drying chamber [53]. Each soil sample weight was measured after the drying process. Equation (1) was used to calculate the gravimetric soil moisture content.

$$W = \frac{W_1 - W_2}{W_2} \times 100 \quad (1)$$

where W is the gravimetric soil moisture content (%), W_1 is the weight of the wet soil, and W_2 is the weight of the dry soil.

Finally, through the screening of the Landsat images during the sampling period, the image data of the cloud cover on 12 June 2014 was only 0.06%. Finally, we used the soil sampling experiment data from 12 June 2014 for index verification.

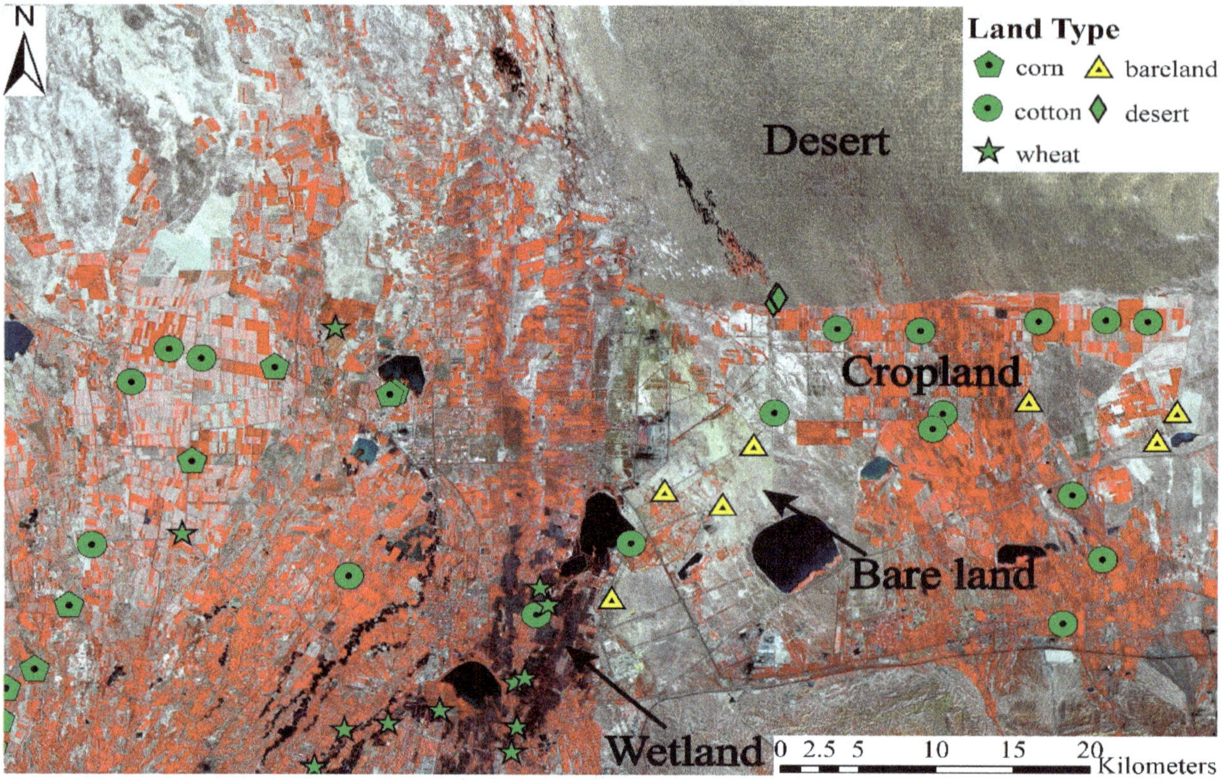

Figure 2. The positions of the fifty-one soil moisture samples with different land cover types in the surrounding areas of the Fukang City. Typical land cover types, such as desert, farmland, bare land, wetlands, are marked with text in the map.

2.3. *Satellite Image Data and Its Pre-Processing*

In order to evaluate the dryness index proposed in this paper and to verify its applicability at different scales, two types of satellite images were used in this study. The red (R) and near-infrared (NIR) bands used in the study were obtained by the Landsat-8, with a medium resolution (30 m), and the MODIS was obtained with a coarser resolution (250 m). The Landsat-8 image was used for the evaluation of the proposed dryness index and dryness mapping in the surrounding areas of Fukang City. In addition, MODIS data were used to generate a dryness map of Xinjiang, demonstrating the applicability of the proposed index, on a large scale.

2.3.1. Landsat 8 Data

The cloud free Landsat-8/Operational Land Imager (OLI) L1T image, from 12 June 2014, was downloaded from the United States Geological Survey (USGS) website to develop the proposed index. The surface reflectance was calculated using the 6S atmospheric correction model. Due to the calibration uncertainties of band 11 of the Landsat-8, the split window method is not recommended to invert LST, as mentioned in Gerace et al., [54]. Band 10 of the Landsat-8 was used for the inversion of LST, through the single-window algorithm in this research, as explained by USGS [55]. Furthermore, the spatial resolution of Landsat-8's thermal infrared bands was resampled to 30 meters, by the product provider. Therefore, the spatial resolution of Landsat-8 imagery data used in this study was 30 meters.

2.3.2. MODIS data

MODIS surface spectral reflectance products were used for mapping the dryness distribution in Xinjiang, China. The Xinjiang Uygur Autonomous Region covers a large space and requires a total of six scenes to achieve regional coverage (Path 23–25, Row 04–05). MOD09A1 contains surface reflectance of Terra MODIS Bands 1–7, and its temporal granularity was eight days. All MODIS products were geometrically and radiometrically corrected and were in a sinusoidal projection system, downloaded from https://search.earthdata.nasa.gov/. The sinusoidal projection of the MODIS products was converted to an Albers equal area projection with a WGS-84 datum, and the products on the same date were mosaiced. All these processes were completed using the MODIS Reprojection Tool (MRT). The georeferencing process was performed in ArcGIS 10.2, with a georeferencing error less than 0.5–1 pixels. The products were extracted using the vector boundary of the Xinjiang Uygur Autonomous Region. The spatial resolution of the MODIS09A1 reflectance product was 500 meters.

We selected 18 tiles of the MODIS image (h23v04, h23v05, h24v04, h24v05, h25v04, h25v05) from 2 June 2013, 2 June 2014, and 21 August 2014, to show the intra- and inter-annual differences in the RDMI index, within the study area.

3. Method

Since the NIR–Red spectral space theory is the basis of the proposed dryness index, as well as other indices, a brief explanation of the NIR–Red spectral space is provided in Section 3.1. Then, the dryness indices based on the 2/3D space, which were used to compare the proposed index in the surrounding areas of Fukang City, are discussed in Section 3.2. Finally, the new dryness monitoring index is proposed in Section 3.3.

3.1. The NIR–Red Spectral Space

Due to vegetation canopies' strong absorption in the red band and strong reflection in the near-infrared band, the red and near-infrared bands in the remote sensing data are used to generate various vegetation indices, such as the Ratio Vegetation Index (RVI), the Difference Vegetation Index (DVI), and the NDVI. The NIR–Red spectral space and the spectral features were originally designed by Richardson and Wiegand to construct the PVI [56]. Based on this theory, Price estimated the vegetation amount from visible and near infrared reflectances [57].

In addition, Ghulam and Zhan found that the near-infrared spectral space contained information on not only vegetation status, but also soil moisture [43]. PDI and MPDI were proposed on the basis of this feature. In the triangular feature space (Figure 3a), because bare soils were not affected by vegetation cover, their reflectivity in red and near-infrared bands are only affected by their moisture content. With the decrease of soil moisture, the reflectivity of the red, and near-infrared bands increase, forming one edge of the triangle in the NIR–Red spectral space, represented by the soil line (Figure 3a). With the increase of vegetation coverage, the rate of increase of reflectance in the near-infrared band is much higher than that in the red band. PVI uses the arbitrary, pixel to soil-line distance, to characterize the degree of vegetation cover. The distribution characteristics of the pixels parallel to the soil line in the triangular feature space are also used by Ghulam and Zhan to describe the connection with soil moisture and to establish their dryness index. Zhan et al. developed the Soil Moisture Monitoring by Remote Sensing (SMMRS) model, based on the distribution characteristics of the near-infrared and red spectral spaces, for dryness monitoring [58]. This model is relatively simple, but the mixed information of soil and vegetation was not considered.

PDI has been verified to be very effective in soil moisture monitoring over bare soils, but similar to SMMRS its performance is greatly reduced in areas with dense vegetation. MPDI has been validated with in-situ soil moisture measurements in the Henan province of China [45]. The characteristics of vegetation and soil moisture in the NIR–Red spectral space were widely used in the studies of Amani and Mobasheri [59,60]. In particular, TVMDI has been verified with Australia's Yanco and

Iran, as research areas. The above-mentioned indices, based on this feature space, are listed in Table 3. From the application of this feature space, it has a good performance in practical applications, such as vegetation conditions and soil moisture monitoring.

Table 3. The dryness monitoring indices based on 2/3D space.

Dryness Index	Indicators	Remote Sensors	Verification Area	References
PDI	Red band (0.63–0.69 μm) Near infrared (0.77–0.90 μm)	Landsat-7 ETM+	The Shunyi Remote Sensing Experimental Base, Beijing, China	[61–63]
MPDI	Red band (0.63–0.69 μm) Near infrared (0.77–0.90 μm)	Landsat-7 ETM+	The Shunyi Remote Sensing Experimental Base, Beijing, China	[64,65]
	Red (0.62–0.67 μm) Near infrared (0.84–0.87 μm)	MODIS	Ningxia Huizu Autonomous Region of China	
TVDI	NDVI and LST	NOAA-AVHRR MODIS Landsat TM/ETM+	Senegal river valley in Senegal Sichuan Basin Fuxin, China Northern China	[66,67]
TVMDI	NDVI, LST and MPDI	Landsat-8 OLI MODIS	Yanco AustraliaIran	[49]

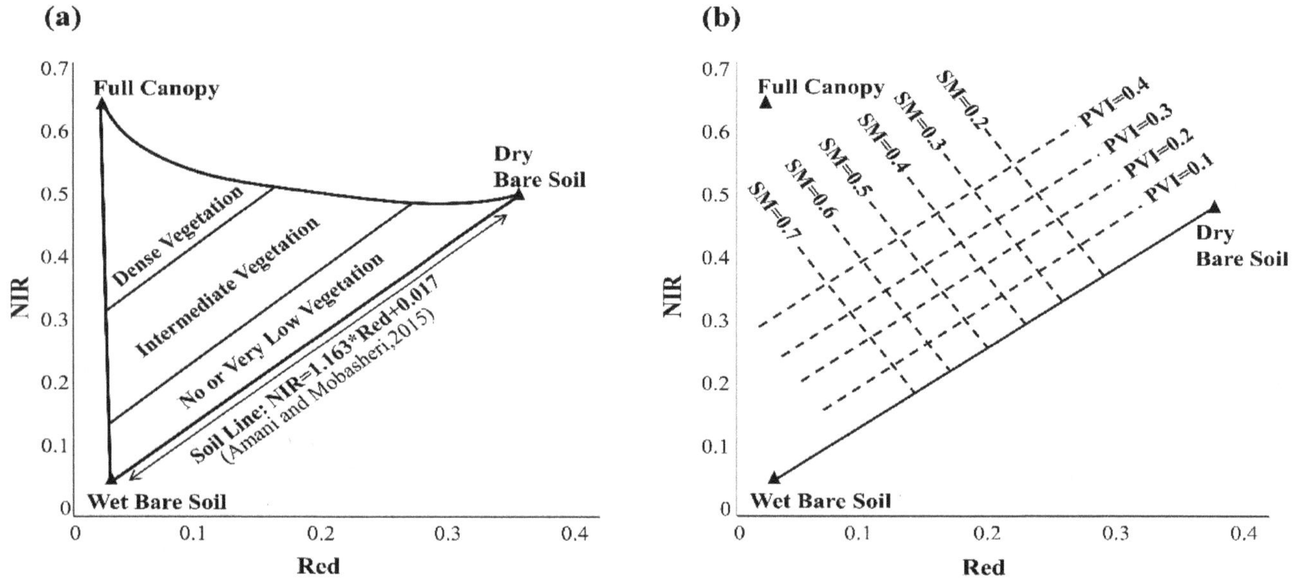

Figure 3. (**a**) The triangle distribution characteristics in the near-infrared (NIR)–Red spectral space. (**b**) The PVI and soil moisture (SM) iso-lines [49]. NIR and Red represent near-infrared and red band reflectance, respectively.

3.2. Satellite-Based Dryness Indices

3.2.1. PDI

PDI is a dryness monitoring index considering the NIR–Red spectral space under different soil moisture conditions, as follows:

$$PDI = \frac{1}{\sqrt{M^2+1}}(R_{red} + MR_{NIR}) \quad (2)$$

where R_{red} and R_{NIR} are the reflectance values of the red and near-infrared bands, respectively, and M indicates the slope of the soil line equation that is defined according to the conditions in the study area. The definition of soil moisture in PDI development is explained by Figure 4. L is a line passing through the origin of the spectral space and is perpendicular to the soil line. The distance from a random pixel to L can be represented as the degree of dryness in the region that the pixel belongs to. Generally speaking, as the distance of a random pixel is closer to L, the dryness value is closer to 0, indicating that the regions are either water bodies or extremely wet areas.

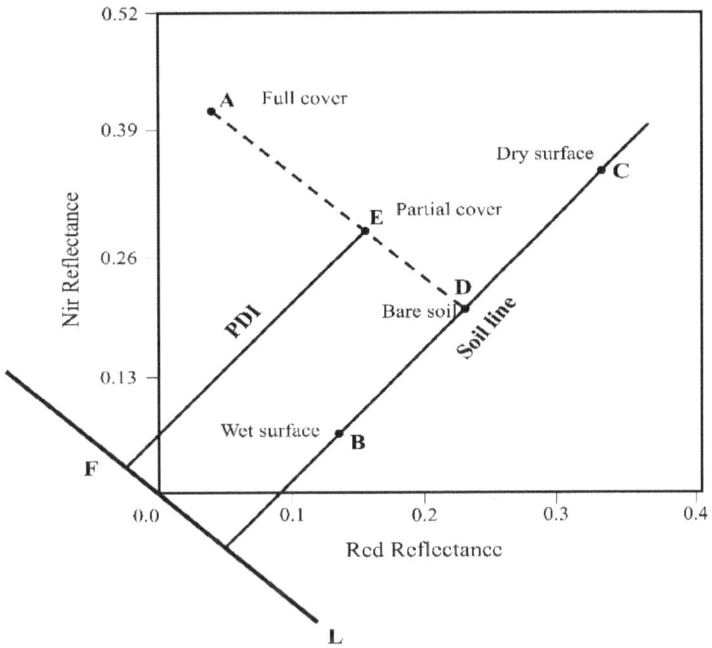

Figure 4. The definition of PDI, the PDI value is defined as the distance of E and F in the NIR–Red spectral space [43].

3.2.2. MPDI

In order to eliminate the influence of vegetation in the PDI, and improve soil moisture monitoring accuracy, Ghulam et al. introduced the vegetation fraction concept and proposed MPDI, on the basis of PDI. This index can effectively reflect soil moisture in areas with different kinds of vegetation coverage. The equation of MPDI is calculated as follows:

$$MPDI = \frac{R_{red} + MR_{NIR} - f_v(R_{v,red} + MR_{v,NIR})}{(1 - f_v)\sqrt{M^2 + 1}}, \quad (3)$$

where R_{red} and R_{NIR} are the reflectance values of red and near-infrared bands, respectively; M indicates the slope of the soil line equation which is defined according to the conditions in the study area; f_v is the vegetation cover fraction, which can be estimated with various methods, such as neural networks [68], linear spectral mixture decomposition [69], and vegetation indices [70]; and $R_{v,red}$ and $R_{v,NIR}$ are the red and near-infrared reflectances of vegetation, which are set to 0.05 and 0.5, by field measurements, respectively [44].

3.2.3. TVDI

TVDI is a dryness index based on the empirical parameterization of the relationship between the surface temperature and vegetation index in the NDVI-LST triangle space, as shown in Figure 5. Its definition is as follows:

$$TVDI = \frac{T_s - T_{s_{min}}}{a + bNDVI - T_{s_{min}}}, \quad (4)$$

where Ts_{min} is the minimum surface temperature in the triangle space that represents the wet edge; T_s and NDVI are the surface temperature and NDVI values for a given pixel, respectively, which are estimated using remote sensing; a and b are the linear fitting parameters of the dry edge, which can be calculated using the maximum surface temperature values, at a given NDVI, as follows:

$$Ts_{max} = a + bNDVI, \quad (5)$$

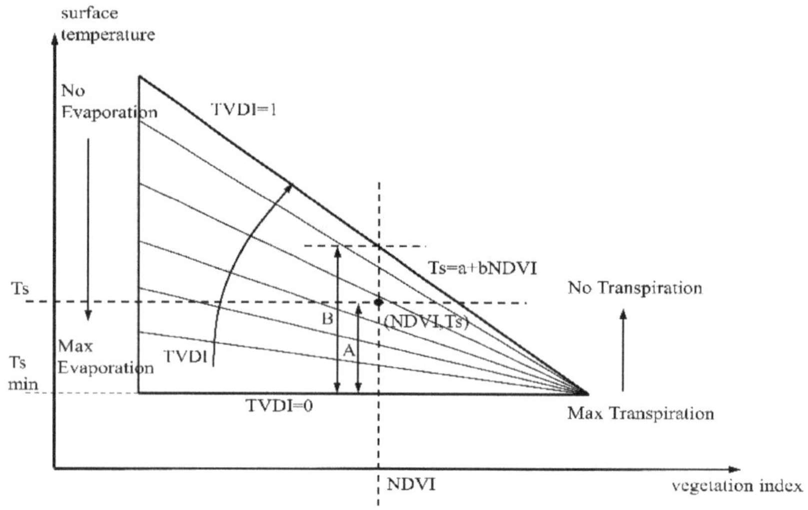

Figure 5. The sketch map of the TVDI. TVDI is established in the two-dimensional space of NDVI and LST, and the TVDI value is defined as the ratio of the distance from the pixel to the wet edge (A) to the distance between the dry and wet edges (B). [46].

3.2.4. TVMDI

TVMDI is constructed in a three-dimensional space integrating normalized land surface temperature (LST_{norm}), normalized soil moisture (SM_{norm}), and vegetation status (PVI), which are commonly used for dryness assessment (Figure 6), as follows:

$$TVMDI = \sqrt{LST_{norm}^2 + SM_{norm}^2 + \left(\frac{\sqrt{3}}{3} - PVI\right)^2}, \quad (6)$$

SM are calculated as follows:

$$SM = \frac{R_{NIR} + \frac{R_{red}}{M} - b}{\sqrt{1 + \frac{1}{M^2}}}, \quad (7)$$

where R_{red} and R_{NIR} are the reflectances of near-infrared and red bands, after atmospheric correction, and M, b, are the slope and the intercept of the linear-fitting formula of the soil line, respectively.

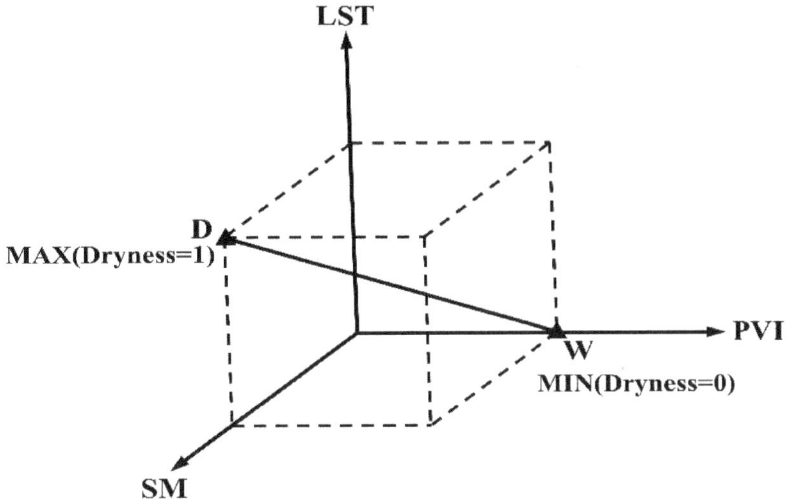

Figure 6. The definition of Temperature–Vegetation–Soil Moisture Dryness Index (TVMDI). TVMDI is based on PVI, LST, and SM as the axis of the three-dimensional space. The TVMDI value is defined as the square root of the square sum of PVI, LST, and SM at the pixel position [49].

All of the above-mentioned dryness indices have their own strengths and weaknesses. PDI and MPDI make full use of the spectral characteristics of soil, under different water conditions, but the PDI index does not take the impact of vegetation cover into account. MPDI introduces the vegetation cover factor to overcome PDI's shortcoming under medium and high vegetation cover conditions. However, in practical applications, the vegetation coverage factor shows uncertainty, due to the use of empirical parameters, in the calculation [62,71]. TVDI introduced land surface temperature and vegetation indices to estimate dryness status, but underutilized the information of red and near infrared bands. In addition, the estimation of TVDI is influenced by the uncertainties of land surface temperature and the NDVI inversion process. TVMDI assesses dryness status by comprehensively considering the land surface temperature, vegetation coverage, and soil moisture. However, the method of TVMDI for soil moisture evaluation uses the theory of PDI, and uncertainties caused by the mixed information of vegetation and soil still exist.

3.3. The Proposed Ratio Dryness Monitoring Index (RDMI)

As shown in Figure 3, the NIR–Red spectral space contains rich information on vegetation coverage and soil moisture. The low reflectance in the red band and high reflectance in the near infrared band is a typical spectral feature of a healthy vegetation. Richardson and Wiegand proposed the PVI, based on the distance of a given pixel to the soil line in the NIR–Red spectral space, to express the vegetation coverage. The distribution of the pixels on the PVI isoline that have the same vegetation coverage condition, is mainly affected by soil moisture and leaf water content of the vegetation [43]. The lower the reflectance in the red and near infrared bands of a given pixel, the closer the pixel is to the lower left in the PVI isoline and the less water stress the vegetation has, and vice versa.

As discussed above, it can be seen from Figure 7 that the line AB is the soil edge and that the line AC represents the smallest water stress areas under various vegetation coverage conditions. AC and BC are defined as the wet edge and the dry edge, respectively.

The position of a random pixel in the approximate triangle, formed by AB, AC, and BC, contains two kinds of information. The first is the vegetation coverage conditions that are expressed as the distance from a given pixel to the soil edge (AB). The second is the water stress status under the same vegetation coverage conditions, which can be expressed as the ratio of the distance from a given pixel to the wet edge (AC) and the distance from the wet edge to the dry edge (BC).

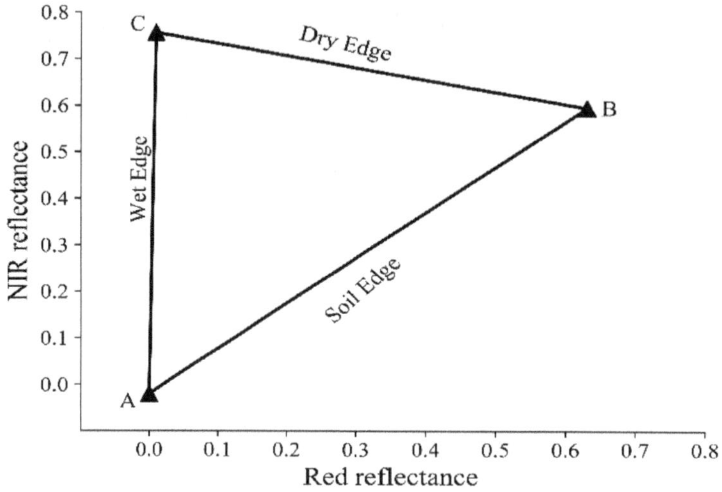

Figure 7. The definition of the triangle edges. AB is the soil edge, AC is the wet edge, and BC is the dry edge. All pixels are distributed in this triangular region in the NIR–Red spectral space.

3.3.1. The Edges of the Triangle Extractions

According to the above discussions, in order to express the water stress state under different vegetation coverage conditions, it is necessary to extract the pixels on the three sides in the scatter plot, for the fitting of the soil edge and the wet edge, respectively. The least squares linear regression was used for the determination of the parameters of the three edges, so as to balance the accuracy of the edge fitting and the performance of the algorithm.

(1) Soil edge:

In the case of the same reflectance in the red band, the reflectance of bare soil in the near infrared band is always lower than that of the vegetation, because of the physical properties of the vegetation leaves [43]. In order to determine the parameters used to describe the soil edge, pixels with the minimum near-infrared reflectance values corresponding to each red reflectance value, were extracted as the bare soil pixel set. Then, the parameters were calculated, using the least squares linear regression method, with the extracted bare soil pixel set. The soil edge equation can be formed as the following:

$$R_{NIR} = S_{soil}R_{red} + I_{soil}, \tag{8}$$

where R_{NIR} and R_{red} are the reflectance values of the near infrared and red bands of the bare soil pixel set, and S_{soil} and I_{soil} are the slope and intercept of the soil edge equation, respectively.

(2) Wet edge:

It can be seen from Figure 7 that, due to the growth of vegetation and the expansion of vegetation coverage in a given pixel, the near infrared band reflectance increases, and the red band reflectance decreases. The reason for this can be analyzed in two aspects. First, the sufficient water content that is required for the growth of vegetation has a strong absorption in the red band, reducing the red band reflectance. Second, the leaf cell structure has a stronger reflection and transmission ability in the near infrared band than the bare soil, increasing the reflectance value in the near infrared band [43]. From the above analysis, we can assume that pixels closer to the near infrared axis are not affected by water stress, and are then defined as the "wet edge". Pixels with the minimum red reflectance values corresponding to each near infrared reflectance value were extracted as the wet pixel set. After that, the least squares linear regression method was introduced to extract the parameters of the wet edge equation, with the wet pixel set. The wet edge can be described as the following:

$$R_{NIR} = S_{wet}R_{red} + I_{wet}, \tag{9}$$

where R_{NIR} and R_{red} are the reflectance values of the near infrared and red band of the wet pixel set, and S_{wet} and I_{wet} are the slope and intercept of the wet edge equation, respectively.

(3) Dry edge:

To determine the dry edge, we must first define the positions of the three points of the triangle. According to the application of the triangle characteristics in the spectral space in previous studies [43,56], the following steps are used to determine the positions of the three vertices of the triangle. First, point A, which represents the wettest bare soil, was the cross-point of the wet edge and the soil edge. Second, point C, which represents the densest healthy vegetation canopies, was calculated using the wet edge equation and the highest near-infrared reflectance in the wet boundary point set. Third, point B, which represents the driest bare soil, was calculated using Equation (7) and the highest red reflectance. According to the positions of B and C, the values of S_{dry} and I_{dry} in Equation (10), which define the dry edge, can then be calculated. Finally, the mathematical description of a triangle space including all pixels was created.

$$R_{NIR} = S_{dry}R_{red} + I_{dry}, \quad (10)$$

3.3.2. Constructing RDMI

Based on the extracted triangular boundary, as shown in Figure 8, we can construct a line passing through a given pixel P (R_{nir}, R_{red}), paralleling the soil line. According to the definition of PVI [56], all the pixels on this line have the same PVI values, which means that they have the same vegetation condition. This line intersects with the wet and the dry edges in D and E, respectively. Based on the above discussion, the position of a point on this line depends on its soil moisture condition. Point D represents the absence of water stress, under this vegetation condition, while point E can represent the maximum water stress situation. The distance DP can represent the difference in the soil moisture situation of point P with point D. Therefore, the degree of dryness, under a certain vegetation coverage condition, can be expressed as the ratio of the distances of DP and DE, as shown in the following formula:

$$RDMI = dist_{DP}/dist_{DE}, \quad (11)$$

Pixels near point D representing the highest soil moisture of this PVI level, will have the minimum dryness values (0), and point E represents the maximum dryness value (1). That is, a larger RDMI value represents, more dryness and less soil moisture, and vice versa.

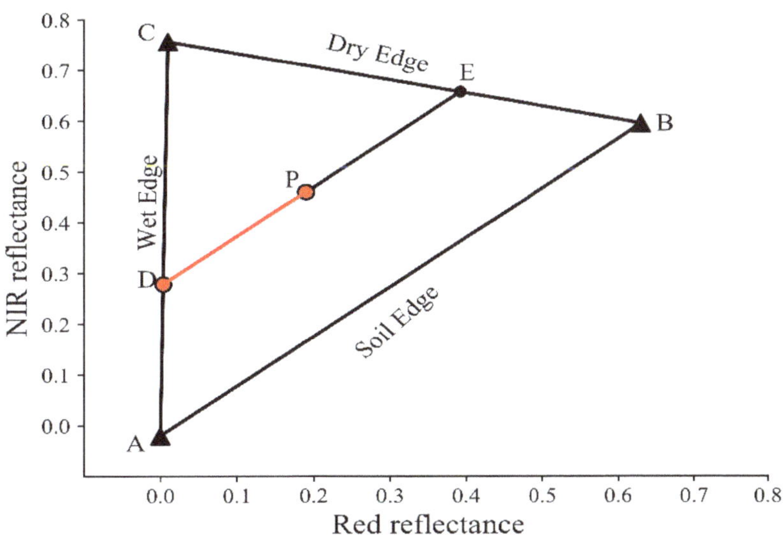

Figure 8. Development of the Ratio Dryness Monitoring Index (RDMI) based on the NIR–Red spectral space. DP and DE are the lines passing through the pixel P that are parallel to soil edge (AB). The RDMI value is the ratio of DP and DE.

3.3.3. Implementing RDMI with Landsat-8

Taking the Landsat-8 data of 12 June 2014 as an example, the RDMI value can be calculated using the band 4 and band 5 reflectance values of the Landsat-8, based on the following steps, as shown in Figure 9.

(a) The pixels on the triangle boundary can be extracted according to the method mentioned in Section 3.3.1. As shown in Figure 10, each pixel in the imagery can be represented by a point which is described by its red and near infrared reflectance values. The pixels of the entire imagery can be abstracted into a n*2 array (n is the number of pixels). According to the value of the red band reflectance, the array is sorted and evenly divided into m groups and, m is decided using the quantity of the imagery and computational efficiency. For each group, the point which has the minimum near-infrared value is selected, and these points form the soil-edge point set. Using the same method, according to the near infrared reflectance values of the points, the array is sorted and divided into groups, and the points with the minimum red reflectance value in each group form the wet-edge point set.

(b) S_{soil}, I_{soil}, S_{wet}, I_{wet} were calculated using the least squares method, based on the point sets extracted. Based on the extracted points sets, S_{dry}, I_{dry} can be calculated using Equation (9).

(c) For a random point P ($R_{NIR,P}$, $R_{red,P}$) in the spectral space, we can define a line with a slope of S_{soil} passing through this point. The intersection points D and E with the dry edge and the wet edge can be obtained by solving the intersection of Equations (9) and (10), respectively. $dist_{DP}$ and $dist_{DE}$ can be calculated using Equation (12).

$$dist_{p1,p2} = \sqrt{\left(R_{red,p1} - R_{red,p2}\right)^2 + \left(R_{NIR,p1} - R_{NIR,p2}\right)^2}, \tag{12}$$

$dist_{p1,p2}$ is the distance of point $p1$ and point $p2$, $R_{red,p1}$ and $R_{red,p2}$ are the red band reflectances of point $p1$ and point $p2$, and $R_{NIR,p1}$ and $R_{NIR,p2}$ are the near infrared band reflectances of $p1$ and point $p2$. The RDMI value of point P can be calculated using Equation (11).

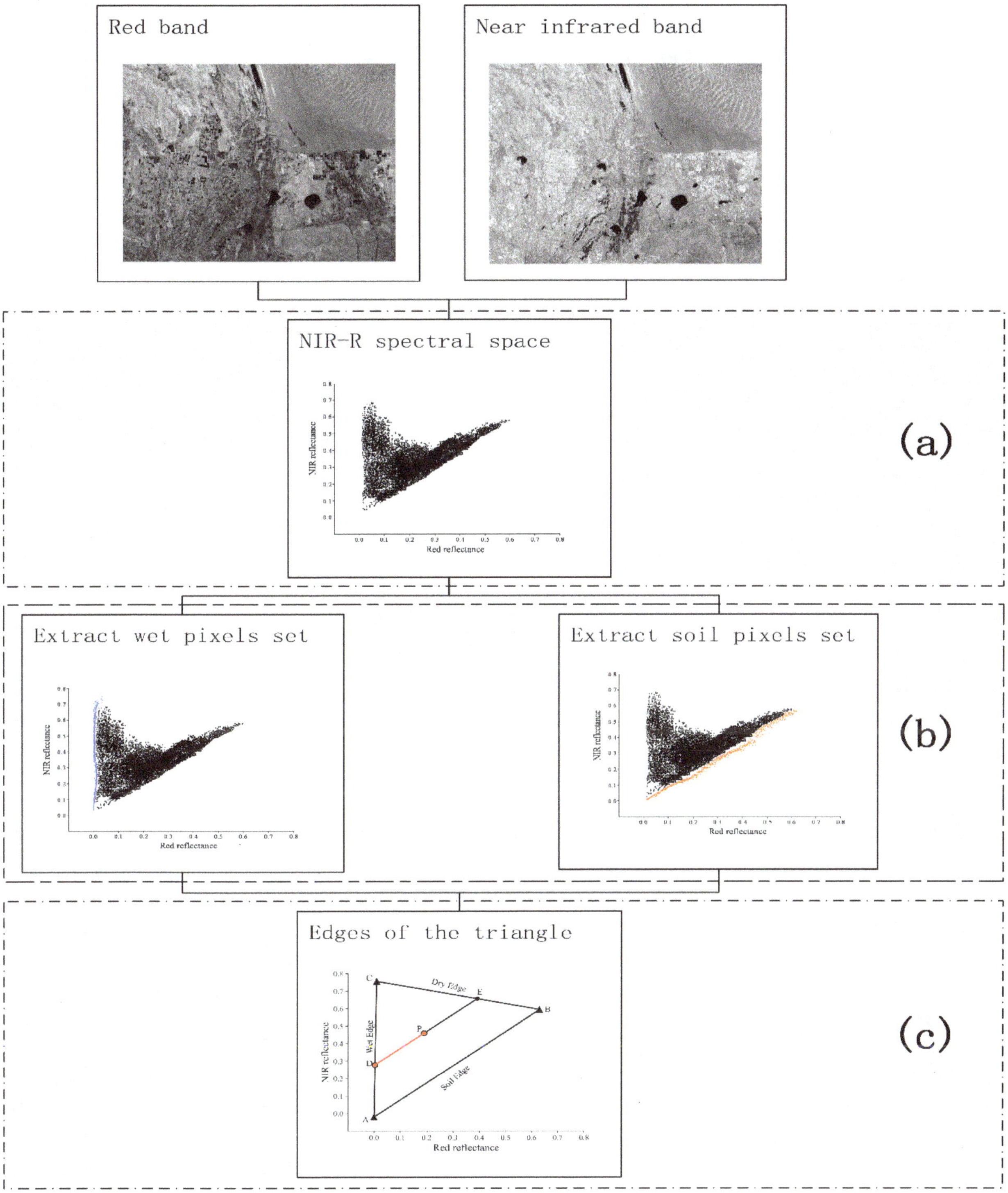

Figure 9. Flowchart for calculating the RDMI. (**a**) Establishment of the NIR–Red spectral space, (**b**) extracting the triangle boundary point set, and (**c**) the triangle boundary fit according to the point set, and calculation of the RDMI.

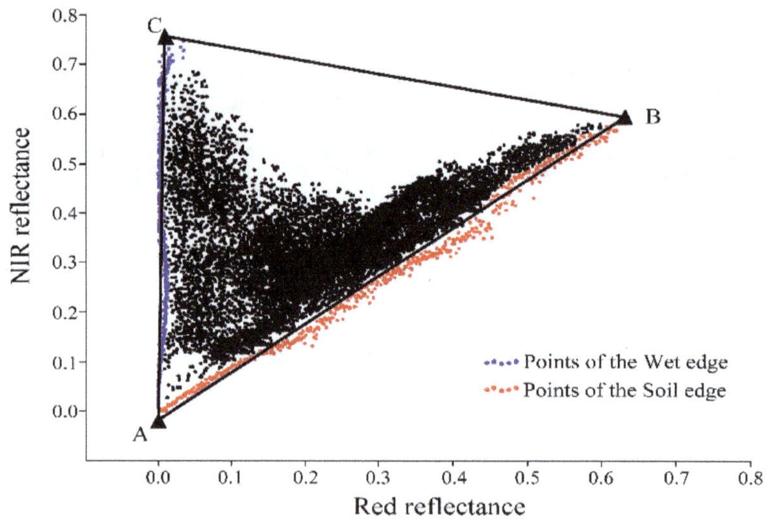

Figure 10. The triangle distribution of the NIR and red bands values in the NIR–Red spectral space generated with real Landsat-8 OLI data. The yellow points represent the soil pixels and the blue points represent the wet pixels.

In summary, PDI, MPDI, and TVMDI are based on the dryness index established by the distribution characteristics of the soil moisture in the NIR–Red spectral space. Based on the same spectral space, this paper discusses the distribution of soil moisture in the feature space and establishes the RDMI index. TVDI is a simple land surface dryness index based on NDVI and LST, which is realized by calculating the distance between a certain pixel and the cold and hot boundaries. The proposed index is also established from the perspective of the distribution characteristics of soil moisture in a two-dimensional space. Therefore, in this paper, PDI, MPDI, TVDI, and TVMDI, as the indices based on multi-spectral data, are used to compare with RDMI, to evaluate the effectiveness of RDMI. To evaluate the results, the Pearson correlation coefficients (r) were chosen to estimate the strength of the linear relationship between in situ data and the estimated dryness index values, the r results were assessed at 0.001 significance level (P indicates the P-value in this paper) [49]. By comparing the correlation coefficients between the different indices and the measured data, the sensitivity to dryness was compared.

4. Results and Discussions

4.1. RDMI Vs. In Situ Measured Data

In order to evaluate the sensitivity of RDMI, with the soil moisture, we used 26 soil moisture samples from the in situ soil moisture data to analyze the correlation between RDMI and soil moisture. The results showed that there was a significant negative correlation between RDMI and soil moisture ($r = -0.89$, $p < 0.001$), as shown in Figure 11a. In other words, as the soil moisture decreases, the RDMI value increases.

To further illustrate the relationship between RDMI and soil moisture, we performed a linear fit between the RDMI and in situ data, to obtain a linear equation. Based on this linear equation, we calculated the simulated soil moisture ($SM_{estimate}$) for the remaining 25 sample locations. We assessed the performance of this relationship between the $SM_{estimate}$ and the in situ soil moisture (SM), using the root mean square error (RMSE) and coefficient of determination (R^2). As shown in Figure 11b, the RMSE was 2.99%, and the R^2 was 0.76. $SM_{estimate}$ showed a strong correlation with in situ soil moisture data, which means that a change in the RDMI value well reflected a change in the soil moisture levels.

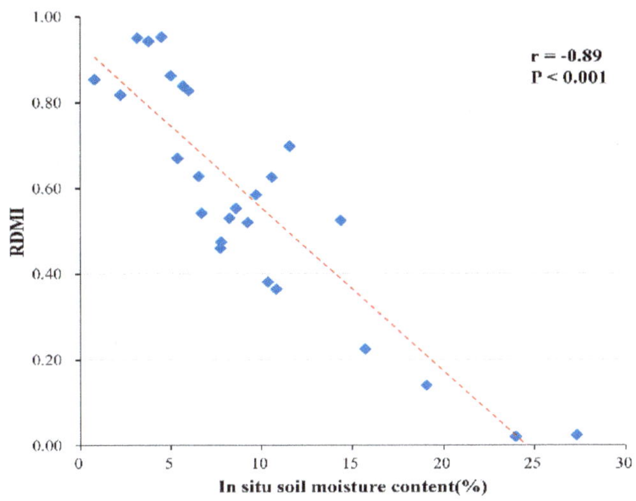

(a)

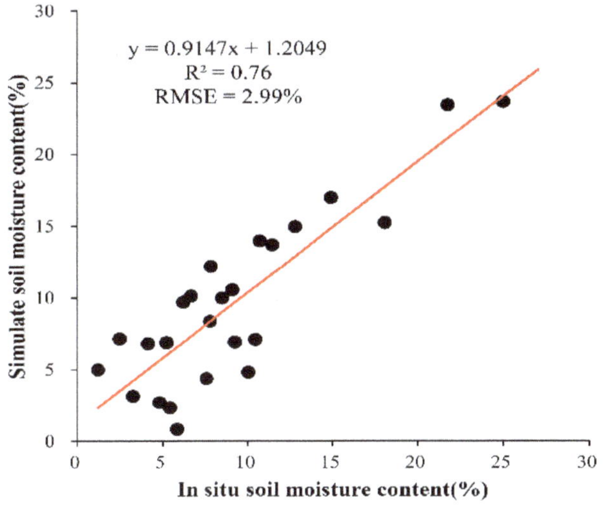

(b)

Figure 11. (**a**) The relationship between the RDMI and in situ soil moisture (n = 26). (**b**) The correlation between the estimated soil moisture content and the in situ soil moisture content.

4.2. RDMI Vs. Other Satellite-Based Dryness Indices

In this section, a comparative study among the RDMI and several existing dryness indices based on remote sensing, including the PDI, MPDI, TVDI, and TVMDI is presented, by calculating each correlation coefficient, slope, intercept, and RMSE of the linear regression, with in situ soil moisture content data (Table 4).

Table 4. The correlation coefficients, slope, intercept, and RMSE of RDMI, Perpendicular Drought Index (PDI), Modified Perpendicular Drought Index (MPDI), TVDI, and TVMDI, under different land cover conditions (n is the amount of in-situ data under this land cover).

		RDMI	PDI	MPDI	TVDI	TVMDI
Total (n = 51)	r	−0.89	−0.72	−0.74	−0.84	−0.87
	Slope	−0.04	−0.03	−0.03	−0.03	−0.03
	Intercept	0.92	0.87	0.89	0.89	0.81
	RMSE	0.09	0.17	0.14	0.11	0.11
Winter wheat (n = 8)	r	−0.76	−0.60	−0.65	−0.64	−0.73
	Slope	−0.03	−0.04	−0.04	−0.02	−0.03
	Intercept	0.79	1.08	1.14	0.72	0.86
	RMSE	0.08	0.19	0.15	0.09	0.09
Cotton (n = 19)	r	−0.64	−0.26	−0.47	−0.61	−0.62
	Slope	−0.05	−0.05	−0.02	−0.03	−0.03
	Intercept	1.01	1.01	0.78	0.91	0.87
	RMSE	0.09	0.17	0.13	0.11	0.10
Spring maize (n = 14)	r	−0.69	−0.50	−0.41	−0.55	−0.65
	Slope	−0.02	−0.02	−0.04	−0.02	−0.03
	Intercept	0.77	0.80	1.09	0.84	0.75
	RMSE	0.09	0.14	0.12	0.10	0.09
Bare land (n = 8)	r	−0.68	−0.56	−0.53	−0.41	−0.50
	Slope	−0.02	−0.01	−0.01	0.01	0.03
	Intercept	0.76	0.81	0.85	0.79	0.87
	RMSE	0.06	0.17	0.08	0.11	0.08

The number of samples with land cover for the desert was only 2, and the correlation coefficient was not calculated.

In order to distinguish the dryness monitoring ability of each index, we analyzed the correlation between the values of each index and the in-situ soil moisture data, under different land cover conditions, as shown in Table 4. The results for all samples indicated that the RDMI shows the strongest linear correlation with soil moisture content data measured in situ, with an r value of -0.89, followed by TVMDI, with an r value of -0.87. The results of the samples under different land cover conditions indicated that RDMI showed a strong correlation with in situ data, under both varied vegetation cover and bare soil conditions. PDI and MPDI showed strong correlation with in situ data under different bare soil conditions. The correlation between TVMDI and in situ data under different vegetation coverages, was second only to RDMI, but, in varied bare soil conditions, the correlation was weaker than in that of other indices. RDMI has a stable response to soil moisture under different land cover conditions, which is better than other dryness indices. This result could be attributed to two aspects. First, the RDMI is constructed on the basis of two variables which reflect the status of vegetation and soil moisture intuitively. Second, the distribution of pixels in the NIR–Red spectral space is mathematically determined.

TVMDI uses the three indicators of LST, NDVI, and SM for a regional dryness assessment. The results showed a high correlation with the measured soil moisture, but there were three issues to be considered. First, there was an inherent correlation between the three variables. A change in one indicator would cause another indicator to change. The impact of this intrinsic correlation on TVMDI should be explored further [49]. Second, TVMDI uses NDVI to characterize vegetation status. A disadvantage of the NDVI, i.e., being saturated in areas with high vegetation coverage, will affect the TVMDI result, over these areas. Third, the LST inversion uses a single-channel algorithm. The real-time atmospheric profile data required in this algorithm is often difficult to obtain for most study areas. Due to the spatial resolution differences between the thermal infrared sensors and visible near infrared sensors, all LST products must be resampled to ensure consistency with other spatial data resolutions [72]. Uncertainty in the LST inversion process can further accumulate in the TVMDI.

Other dryness indices, such as the TVDI, the PDI, and the MPDI showed a relatively weaker correlation with the measured soil moisture data (Table 4). TVDI uses the vegetation index and surface temperature as variables for constructing the LST-NDVI space.Thus, it is more sensitive to the uncertainty of remote sensing-based inversion, as compared with TVMDI. Among all the indices examined, PDI showed the least significant correlation with the measured soil moisture data ($r = -0.72$, $p < 0.001$), and its application for vegetated surfaces is limited, due to the mixed pixel phenomenon [73]. Compared with PDI, MPDI showed a slightly stronger correlation ($r = -0.74$, $p < 0.001$) with measured soil moisture data, which could be attributed to the mixed pixel decomposition elimination of the vegetation component [45]. Soil types and properties are diverse in different areas, as are vegetation types and conditions, but most studies using the MPDI method used fixed parameters, such as the slope of the soil line and the vegetation reflectance of the red and near infrared bands [65]. Such treatment greatly affects the practical applications of MPDI to different regions and at different scales.

Compared with the existing remote sensing-based dryness monitoring index, RDMI is calculated based on the red and near-infrared band reflectance values, directly acquired using satellite sensors, which avoids the accumulation of errors in LST, NDVI, and other inversion processes. Through the mathematical description of the triangular feature space, RDMI explained the variation of soil moisture, under different vegetation conditions.

4.3. RDMI Dryness Maps with Landsat 8 and MODIS Imagery

4.3.1. Dryness Indices Map with Landsat 8 Data

In order to analyze the difference in the dryness monitoring ability between RDMI and other dryness indices, including PDI, MPDI, TVDI, and TVMDI, in the area around Fukang, we used the Landsat-8 data to map the dryness indices in this area (Figure 12). In order to facilitate comparison among these indices, all the mapping results were normalized using Equation (13).

$$I_{normalized} = (X_i - X_{min})/(X_{max} - X_{min}), \tag{13}$$

where $I_{normalized}$ is the normalized dryness index value, is the dryness index value, and X_{max} and X_{min} are the maximum and minimum of the dryness index values.

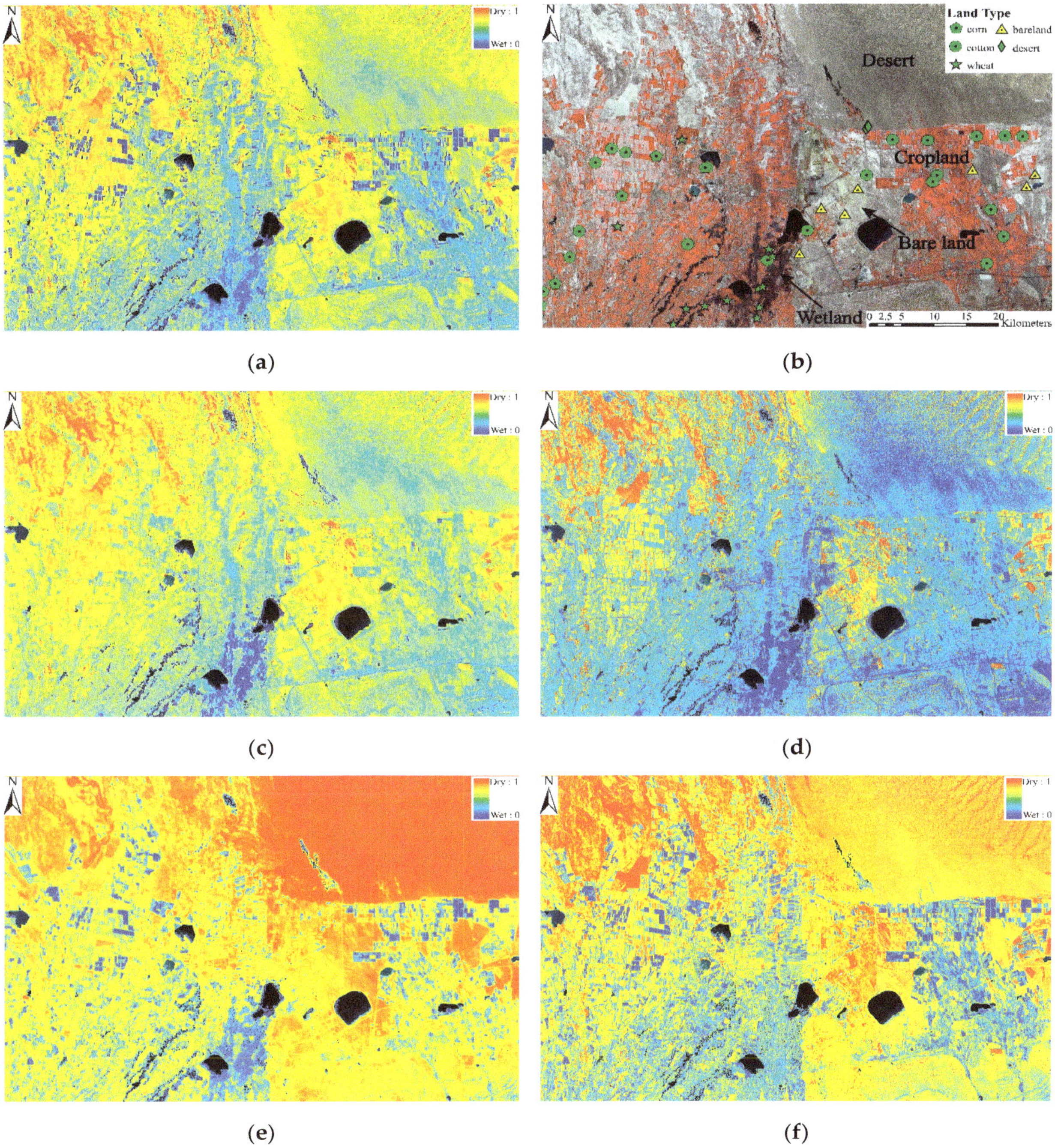

Figure 12. The dryness maps of the surrounding area of Fukang City using Landsat-8 imagery on the 12 June 2014. (**a**) RDMI, (**b**) the false color composite image, (**c**) PDI, (**d**) MPDI, (**e**) TVDI, and (**f**) TVMDI. (All the dryness values in these maps were normalized. The spatial resolution of these maps is 30 m).

Figure 12 shows the dryness of each index in the area around Fukang. The larger the index value in these figures, the higher the surface dryness. Among them, RDMI, TVDI, and TVMDI can clearly distinguish between different land cover types. As irrigation is the main supplier of soil moisture under arid conditions [74], compared with bare land, desert, the cropland has been effectively replenished with water. Therefore, its dryness value is low, and the soil is moist. The dryness values of PDI and MPDI are not significantly different, under different land covers. Figure 13 shows the frequency distributions of the dryness map of these indices. From the statistical distribution of the frequency, RDMI pixels were normally distributed in the range of 0.1–0.7, and the number of pixels with a pixel value of 0.4 was the greatest. RDMI could describe the spatial difference of regional dryness distribution. The pixel values of TVMDI were distributed between 0.1–1, but a large number of pixels were concentrated in the high value range. There were two peaks in the frequency distribution of TVDI. In combination with the information in Figure 12e, the TVDI value in the desert area was abnormally high. PDI pixels were distributed in the value range of 0.5–0.9, and the dryness of this region was significantly overestimated. MPDI was mainly concentrated between the range of 0.15 and 0.4, and the dryness of this region was significantly underestimated. The MPDI's ability to distinguish dryness under different land cover conditions, was limited.

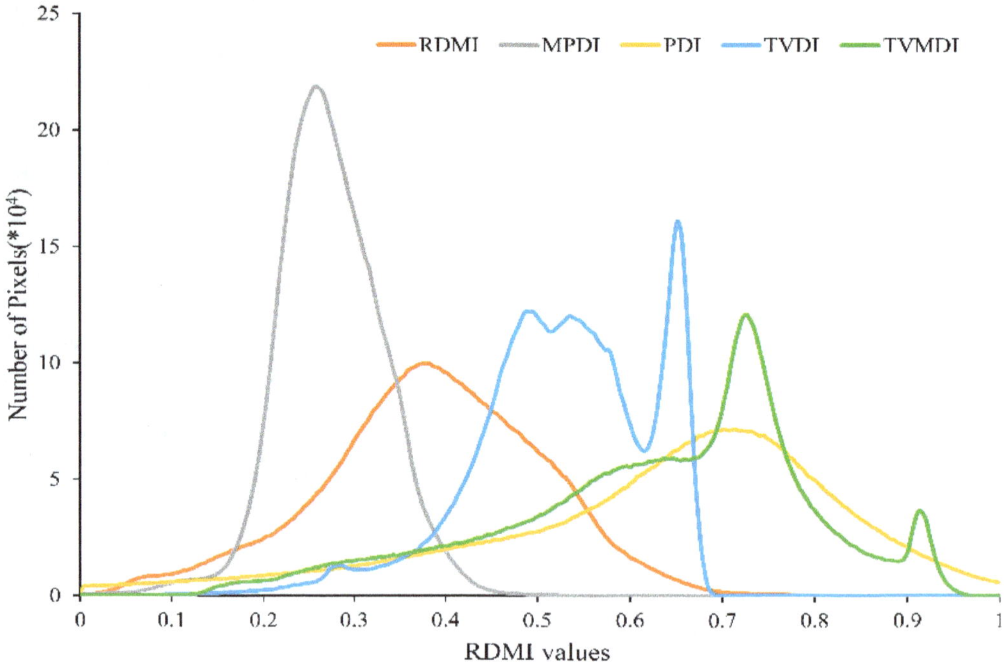

Figure 13. A comparison of the frequency distributions of the RDMI, MPDI, PDI, TVDI, and the TVMDI dryness maps.

It is worth noting that all dryness values of the TVDI in the Gurbantunggut desert, were high. In fact, the soil moisture on the southern edge of the Gurbantunggut Desert was not as low as expected. On one hand, due to the penetration of the snowmelt water in spring, the soil moisture content of sand dunes might have been relatively high during this season. On the other hand, ephemeral plants are widely distributed in this area, which is an important part of desert vegetation as it has a positive effect on maintaining soil moisture [75]. This was faithfully displayed in the RDMI map. In summary, the RDMI exhibited good dryness monitoring capabilities under different land covers, as compared to other drought indices.

4.3.2. The RDMI Map with MODIS Data

To evaluate the applicability of the proposed RDMI over large-scale areas, two RDMI maps were produced for the entire Xinjiang province, using two MODIS images that were acquired on 2 June 2013 and 2 June 2014.

Figure 14a shows the spatial distribution of RDMI on 2 June 2013. It can be seen that the RDMI values are higher in the central area of the Tarim Basin and in the Northern Xinjiang Junggar Basin, where the land surface is much drier than in other areas. Furthermore, the values are lower in the Tianshan mountainous area and West of the Ili River Valley, which are relatively wetter than other areas.

As shown in Figure 15, the soil edge parameters of the three different dates were very stable, the slopes of the soil edges were maintained between 0.91 and 0.92, and the intercept changes were negligible. The slopes of the wet edges had tiny variations among these dates, but it is intuitively seen from Figure 16 that this change had an insignificant effect on the characteristic of the wet edges in the spectral space.

The RDMI values in North of the Altay region, Tacheng, Yumin, and the Northern area of the Kunlun Mountains were relatively low, which means that the soil in these areas was moist. Through the analysis of the spatial distribution characteristics of RDMI in Xinjiang, it could be seen that the RDMI values in the growing season of Xinjiang's mountain forests and the main cultivated areas, maintained a low level, indicating that the soil moisture remained relatively high. As the supply of water resource in Xinjiang mainly comes from precipitation, snow, glaciers, and groundwater in mountain areas, the spatial distribution of water resources in Xinjiang is extremely uneven [76]. The vertical zonal distribution of dry and wet conditions in Xinjiang is relatively obvious [77]. The northern slope of the Tianshan Mountains, the Ili River Valley, and the Southern slope of Altay were relatively humid, while the Taklamakan Desert and the Gurbantunggut Desert were dry [78]. The dry and wet conditions reflected by the RDMI mapping, based on the MODIS images were consistent with the actual dry and wet patterns in Xinjiang.

Figure 14c shows the RDMI distribution on the 2 June 2014. In the spring and summer of 2014, most areas of Xinjiang were dry, because the temperatures in Northern Xinjiang, the Tianshan Mountain area, and the Southern Xinjiang, were higher than that in the previous years, and precipitation in the whole region was lower in the spring of 2014. The Ili River Valley suffered the most severe drought in the past 60 years, and the southern Xinjiang had the least precipitation for the same period of six consecutive months [79]. As can be seen clearly, the RDMI values of 2014 in the Ili River Valley, Aksu, Kashgar, and Hotan areas were higher than that on the same date of 2013. Figure 14d shows the RDMI difference results, for the same period, between the two years. The RDMI spatial distribution of the entire Xinjiang province in 2014, compared to 2013, Tacheng, Ili River Valley, the Northeastern Tarim Basin, and the Eastern section of the Northern slope of the Tianshan Mountains, was significantly dry. In contrast, the Northern part of the Tarim Basin obviously became wetter.

As shown in Figure 14e, since the end of August, the RDMI values in the Altay, Tacheng, Tianshan North Slope Economic Zone, Ili River Valley, Hotan, and Aksu regions were low, indicating that the soil moisture was high. The RDMI values were higher in the Gurbantunggut Desert and in the Taklimakan Desert. Compared with June 2 of the same year (Figure 14f), the RDMI values in most areas of the Xinjiang region did not change much. In the major agricultural areas, such as the Tacheng area, the Northern slope of the Tianshan Mountains, the Ili River valley, the Hotan area, and the Aksu area, the RDMI values became lower, which means the soil became wetter. This is related to irrigation in Xinjiang, in mid-August.

To further illustrate the ability of RDMI for dryness monitoring, we performed a comparison of the frequency distribution of RDMI maps, for the three dates. As shown in Figure 16a, the frequency distribution of the RDMI values shifted to the high-value area on 2 June 2014, compared with 2 June 2013, and the number of high-value pixels was significantly higher than that on 2 June 2013. This shows that Xinjiang, on 2 June 2014, was drier than that on 2 June 2013. As shown in Figure 16b, the frequency distribution on 21 August 2014, was more evenly distributed between 0.2 and 0.6, compared to 2 June 2014, and the peak was significantly lower. This shows that Xinjiang's dryness on 31 August 2014 had, eased as compared to the dryness levels of 2 June 2014.

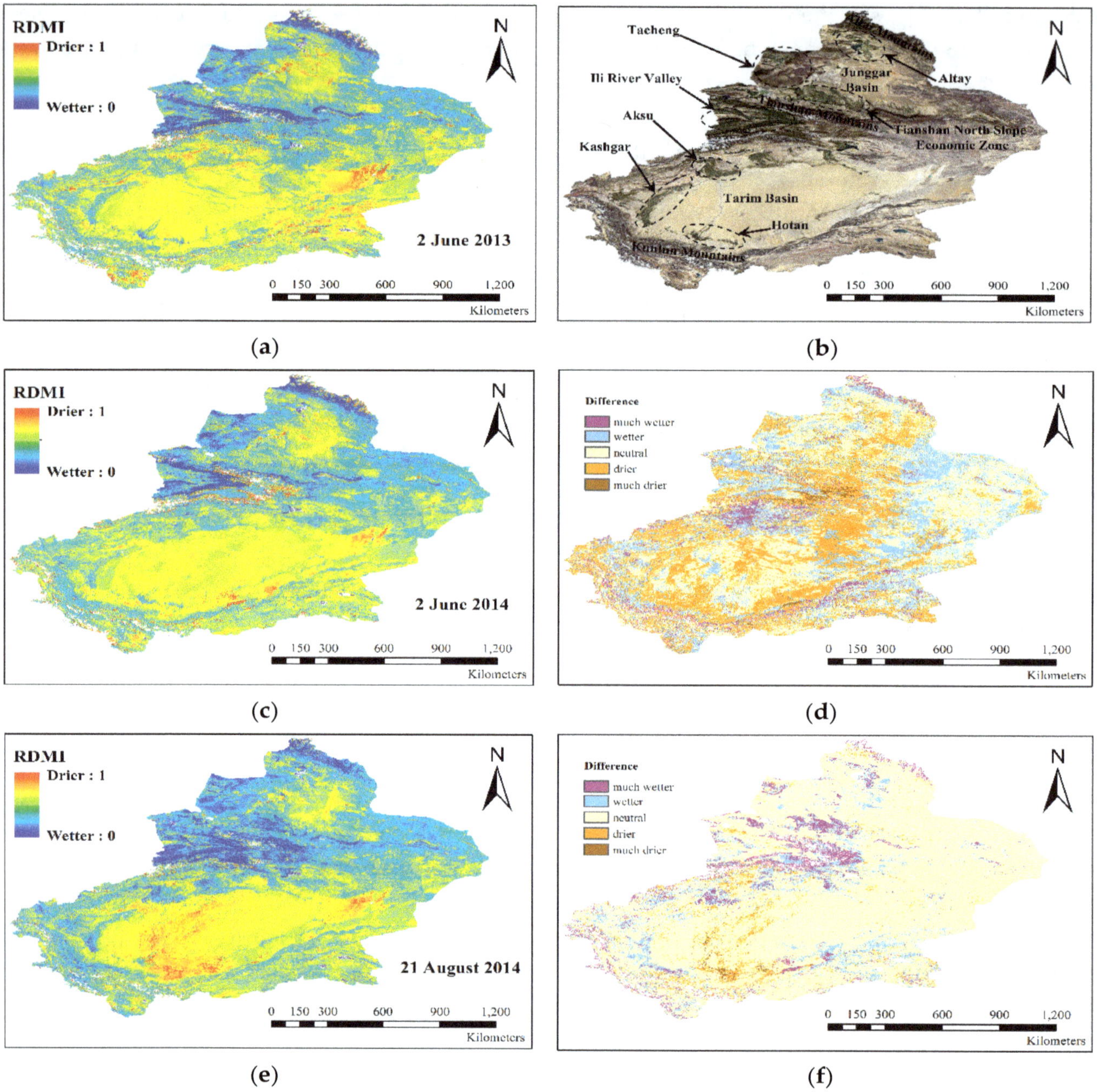

Figure 14. The RDMI maps of the Xinjiang province and the differences in dryness in spatial distributions. (**a**) RDMI map for June 2, 2013; (**b**) illustration of the Xinjiang area; (**c**) RDMI map for June 2, 2014; (**d**) the RDMI difference on the same date of different years; (**e**) RDMI map for August 21, 2014; (**f**) the RDMI difference on the different date of the same year.

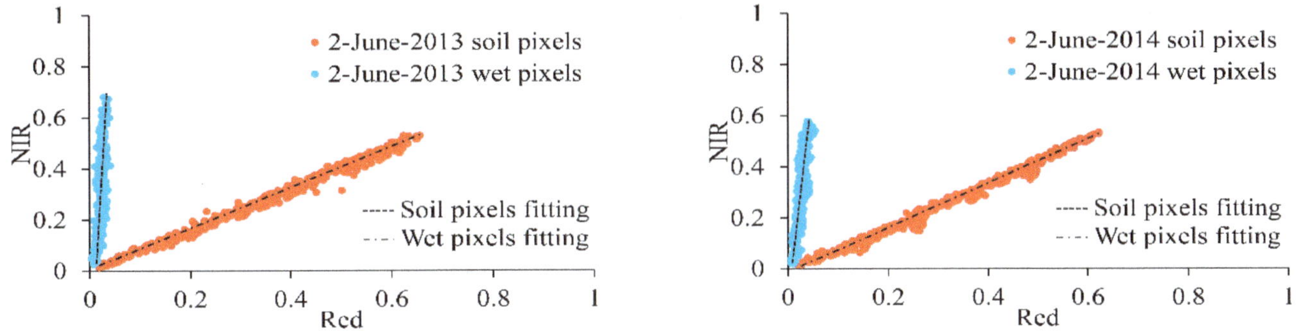

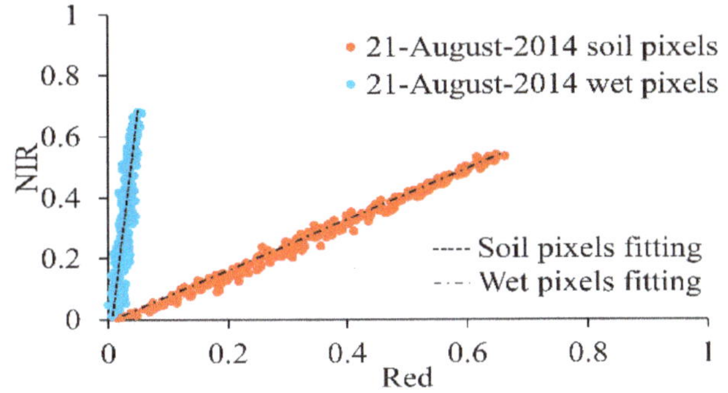

Time	Soil Edge		Wet Edge	
	S_{soil}	I_{soil}	S_{wet}	I_{wet}
2-June-2013	0.91	-0.02	19.76	-0.03
2-June-2014	0.91	0.00	11.06	0.16
21-August-2014	0.92	-0.02	14.61	-0.06

Figure 15. The soil, wet edge pixels, and fitting parameters of the NIR–Red spectral spaces in Xinjiang on three different dates (2 June 2013, 2 June 2014, and 21 August 2014). S_{soil}, I_{soil} are the slopes and intercepts of soil edge, S_{wet}, I_{wet} are the slopes and intercepts of the wet edge.

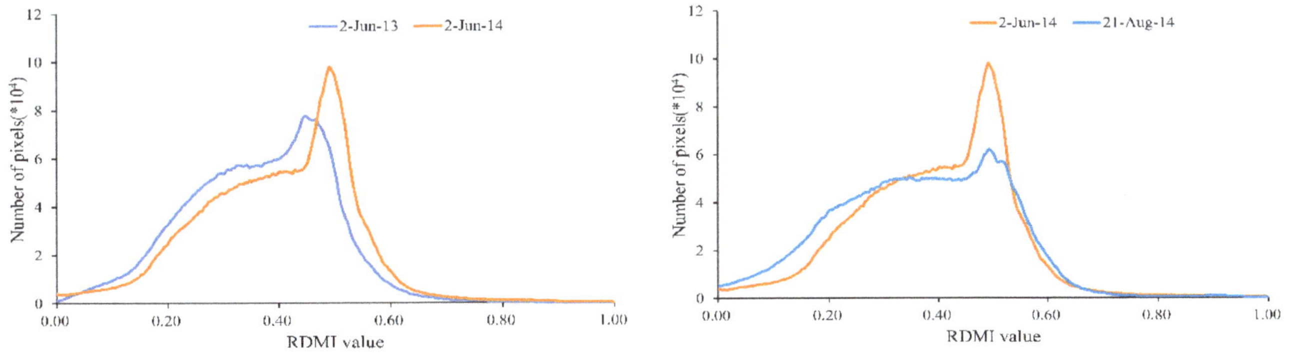

(a) (b)

Figure 16. A comparison of the RDMI frequency distribution on different dates. (**a**) A comparison between 2 June 2013 and 2 June 2014; and (**b**) a comparison between 2 June 2014 and 21 August 2014.

Land cover, regional climate, and soil properties affected soil moisture, and land cover might be the most crucial factor determining the short-term, and eventually long-term, evolution of the soil moisture fields [80]. Different land cover has different evapotranspiration conditions, which affect soil moisture over time [81]. Figure 17 shows the distribution of RDMI mean values in different land cover conditions in Xinjiang. According to the average RDMI distribution under different land cover conditions, the average RDMI of agricultural land, including dry cropland and irrigated cropland, was between 0.2 and 0.4, indicating that cropland soil was wetter than other land covers. With the decrease of grassland vegetation coverage, RDMI gradually increased, indicating that the lower the grassland coverage, the drier the soil. The RDMI of bare land and desert was high, indicating that the soil moisture content was extremely low in these areas. According to the RDMI changes at different times, under the same land cover conditions, the RDMI average values on 2 June 2014 increased for various land cover regions, as compared to 2 June 2013, which showed that the soil moisture in Xinjiang on

2 June 2014 was higher than it was on 2 June 2013. Similarly, the RDMI value on 21 August 2014 was lower than that on 2 June 2014.

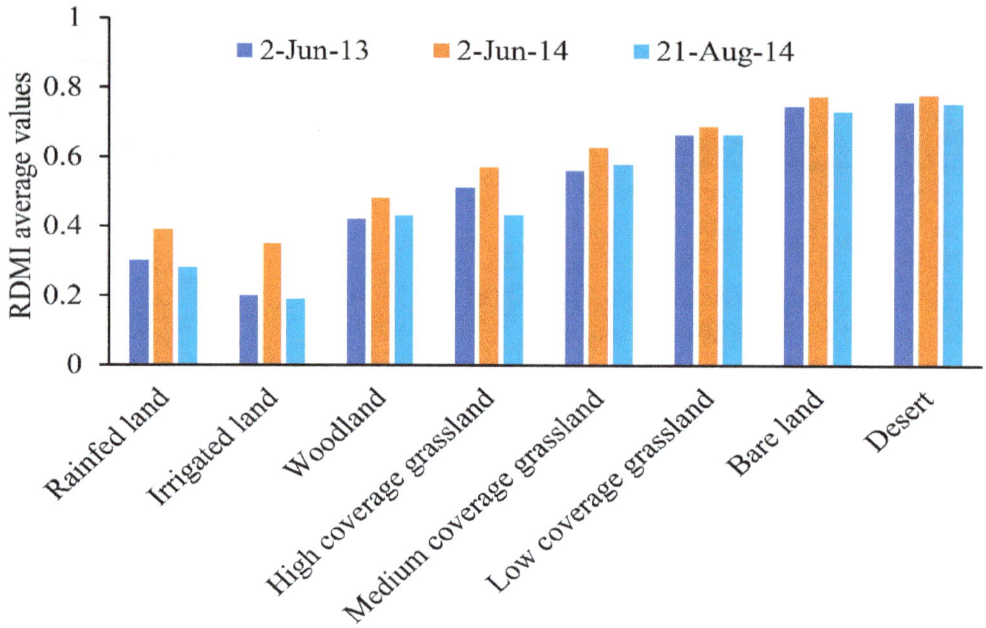

Figure 17. The RDMI average values on different dates for different land cover conditions in Xinjiang.

5. Conclusions

Soil dryness poses serious health, social, economic, and political issues with far-reaching consequences. Therefore, it is crucial to understand how soil dryness affects the environment. In this regard, remote sensing methods have gained increasing attention, due to their larger spatial coverage and frequent temporal sampling abilities. The combination of different bands of spectral reflectance values can reflect the vegetation growth and soil moisture for a given region. Based on previous studies on the spectral space, this paper proposed a new dryness monitoring index, named RDMI, and verified it in the arid area. First, based on the NIR–Red spectral space, we analyzed the pixel triangular distribution characteristics, proposed a fitting method for establishing triangle boundaries, and then constructed a new dryness monitoring index. Second, we validated the proposed index using in-situ soil moisture data. The results showed that the RDMI values had a strong correlation with the in situ soil moisture values ($r = -0.89$, $p < 0.001$). Third, we used RDMI and other dryness indices, such as PDI, MPDI, TVDI, and TVMDI to map the dryness spatial distribution of the surrounding areas of Fukang City, with the Landsat-8 image. By comparing the correlation between the values of these indices and the in-situ soil moisture data under different land cover conditions, we found that the RDMI had a stable performance under various land covers. Finally, in order to verify the application of this index on a large scale, MODIS data acquired from three dates were used to map and compare the dryness spatial distribution in Xinjiang. The results showed that this index could reflect the changes in dryness spatial distributions in Xinjiang, at different times.

The index proposed in this paper expressed the dryness state at a certain moment in a specific region, and is an estimate of the dryness and wetness pattern at a specific time, taken with "snapshot" characteristics. The comparative analysis of the dryness spatial distribution changes in Xinjiang, for the three periods, showed that this index could reflect the change of regional dryness. In the future, we will conduct a comparative study of the long-term sequence dryness spatial distribution, based on this method, and will try to use this method to monitor the regional drought events. We can compare this method with the widely used drought indices, such as PDSI and SPI, to expand the application scope of this method. In addition, this paper used Landsat and MODIS data sources to verify the feasibility of this method for medium resolution and coarse resolution remote sensing data. The results showed that this method had a good performance in dryness monitoring for different spatial resolutions. However, because the acquisition times of these two data sources were different, and the soil moisture might have changed in a short time, the correlation of the RDMI values on these two scales were not analyzed. We will discuss this issue in a future long-term sequence study of RDMI.

In summary, the RDMI index could provide accurate surface dryness information and was robust, for different scales and different surface cover conditions.

Author Contributions: Data curation, J.Z.; Funding acquisition, Q.Z.; Investigation, J.Z.; Methodology, J.Z.; Project administration, Q.Z.; Resources, A.B.; Writing–original draft, J.Z.; Writing–review & editing, Q.Z., Y.W., and A.B.

Acknowledgments: We are very grateful to Yuankang Xiong and Jin Zhao for their help during the soil sample collection. We thank very much the editors and the anonymous reviewers for their insightful comments and suggestions, which helped to clarify the text and improve the manuscript significantly.

References

1. Wilhite, D.A. Drought as a natural hazard: Concepts and definitions. *Drought A Glob. Assess.* **2000**, *1*, 3–18.
2. Mishra, A.K.; Singh, V.P. A review of drought concepts. *J. Hydrol.* **2010**, *391*, 202–216. [CrossRef]
3. Jiang, L.; Jiapaer, G.; Bao, A.; Guo, H.; Ndayisaba, F. Vegetation dynamics and responses to climate change and human activities in central asia. *Sci. Total Environ.* **2017**, *599*, 967–980. [CrossRef] [PubMed]
4. Makundi, W.R.; Shemdoe, R.; Kibassa, D.; Ravilious, C.; Bertzky, M.; Miles, L.; Angelsen, A.; Wertzkanounnikoff, S.; Hatcher, J. *Climate Change and Water*; Intergovernmental Panel on Climate Change Secretariat: Geneva, Switzerland, 2008.
5. Hao, L.; Liu, S. Risk assessment to china's agricultural drought disaster in county unit. *Nat. Hazards* **2012**, *61*, 785–801. [CrossRef]
6. Li, Y.; Chen, C.Y.; Sun, C.F. Drought severity and change in xinjiang, china, over 1961–2013. *Hydrol. Res.* **2017**, *48*, 1343–1362. [CrossRef]
7. Santos, J.F.; Pulido-Calvo, I.; Portela, M.M. Spatial and temporal variability of droughts in portugal. *Water Resour. Res.* **2010**, *46*. [CrossRef]
8. Palmer, W. Meteorological drought. In *U.S. Department of Commerce Weather Bureau Research Paper*; U.S. Government Printing Office: Washington, DC, USA, 1965.
9. Van Rooy, M. A rainfall anomaly index independent of time and space. *Notos* **1965**, *14*, 6.
10. Palmer, W.C. Keeping track of crop moisture conditions, nationwide: The new crop moisture index. *Weatherwise* **1968**, *21*, 156–161. [CrossRef]
11. Bhalme, H.N.; Mooley, D.A. Large-scale droughts/floods and monsoon circulation. *Mon. Weather Rev.* **1980**, *108*, 1197. [CrossRef]
12. Shafer, B.A.; Dezman, L.E. Development of a surface water supply index (swsi) to assess the severity of drought conditions in snowpack runoff areas. In Proceedings of the 50th Annual Western Snow Conference, Reno, NV, USA, 19–23 April 1982.
13. Katz, R.W.; Glantz, M.H. Anatomy of a rainfall index. *Mon. Weather Rev.* **1986**, *114*, 764–771. [CrossRef]
14. McKee, T.B. Drought monitoring with multiple time scales. In Proceedings of the 9th Conference on Applied Climatology, Dallas, TX, USA, 15–20 January 1995.
15. Easterling, D.R. *Global Data Sets for Analysis of Climate Extremes*; Springer: Berlin, Germany, 2013; pp. 347–361.
16. Lessel, J.; Sweeney, A.; Ceccato, P. An agricultural drought severity index using quasi-climatological anomalies of remotely sensed data. *Int. J. Remote Sens.* **2016**, *37*, 913–925. [CrossRef]
17. Atzberger, C. Advances in remote sensing of agriculture: Context description, existing operational monitoring systems and major information needs. *Remote Sens.* **2013**, *5*, 949–981. [CrossRef]
18. Yin, J.; Zhan, X.; Hain, C.R.; Liu, J.; Anderson, M.C. A method for objectively integrating soil moisture satellite observations and model simulations toward a blended drought index. *Water Resour. Res.* **2018**, *54*, 6772–6791. [CrossRef]
19. Hao, Z.; AghaKouchak, A.; Nakhjiri, N.; Farahmand, A. Global integrated drought monitoring and prediction system. *Sci. Data* **2014**, *1*, 140001. [CrossRef] [PubMed]
20. Paridal, B.R.; Collado, W.B.; Borah, R.; Hazarika, M.K.; Sarnarakoon, L. Detecting drought-prone areas of rice agriculture using a modis-derived soil moisture index. *Gisci. Remote Sens.* **2008**, *45*, 109–129. [CrossRef]

21. Damberg, L.; AghaKouchak, A. Global trends and patterns of drought from space. *Theor. Appl. Climatol.* **2014**, *117*, 441–448. [CrossRef]
22. AghaKouchak, A.; Cheng, L.; Mazdiyasni, O.; Farahmand, A. Global warming and changes in risk of concurrent climate extremes: Insights from the 2014 california drought. *Geophys. Res. Lett.* **2014**, *41*, 8847–8852. [CrossRef]
23. Barré, H.M.; Duesmann, B.; Kerr, Y.H. SMOS: The mission and the system. *IEEE Trans. Geosci. Remote Sens.* **2008**, *46*, 587–593. [CrossRef]
24. Entekhabi, D.; Njoku, E.G.; O'Neill, P.E.; Kellogg, K.H.; Crow, W.T.; Edelstein, W.N.; Entin, J.K.; Goodman, S.D.; Jackson, T.J.; Johnson, J. The soil moisture active passive (smap) mission. *Proc. IEEE* **2010**, *98*, 704–716. [CrossRef]
25. Njoku, E.G.; Jackson, T.J.; Lakshmi, V.; Chan, T.K.; Nghiem, S.V. Soil moisture retrieval from amsr-e. *IEEE Trans. Geosci. Remote Sens.* **2003**, *41*, 215–229. [CrossRef]
26. Chen, N.C.; He, Y.Q.; Zhang, X. Nir-red spectra-based disaggregation of smap soil moisture to 250 m resolution based on oznet in southeastern australia. *Remote Sens.* **2017**, *9*, 51. [CrossRef]
27. Liu, D.; Mishra, A.K.; Yu, Z.B.; Yang, C.G.; Konapala, G.; Vu, T. Performance of smap, amsr-e and lai for weekly agricultural drought forecasting over continental united states. *J. Hydrol.* **2017**, *553*, 88–104. [CrossRef]
28. Martinez-Fernandez, J.; Gonzalez-Zamora, A.; Sanchez, N.; Gumuzzio, A.; Herrero-Jimenez, C.M. Satellite soil moisture for agricultural drought monitoring: Assessment of the smos derived soil water deficit index. *Remote Sens. Environ.* **2016**, *177*, 277–286. [CrossRef]
29. Park, S.; Im, J.; Park, S.; Rhee, J. Drought monitoring using high resolution soil moisture through multi-sensor satellite data fusion over the korean peninsula. *Agric. For. Meteorol.* **2017**, *237*, 257–269. [CrossRef]
30. Du, L.; Tian, Q.; Yu, T.; Meng, Q.; Jancso, T.; Udvardy, P.; Huang, Y. A comprehensive drought monitoring method integrating modis and trmm data. *Int. J. Appl. Earth Obs.* **2013**, *23*, 245–253. [CrossRef]
31. Naumann, G.; Barbosa, P.; Carrao, H.; Singleton, A.; Vogt, J. Monitoring drought conditions and their uncertainties in africa using trmm data. *J. Appl. Meteorol. Climatol.* **2012**, *51*, 1867–1874. [CrossRef]
32. Joyce, R.J.; Janowiak, J.E.; Arkin, P.A.; Xie, P.P. Cmorph: A method that produces global precipitation estimates from passive microwave and infrared data at high spatial and temporal resolution. *J. Hydrometeorol.* **2004**, *5*, 487–503. [CrossRef]
33. Janowiak, J.E.; Joyce, R.J.; Yarosh, Y. A real-time global half-hourly pixel-resolution infrared dataset and its applications. *Bull. Am. Meteorol. Soc.* **2001**, *82*, 205–217. [CrossRef]
34. Guo, H.; Bao, A.; Liu, T.; Chen, S.; Ndayisaba, F. Evaluation of persiann-cdr for meteorological drought monitoring over china. *Remote Sens.* **2016**, *8*, 379. [CrossRef]
35. Adler, R.F.; Huffman, G.J.; Chang, A.; Ferraro, R.; Xie, P.P.; Janowiak, J.; Rudolf, B.; Schneider, U.; Curtis, S.; Bolvin, D.; et al. The version-2 global precipitation climatology project (gpcp) monthly precipitation analysis (1979-present). *J. Hydrometeorol.* **2003**, *4*, 1147–1167. [CrossRef]
36. Price, J.C. On the information content of soil reflectance spectra. *Remote Sens. Environ.* **1990**, *33*, 113–121. [CrossRef]
37. Lobell, D.B.; Asner, G.P. Moisture effects on soil reflectance. *Soil Sci. Soc. Am. J.* **2002**, *66*, 722–727. [CrossRef]
38. Bablet, A.; Vu, P.V.H.; Jacquemoud, S.; Viallefont-Robinet, F.; Fabre, S.; Briottet, X.; Sadeghi, M.; Whiting, M.L.; Baret, F.; Tian, J. Marmit: A multilayer radiative transfer model of soil reflectance to estimate surface soil moisture content in the solar domain (400–2500 nm). *Remote Sens. Environ.* **2018**, *217*, 1–7. [CrossRef]
39. Whalley, W.R.; Leedsharrison, P.B.; Bowman, G.E. Estimation of soil-moisture status using near-infrared reflectance. *Hydrol. Process.* **1991**, *5*, 321–327. [CrossRef]
40. Fabre, S.; Briottet, X.; Lesaignoux, A. Estimation of soil moisture content from the spectral reflectance of bare soils in the 0.4–2.5 mu m domain. *Sensors* **2015**, *15*, 3262–3281. [CrossRef] [PubMed]
41. Oltra-Carrio, R.; Baup, F.; Fabre, S.; Fieuzal, R.; Briottet, X. Improvement of soil moisture retrieval from hyperspectral vnir-swir data using clay content information: From laboratory to field experiments. *Remote Sens.* **2015**, *7*, 3184–3205. [CrossRef]
42. Gao, B.C. Ndwi—A normalized difference water index for remote sensing of vegetation liquid water from space. *Remote Sens. Environ.* **1996**, *58*, 257–266. [CrossRef]
43. Ghulam, A.; Qin, Q.; Zhan, Z. Designing of the perpendicular drought index. *Environ. Geol.* **2007**, *52*, 1045–1052. [CrossRef]

44. Ghulam, A.; Qin, Q.M.; Teyip, T.; Li, Z.L. Modified perpendicular drought index (mpdi): A real-time drought monitoring method. *ISPRS J. Photogramm. Remote Sens.* **2007**, *62*, 150–164. [CrossRef]
45. Zhang, J.; Zhou, Z.; Yao, F.; Yang, L.; Hao, C. Validating the modified perpendicular drought index in the north china region using in situ soil moisture measurement. *IEEE Geosci. Remote Sens. Lett.* **2015**, *12*, 542–546. [CrossRef]
46. Sandholt, I.; Rasmussen, K.; Andersen, J. A simple interpretation of the surface temperature/vegetation index space for assessment of surface moisture status. *Remote Sens. Environ.* **2002**, *79*, 213–224. [CrossRef]
47. Kogan, F.N. Application of vegetation index and brightness temperature for drought detection. *Adv. Space Res.* **1995**, *15*, 91–100. [CrossRef]
48. Ghulam, A. Remote Monitoring of Farmland Drought Based n-Dimensional Spectral Feature Space. Ph.D. Thesis, Peking University, Beijing, China, 2006. (In Chinese)
49. Amani, M.; Salehi, B.; Mahdavi, S.; Masjedi, A.; Dehnavi, S. Temperature-vegetation-soil moisture dryness index (tvmdi). *Remote Sens. Environ.* **2017**, *197*, 1–14. [CrossRef]
50. Dash, P.; Gottsche, F.M.; Olesen, F.S.; Fischer, H. Land surface temperature and emissivity estimation from passive sensor data: Theory and practice-current trends. *Int. J. Remote Sens.* **2002**, *23*, 2563–2594. [CrossRef]
51. Yao, J.; Chen, Y.; Zhao, Y.; Mao, W.; Xu, X.; Liu, Y.; Yang, Q. Response of vegetation ndvi to climatic extremes in the arid region of central asia: A case study in xinjiang, china. *Theor. Appl. Climatol.* **2017**, *131*, 1503–1515. [CrossRef]
52. Zhang, Q.; Singh, V.P.; Bai, Y. Spi-based evaluation of drought events in xinjiang, china. *Nat. Hazards* **2012**, *64*, 481–492. [CrossRef]
53. Li, H.X.; Yang, J.; Hao, X.M. Retrieval of soil moisture information in xinjiang using modis. *Acta Prataculturae Sin.* **2017**, *26*, 16–27. (In Chinese)
54. Gerace, A.; Montanaro, M. Derivation and validation of the stray light correction algorithm for the thermal infrared sensor onboard landsat 8. *Remote Sens. Environ.* **2017**, *191*, 246–257. [CrossRef]
55. Available online: https://landsat.usgs.gov/2014 (accessed on 24 December 2014).
56. Richardson, A.J. Distinguishing vegetation from soil background information. *Photogramm. Eng. Remote Sens.* **1977**, *43*, 1541–1552.
57. Price, J.C. Estimating vegetation amount from visible and near-infrared reflectances. *Remote Sens. Environ.* **1992**, *41*, 29–34. [CrossRef]
58. Zhan, Z.; Qin, Q.; Ghulan, A.; Wang, D. Nir-red spectral space based new method for soil moisture monitoring. *Sci. China Ser. D Earth Sci.* **2007**, *50*, 283–289. [CrossRef]
59. Amani, M.; Parsian, S.; MirMazloumi, S.M.; Aieneh, O. Two new soil moisture indices based on the nir-red triangle space of landsat-8 data. *Int. J. Appl. Earth Obs.* **2016**, *50*, 176–186. [CrossRef]
60. Mobasheri, M.R.; Bidkhan, N.G. Development of new hyperspectral angle index for estimation of soil moisture using in situ spectral measurments. *ISPRS Int. Arch. Photogramm. Remote Sens. Spat. Inf. Sci.* **2013**, *40*, 481–486. [CrossRef]
61. Qin, Q.; Ghulam, A.; Zhu, L.; Wang, L.; Li, J. Application of perpendicular drought index in the drought assessment: A case study in ningxia huizu autonomous region of china using modis data. In Proceedings of the 2006 IEEE International Geoscience and Remote Sensing Symposium, Denver, CO, USA, 31 July–4 August 2006; Volumes 1–8, pp. 3078–3081.
62. Shahabfar, A.; Eitzinger, J. Agricultural drought monitoring in semi-arid and arid areas using modis data. *J. Agric. Sci.* **2011**, *149*, 403–414. [CrossRef]
63. Liu, Y.; Yue, H.; Wang, H.; Zhang, W.; Iop. Comparison of smmi, pdi and its applications in shendong mining area. In Proceedings of the International Symposium on Earth Observation for One Belt and One Road, Sanya, China, 25–27 November 2017; Volume 57.
64. Zormand, S.; Jafari, R.; Koupaei, S.S. Assessment of pdi, mpdi and tvdi drought indices derived from modis aqua/terra level 1b data in natural lands. *Nat. Hazards* **2017**, *86*, 757–777. [CrossRef]
65. Sheng, C.; Bingfang, W.; Nana, Y.; Jianjun, Z.; Qi, W.; Feng, X. A refined crop drought monitoring method based on the chinese gf-1 wide field view. *Sensors* **2018**, *18*, 1297.
66. Gao, Z.; Gao, W.; Chang, N.-B. Integrating temperature vegetation dryness index (tvdi) and regional water stress index (rwsi) for drought assessment with the aid of landsat tm/etm plus images. *Int. J. Appl. Earth Obs.* **2011**, *13*, 495–503. [CrossRef]
67. Sun, L.; Wu, Q.; Pei, Z.; Li, B.; Chen, X. Study on drought index in major planting area of winter wheat of china. *Sens. Lett.* **2012**, *10*, 453–458. [CrossRef]

68. Carpenter, G.A.; Gopal, S.; Macomber, S.; Martens, S.; Woodcock, C.E.; Franklin, J. A neural network method for efficient vegetation mapping. *Remote Sens. Environ.* **1999**, *70*, 326–338. [CrossRef]
69. Elmore, A.J.; Mustard, J.F.; Manning, S.J.; Lobell, D.B. Quantifying vegetation change in semiarid environments: Precision and accuracy of spectral mixture analysis and the normalized difference vegetation index. *Remote Sens. Environ.* **2000**, *73*, 87–102. [CrossRef]
70. Baret, F.; Clevers, J.G.; Steven, M.D. The robusteness of canopy gap fraction estimates from red and near-infrared reflectances: A comparison of approaches. *Remote Sens. Environ. Remote Sens. Environ.* **1995**, *54*, 141–151. [CrossRef]
71. Shahabfar, A.; Ghulam, A.; Eitzinger, J. Drought monitoring in iran using the perpendicular drought indices. *Int. J. Appl. Earth Obs.* **2012**, *18*, 119–127. [CrossRef]
72. Qin, Z.; Karnieli, A.; Berliner, P. A mono-window algorithm for retrieving land surface temperature from landsat tm data and its application to the israel-egypt border region. *Int. J. Remote Sens.* **2001**, *22*, 3719–3746. [CrossRef]
73. Shahabfar, A.; Reinwand, M.; Conrad, C.; Schorcht, G. A re-examination of perpendicular drought indices over central and south-west asia. In *Remote Sensing for Agriculture, Ecosystems, and Hydrology Xiv*; Neale, C.M.U., Maltese, A., Eds.; Spie-International Socity Optical Engineering: Bellingham, WA, USA, 2012; Volume 8531.
74. Zhao, C.Y.; Sheng, Y.; Yimam, Y. Quantifying the impacts of soil water stress on the winter wheat growth in an arid region, xinjiang. *J. Arid Land* **2009**, *1*, 34–42.
75. Li, J.; Zhao, C.Y.; Song, Y.J.; Sheng, Y.; Zhu, H. Spatial patterns of desert annuals in relation to shrub effects on soil moisture. *J. Veg. Sci.* **2010**, *21*, 221–232. [CrossRef]
76. Li, S.Y.; Tang, Q.L.; Lei, J.Q.; Xu, X.W.; Jiang, J.; Wang, Y.D. An overview of non-conventional water resource utilization technologies for biological sand control in xinjiang, northwest china. *Environ. Earth Sci.* **2015**, *73*, 873–885. [CrossRef]
77. Hu, R.J.; Ma, H.; Fan, Z.L. The climate trend demonstrated by changes of the lakes in xinjiang since recent years. *J. Arid Land Resour. Environ.* **2002**, *16*, 20–27. (In Chinese)
78. Jiang, L. Analysis on the trend of climate and runoff changes of manas river basin upstream. *Environ. Prot. Xinjiang* **2016**, *38*, 44–48. (In Chinese)
79. Xinhua. Drought Affects Half a Million in Xinjiang. Available online: http://www.chinadaily.com.cn/china/2014-08/23/content_18475075.htm (accessed on 23 August 2014).
80. Yoo, C.; Kim, S.J.; Lee, J.S. Land cover change and its impact on soil-moisture-field evolution. *J. Hydrol. Eng.* **2001**, *6*, 436–441. [CrossRef]
81. Chen, X.; Su, Y.; Liao, J.; Shang, J.; Dong, T.; Wang, C.; Liu, W.; Zhou, G.; Liu, L. Detecting significant decreasing trends of land surface soil moisture in eastern china during the past three decades (1979–2010). *J. Geophys. Res. Atmos.* **2016**, *121*, 5177–5192. [CrossRef]

8

An Efficient In-Situ Debris Flow Monitoring System over a Wireless Accelerometer Network

Jiaxing Ye [1,*], Yuichi Kurashima [2], Takeshi Kobayashi [2], Hiroshi Tsuda [1], Teruyoshi Takahara [3] and Wataru Sakurai [3]

1 National Metrology Institute of Japan (NMIJ), The National Institute of Advanced Industrial Science and Technology (AIST), Tsukuba 305-0045, Japan
2 Department of Electronics and Manufacturing, The National Institute of Advanced Industrial Science and Technology (AIST), Tsukuba 305-8564, Japan
3 Sabo Department, National Institute for Land and Infrastructure Management(NILIM), Tsukuba 305-0804, Japan
* Correspondence: jiaxing.you@aist.go.jp

Abstract: Debris flow disasters pose a serious threat to public safety in many areas all over the world, and it may cause severe consequences, including losses, injuries, and fatalities. With the emergence of deep learning and increased computation powers, nowadays, machine learning methods are being broadly acknowledged as a feasible solution to tackle the massive data generated from geo-informatics and sensing platforms to distill adequate information in the context of disaster monitoring. Aiming at detection of debris flow occurrences in a mountainous area of Sakurajima, Japan, this study demonstrates an efficient in-situ monitoring system which employs state-of-the-art machine learning techniques to exploit continuous monitoring data collected by a wireless accelerometer sensor network. Concretely, a two-stage data analysis process had been adopted, which consists of anomaly detection and debris flow event identification. The system had been validated with real data and generated favorable detection precision. Compared to other debris flow monitoring system, the proposed solution renders a batch of substantive merits, such as low-cost, high accuracy, and fewer maintenance efforts. Moreover, the presented data investigation scheme can be readily extended to deal with multi-modal data for more accurate debris monitoring, and we expect to expend addition sensory measurements shortly.

Keywords: disaster monitoring; wireless sensor network; debris flow; anomaly detection; machine learning; deep learning; accelerometer sensor

---

1. Introduction

Debris flow is a generic term describing the geological hazard that a large volume of a highly concentrated viscous water–debris mixture rapidly flows downward of the hillslope. The phenomena cause considerable damage throughout the world because it happens all of a sudden, destroys all objects in the paths, and often strikes without a sign. Extensive research efforts have been carried out to investigate debris flow from various aspects, such as material properties, movement mechanism, and velocity [1], to achieve better disaster prevention and management. As a result, the most widely accepted classification of debris flow has been developed [2], which expresses debris flow as a sediment mixture of rocks, mud, and water flows rapidly flush a gully bed. Mountainous regions in Japan are susceptible to debris flows, owing to the geographical features [3]. According to the report of Ministry of Land, Infrastructure, Transport, and Tourism of Japan (MLIT), debris flow counts had almost doubled over the past decade, i.e., jumped from 154 to 305 counts per year, which has caused considerable loss of life and property [4]. Moreover, this trend is anticipated to continue in the future attribute to the urbanization process and abrupt climate change. As a result, in line with the demand for disaster risk management, development of debris flow monitoring and early warning systems had been a long-standing and challenging research theme through decades in Japan [5].

The ultimate objective of debris flow monitoring is to make time for taking practical actions such as road closures and evacuations before disaster strikes the hazardous areas. Nowadays, debris flow monitoring systems can be generally subdivided into two classes: 1. Pre-event warning systems and in-progress early detection system.s The first category commonly performs statistical modeling over geographic information system (GIS) data, including air, land and many other climate related observations to predict the debris flow occurrence [6]. A warning would be issued while the estimate exceeds a certain threshold before debris flow arrival. 2. On the other hand, it is commonly based on another group of sensors installed onsite; and the system is designated to process the streaming data in (near) real-time and raise the alarm immediately once debris flow being detected [7]. Clearly, a pre-event warning system can win more time to avert the disaster. Nevertheless, the latter in-situ monitoring system is superior due to lower false alarm rates.

Currently, with the aid of advanced machine learning and the Internet-of-Things (IoT), more attention has been drawn to in-progress monitoring which allows immediate/dynamic disaster information reporting to public safety officials and local inhabitant. An overview diagram of *in-situ* debris flow monitoring system is shown in Figure 1, which commonly includes monitoring data collection/transmission hardware and a data analysis unit for debris flow event detection. A comprehensive set of sensors, such as rain gauge [8], X-band MP Rader, geophone [9], ultrasonic water level transmitter [10], wired sensor [11], had been evaluated for the task. The sensors are usually installed along the mountain slopes to capture debris flow-induced variations exhibited in the measurements. Robust communications is another critical part that enables the collection of streaming data. Wireless sensor network (WSN) is acknowledged as a viable solution according to the latest literature [11,12]. Subsequently, statistical machine learning algorithms are employed to analyze the captured sensory data to discern specific pattern arises by debris flow. The storage and processing of large volumes of monitoring data are perhaps the biggest challenges throughout system development [13]. Advancements in computing hardware and statistical machine learning algorithms have given rise to possible solutions to characterize the massive sensory data and generate decision supports in (near) real-time [14]. Concretely, data mining tools and techniques commonly convert the raw sensor data into a compressed representation with significant information well retained, such as summary statistics derived from time-series data [15], and Fourier and Wavelet spectrograms extraction for image/audio data [16]. At debris flow detection stage, various statistical machine learning techniques have been adopted to exploit the particular debris flow-induced patterns with the extracted feature representations, including support vector machine (SVM) [17], decision trees [18], and artificial neural networks (ANN) [19]. According to empirical studies, advanced machine learning methods are deemed to be well suited to handle the multivariate sensor data, which usually exhibits wide variability and complex non-linear distributions [20].

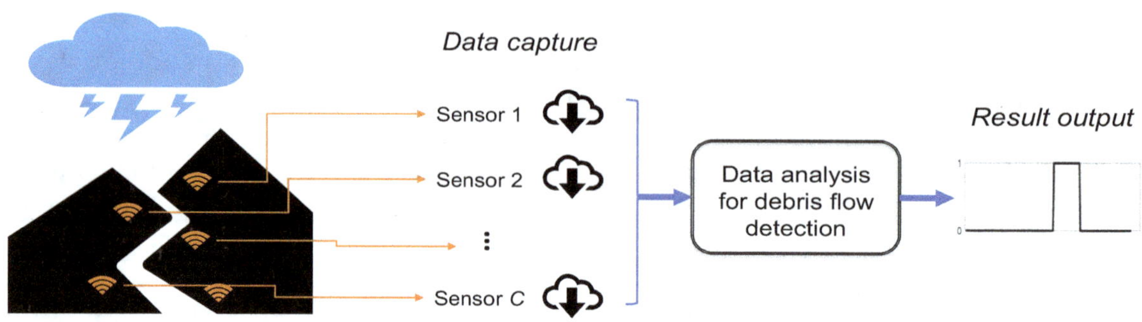

Figure 1. Overview diagram of in-situ debris flow monitoring system.

This study aims at developing an in-situ debris flow monitoring system which integrates advanced machine learning techniques to process massive data captured by the wireless accelerometer network. We summarize the key features of the proposed debris flow monitoring network as follows.

† **Hardware design**. Sensors play a vital role in debris flow monitoring. This study presents a novel design of vibration sensing device which equipped with a triaxial sensor, self-contained power unit with a rechargeable battery and solar panel, and the wireless communication unit. All components are available on the open market, and thus, the total price of one sensing unit is lower than 100 dollars. In the debris monitoring network design, we deployed 20 sensors covering a hazardous area of Nojiri river No.7 dam at Sakurajima, Japan. Through a wireless network, the sensing data were transmitted to a remote computer for further event-based investigation. The total hardware cost is substantially lower compared to other existed methods, such as wire sensor which demands high expense on wiring installation and maintenance. Section 3.1 presents the details related to hardware development.

† **Data investigation framework**. To deal with massive sensor data efficiently, we proposed a two-stage data processing flow, including anomaly detection phase and debris flow induced pattern identification phase. The first stage is designated to screen out a large volume of monitoring data contains no intensive dynamics induced by environmental hazards. It is noteworthy that there exists one critical issue that heavy rain and strong wind can arise significant displacement of accelerometers in addition to debris. To suppress false alarms, we perform a further investigation on the 'suspicious' data by using a deep learning algorithm to characterize the specific data patterns generated by debris flow. To our best knowledge, it is the first attempt to introduce deep learning technique to debris flow monitoring.

† **Accelerometer data fusion**. The above-introduced monitoring data gathering and analysis process work in parallel on every channel of sensor output. Aggregation of multi-sensor data turns out to be an efficient way to achieve more accurate and robust disaster monitoring. To this end, we proposed a multi-sensor information fusion scheme, and the details are shown in Section 3.2.4.

The paper is structured as follows: Section 2 provides an introduction to studied area and disaster background. Section 3 introduces sensing hardware design and data investigation framework for debris monitoring. Next, Sections 4 and 5 outline the results of the case study with discussions. Finally, Section 6 presents the concluding remarks.

2. Overview of the Study Area

The study area is located in a mountainous region near Nojiri river No.7 dam of Sakurajima, which is an active volcano situated at 31°35′ N, 130°30′ E in the southern tip of Kyushu Island, Japan and we present an overview of the area in Figure 2. Sakurajima is 12 km long in the east–west direction and 9 km wide from north to south. It is a land-connected island with an area of about 80 km^2 and the circumference of 50 km. A particular characteristic of Sakurajima's debris flows is that the presence of large volumes of accumulated volcanic ash on the steep mountain slopes results in many debris-flow disasters even when only a small amount of rainfall occurs. There had been plenty of efforts devoted to managing the risk of debris flow at Sakurajima, such as deployment of multiple monitoring sensors including wire sensor, optical sensor, and the acoustic sensor for the disaster detection. Meanwhile, surveillance camera monitoring and mud-sampling are carried out as routine operations [5]. This study attempts to devise a low-cost monitoring system while retaining high accuracy in debris flow detection. Figure 3 shows two surveillance camera snapshots depicting the situation of non-hazardous and debris flow strikes, respectively. It can be seen that even the rain is not very heavy, volcanic ash piled on the steep mountain slopes is easily tuned to debris flow and traveled downhill to small basin areas.

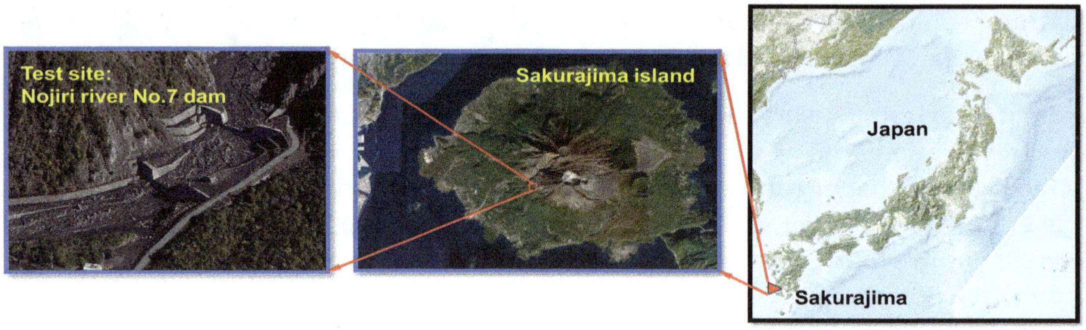

Figure 2. The map of study area: the debris monitoring site at Nojiri river No.7 dam (**left**). Sakurajima island (**middle**). Location of Sakurajima in Japan (**right**).

Figure 3. Snapshots of monitoring camera: before (**left**) and during (**right**) debris flow.

3. Materials and Methods

3.1. Design and Implementation of the Debris Flow Monitoring System Hardware

In this section, we demonstrate the sensing hardware system which lays the fundamentals of this research, which consists of three major parts: triaxial vibration sensor design, sensor deployment scheme over the monitoring area of Nojiri river No.7 high dam, and, wireless communication system settings.

3.1.1. Triaxial Accelerometer Sensing Unit

Figure 4 illustrates the internal design of accelerometer implemented through this study, which is composed of three major components. We introduce those the detail of each part as follows. The power supply has been a long-lasting issue, especially for monitoring systems. In the current design, we expect the system to operate in a self-contained manner with the battery and solar power generator equipped. To this end, three standard 18,650 lithium batteries were used, which reach up to 40 Wh total capacity; meanwhile, we equipped a solar panel rated at 1.5 kWh per day. The operation power consumption is 10 mW, and thus, a vibration sensing unit can work continuously up to two weeks after fully charged. The selection of triaxial vibration sensor is another critical matter to the hardware system. We choose ADXL312 accelerometer developed at Analog Device Inc. At the wireless data transmission stage, we set the carrier band to 920 MHz with GFSK modulation mode. The sampling frequency was fixed to 50 Hz, which had been proved to be reliable for real-time data transmission even under hostile environment conditions.

Figure 4. Internal design (**left**) and exterior look (**right**) of accelerometer unit.

3.1.2. Sensor Deployment Map

In the system design, debris flow monitoring is performed over an accelerometer network with a series of distributed sensing nodes, each of which has data capture and communication capabilities. Since the accelerometer unit dedicated to this task is not pricey, we are able to distribute a batch of sensor nodes to achieve full coverage of the terrain monitored for regional vibration information characterization. It is noteworthy that the positions of nodes can significantly affect vibration observation, such as sensors located in a high place are more vulnerable to wind and rain effects. Based on this consideration, we placed sensing units at both high or low places along the dam tunnel to capture specific vibrations arise by debris flow progress and suppress environmental noises as well. The detailed installation map is shown in Figure 5.

Figure 5. The deployment map of monitoring sensors, in which accelerometer sensors are marked with white circles.

Figure 6 shows examples of 5 min vibration data clips collected by monitoring accelerometer networks regarding environmental conditions. It is evident that the patterns, i.e., waveforms (first column, red/green/blue colors represent x-/y-/z-axes) and time-frequency representations (spectrogram) (2nd to 4th column for the triaxial data, respectively), are different—the first row of normal data contains no environmental impact but only accelerometer sensor noise. In the middle row, we can see raindrops would lead to active impulsive variations to monitoring signal. The last row demonstrates debris flow induced data pattern. As the watery and rocky mud flowing down hills and through the dam tunnel, intense vibration can be sensed. Those differences laid the foundations for automatic disaster identification. Besides, to cope with environmental interferences, we devise an efficient fusion method to aggregate critical debris flow-related information from all sensing nodes. The sensory data analysis system is introduced in the following section.

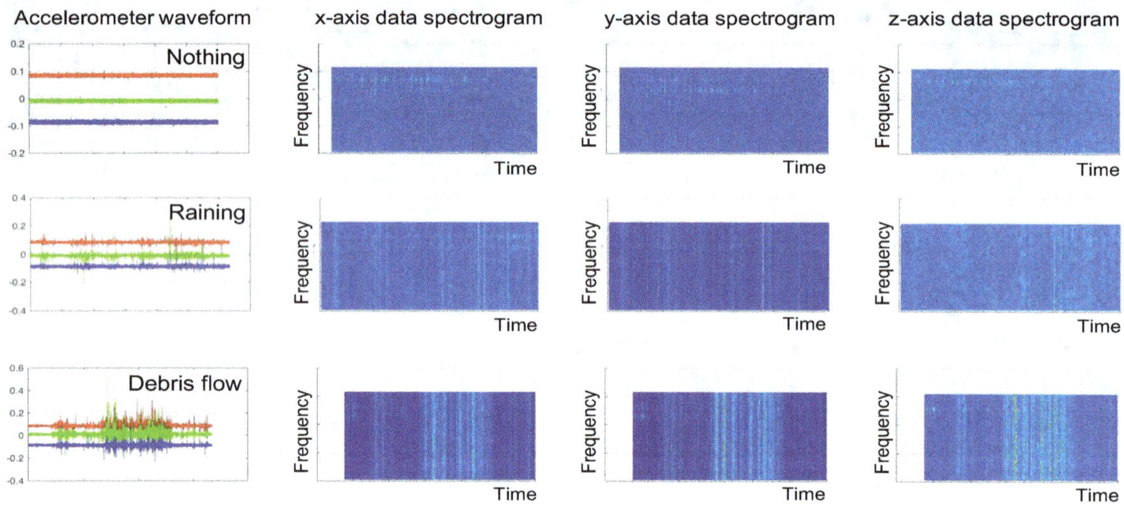

Figure 6. Examples of sensory data with respect to different environmental stimuli. The leftmost column shows raw waveforms with red, green and blue indicating x-/y-/z-axis, respectively. Also spectrogram (time-frequency distribution) plots of triaxial data are presented from 2nd to 4th column.

3.2. The Proposed Debris Flow Detection Algorithm

Within the context of debris flow monitoring, efficient content-level interpretation of input vibration data is the most critical step. Content/event retrieval from accelerometer data series had been a long-standing research theme in the field of signal processing and machine learning, such as in health care [21] and security fields [22]. We demonstrate the vibration data mining approach developed for debris flow monitoring in the following content.

Current accelerometer data investigation systems commonly have two major components: 1. Efficient feature extraction which converts the raw (dense) data to a compact form with discriminant information well retained to the task. 2. Statistical machine learning algorithms which are employed to perform content-based retrieval using extracted features. The latest advancement in machine learning techniques, which enables to characterize nonlinear relationships and interactions between multivariate data adequately, had been successfully adopted for modeling observations and measurements in various environmental fields such as hydrological forecasting [23] and satellite data processing [24]. In this study, we develop state-of-the-art machine learning algorithms to deal with multi-channel vibration data streams, and the overview diagram is presented in Figure 7. Concretely, first, we perform anomaly detection of the overall sensory data to get rid of a large volume of normal data that carry non-hazardous information in an unsupervised manner. Notably, there exists one major drawback in the first-round screening which cannot distinguish debris flow induced sensory data patterns from the ones generated by strong wind and heavy rain; and thus, we further devise a second-stage classification that identifies debris flow induced patterns among others. At this stage, we conducted an extensive comparative evaluation on a series of efficient machine learning algorithms, including logistic regression, support vector machine (SVM) [25] and deep neural networks [26], on real data and the results are exhibited in the validation section.

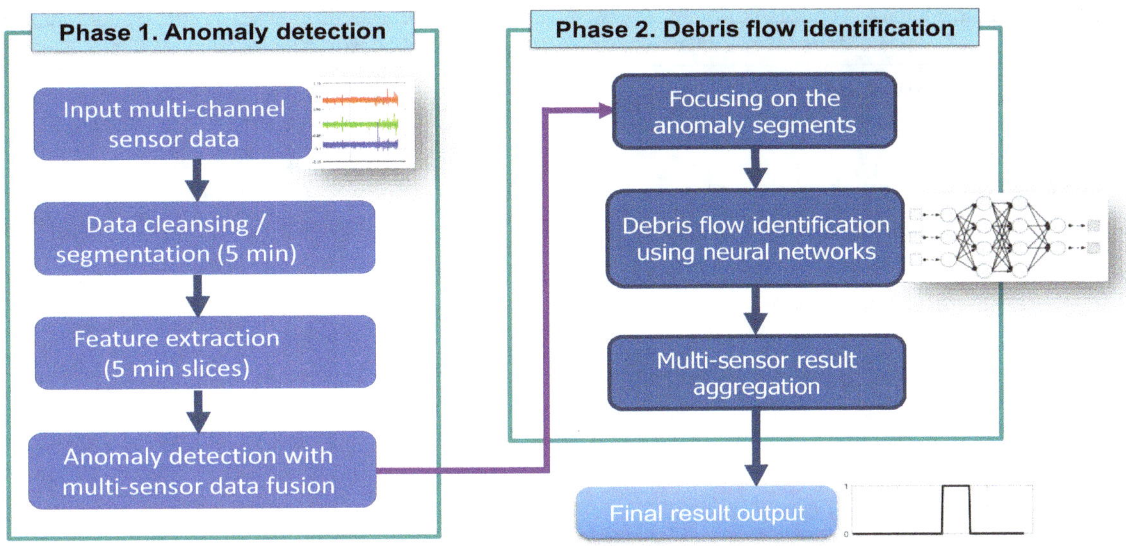

Figure 7. The proposed two-phase accelerometer monitoring data investigation framework.

Before entering into the details of machine learning algorithm design, let us explain mathematical notions. The triaxial data gathered from c-th accelerometer sensor at t time frame is noted as $\mathbf{s}_t^c = [s^a, s^b, s^c]'_{t,c}$, where a, b, c denotes the triaxial measurements, respectively. To characterize the continuous time-series data, we segment the data by using sliding windows with T data points. The resultant i-th clip can be expressed as $\mathbf{x}_i^c = [\mathbf{s}_1^c, \mathbf{s}_2^c, ..., \mathbf{s}_T^c]$ for the sake of brevity. Meanwhile, we perform error checking on sensor data to eliminate sensor faults. The resultant data clips are fed to the above-mentioned two-stage data investigation process for debris flow event identification.

3.2.1. Feature Extraction from Accelerometer Data

Feature representation plays a critical role in event-based accelerometer data interpretation, i.e., discerning debris flow-induced patterns, among others. As shown in Figure 6, different environmental stimuli, such as rain and debris flow, induce specific signal patterns, and feature extraction methods are employed to characterize the signal by using statistics and probabilities. Concretely, the primary purpose of accelerometer feature extraction is to find a set of characteristics such as, stationary property, entropy as a measure of uncertainty, summary statistical attributes of an interquartile range, which can adequately capture discriminant information presented in the observation window. Based on previous studies of accelerometer data mining [27], we established a vibration data feature library dedicated to the debris flow identification task. Table 1 demonstrates the detail list of the features and we extract 12 types of time domain features from triaxial data waveforms. Moreover, to exploit the physical displacement patterns of the sensor, we examine the cross-relational effects of different motion axes, i.e., taking cross-correlation of binary combinations of a-, b-, and c- into account as features. Time–frequency analysis is another efficient method to accelerometer pattern analysis [28], which is preferable to get rid of noises stay at a specific band. In this study, we extracted two types of spectral features for vibration data analysis. As a result, given one segmented data clip noted as $\mathbf{x}_i^c$, we obtain a series of characteristic features through the methods shown in Table 1. Then, we concatenate all feature values as the vector. In this study, the extracted feature is denoted as $\mathbf{f}_i^c$ where c is an index of sensor channel, and i represents ordinal number along time, respectively.

Table 1. Feature list for accelerometer data characterization.

Accelerometer Features for Debris Flow Monitoring		
Feature Type	**Feature Name**	**Description**
Time domain	1. Root mean square (RMS)	$RMS^c = \sqrt{\frac{1}{T}\sum \mathbf{s}_{t,c}^2}, t \in [1,T]$
	2. Mean absolute deviation (MAD)	$MAD^c = \frac{1}{T}\sum_1^T \mathbf{s}_{t,c} - \mu_c, \mu_c = \frac{1}{T}\sum \mathbf{s}_{t,c}$
	3. Interquartile range (IQR)	Descriptive statistics as the difference between 75th and 25th percentiles
	4. Tilt of the sensor	$tilt_c^1 = \frac{1}{T}\sum \lvert s_t^a\rvert + \lvert s_t^b\rvert$ $tilt_c^2 = \frac{1}{T}\sum \lvert s_t^a\rvert + \lvert s_t^c\rvert$ $tilt_c^3 = \frac{1}{T}\sum \lvert s_t^b\rvert + \lvert s_t^c\rvert$
	5. Tilt ratio (TR) of the sensor	$TR_c^1 = tilt_c^1/\lvert s_t^c\rvert$ $TR_c^2 = tilt_c^2/\lvert s_t^b\rvert$ $TR_c^3 = tilt_c^3/\lvert s_t^a\rvert$
	6. Magnitude area (MA)	$MA_c = \frac{1}{T}\sum \lvert\mathbf{s}_t^a\rvert + \lvert\mathbf{s}_t^b\rvert + \lvert\mathbf{s}_t^c\rvert$
	7. Motion intensity (MI)	$MI_c = \frac{1}{T-1}\sum_{t=1}^{T-1} \lvert\mathbf{s}_{t+1}^c - \mathbf{s}_t^c\rvert$
	8. Maxima/Minima (M2M)	$M2M = \max(\mathbf{s}_{t,c}^2)/\min(\mathbf{s}_{t,c}^2), t \in [1,T]$
	9. Binned distribution (BD)	For input data, first calculate the range (R) as maximum–minimium; then, R is divided into 15 equal size bins which records the fraction of data values falls in.
	10. Zero cross rate (ZCR)	Zero-crossing count of the waveform
	11. Cross-axes correlation (CC)	Calculated for each pair of axes as the ratio of the covariance and the product of the standard deviations.
	12. Descriptive statistics	entropy, skewness and kurtosis
Spectral domain	13. Average band power (ABP)	Compute time-average of spectrogram of data
	14. Band standard deviation (BSD)	Compute standard deviation of each band along within observation window

3.2.2. Data Analysis Phase 1: Anomaly Detection

In general, debris flow monitoring is to find a particular event of interest that sparsely superimposed in the continuous observation context. From this aspect, the monitoring data is anticipated to be highly redundant since the majority renders no hazardous information at all. To eliminate the data irrelevant to disaster monitoring with high efficiency and low computation cost, we employ the subspace method that is favorable for various anomaly detection tasks [29].

Subspace method characterizes the highest variance of a multi-dimensional dataset by using a lower dimensional linear subspace defined by a set of orthogonal eigenvectors. According to Figure 6, the normal data that contains no environmental stimuli commonly exhibit stationary characteristic, which can be effectively modeled by a subspace. On the contrary, heavy rain and debris flow can lead to a higher variability signal that may not reside in the subspace. Anomaly detection can be carried out by investigating deviation from input signal to the predominant pattern subspace. That is, the first k principal components returned by eigendecomposition of the data covariance matrix are used to form a "predominant pattern subspace," since it captures the major patterns of normal data. During anomaly detection, all normal data tends to have almost zero length projection on the normal subspace. On the other hand, abnormal data induced by environmental impacts will exhibit significant deviations. We introduce the computational procedure as follows, and it begins with the computing correlation matrix of the input feature series $\mathbf{F}^c = [\mathbf{f}_1^c, ..., \mathbf{f}_n^c, ..., \mathbf{f}_N^c]$:

$$C_{\mathbf{D}} = \frac{1}{N}\sum_{n=1}^{N} \mathbf{f}_n \mathbf{f}_n^{\top}, \tag{1}$$

then, eigen decomposition is performed:

$$\lambda \mathbf{v} = C_{\mathbf{D}} \mathbf{v}. \tag{2}$$

Let $\mathbf{P} = [\mathbf{v}_1, ..., \mathbf{v}_{K'}]$ denotes subspace accommodating predominant textures, which is composed of K'-th eigen vectors with highest eigenvalues. K' is determined by contribution rate which is defined as $\eta'_K = \sum_{k=1}^{K'} \lambda_k / \sum_{k=1}^{K} \lambda_k$. Given input feature vector $\mathbf{f_{in}}$The deviation distance to subspace can be computed by:

$$h_{in} = \mathbf{f_{in}}^{\top}\mathbf{f_{in}} - \mathbf{f_{in}}^{\top}\mathbf{P}\mathbf{P}^{\top}\mathbf{f_{in}}. \tag{3}$$

By examining the deviation distance h_{in}, we are able to detect outliers in the accelerometer data. Concretely, normal data, due to stationary characteristic, will generate quite low h_{in}. In contrast, intense vibrations aroused by environmental impacts will introduce distinct h_{in} values; therefore, h_{in} can be treated as an anomalous score indicating some event may happen regarded to monitoring area. By performing simple thresholding with a defined threshold τ and a gating function $g(\cdot)$, we can convert continuous series of h_{in} to discrete indexes, where 0 and 1 indicate normal and abnormal, respectively:

$$g(h_{in}) = \begin{cases} 1 & h_{in} \geq \tau \\ 0 & \text{otherwise} \end{cases} \tag{4}$$

Above processing can be performed on each channel of sensor measurements, and then we further integrate individual judgment to produce overall output. The fusion rule is straightforward here that we employ logical AND overall channel's anomaly detection results. In other words, we adopt a hypothesis that debris flow would generate significant displacements to all monitoring sensors installed along banks of the dam; and we had validated such assumption on real data. As a result, the anomaly detection algorithm screens out a large portion of non-hazardous data.

However, one significant issue remained in the current result, that is, by using subspace method, both debris flow and other extreme weather conditions such as heavy rain and strong wind will lead to high h_{in} values, and hence, a further distinction can not be made. To tackle debris flow identification problem, we devised a second-stage pattern analysis process.

3.2.3. Data Analysis Phase 2: Debris Flow Identification

This section demonstrates the machine learning algorithm employed to discern the particular debris flow-induced signal patterns among others. Recently, deep neural networks have emerged as a series of learning models that are quite efficient to characterize complex/high-level abstractions from raw data [26]. The architecture is composed of multiple layers which perform the nonlinear transformation on the outputs of previous ones so that a hierarchy of computation interprets input data from low-level raw data values to high-level concepts. Deep learning models had achieved remarkable results in computer vision, and speech recognition [30], it has also been exploited for accelerometer data investigation under the context of human activity recognition [31]. In this study, we introduce deep convolution neural networks (CNN) to investigate the accelerometer signal for debris flow identification, by which convolution operators are designated to deal with two-dimensional data, such as vibration data spectrogram (time-frequency distribution) shown in Figure 6. Moreover, Figure 8 shows detail architecture of the proposed learning model which is composed of 5 convolution layers (conv1 $\sim$ conv5), one full connected layer (fc6) and a classification layer (cl7). First, we denote the computation model of deep neural networks as function $H(\cdot)$, which encodes hierarchical structures. The mechanism of information propagation between layers, e.g., from $k-1$ layer to k, complies with the same principle that performs convolution operation to obtain a convolved feature map. The process can be expressed as follows:

$$\mathbf{h}^{(k)} = g(\mathbf{b}^{(k)} + \mathbf{W}^{(k)}\mathbf{h}^{(k-1)}) \tag{5}$$

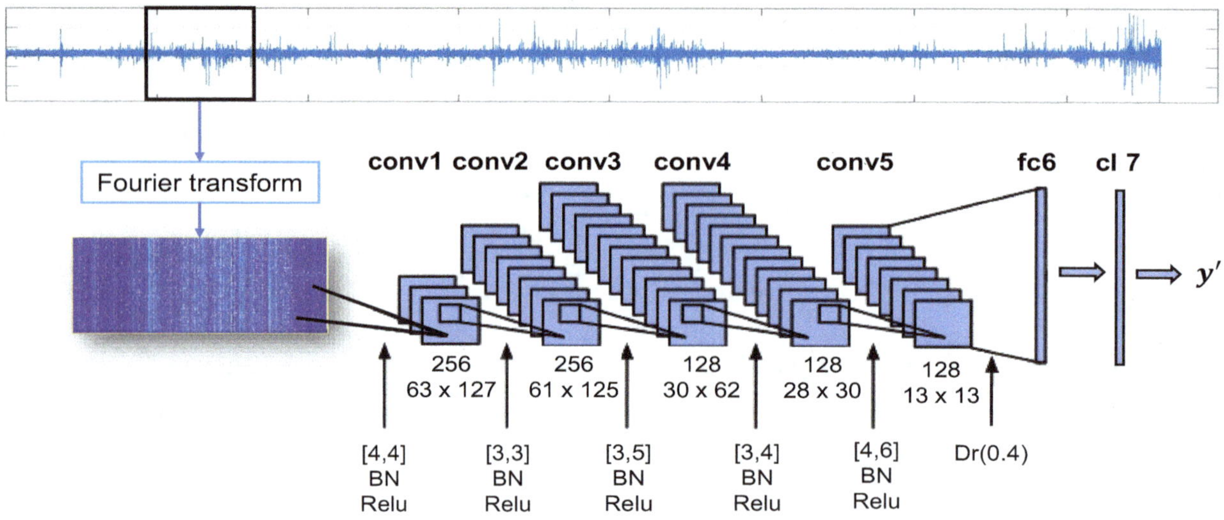

Figure 8. Flow chart of the deep convolution neural network applied for accelerometer data investigation.

It is noteworthy that the $g(\cdot)$, named as activation function, plays a critical role. We employ Rectified linear unit (ReLU), defined by $g(\rho, a(k)) = \max(0, a(k)) + a(k)min(0, \rho)$, owing to two facts: 1. the embedded linear transform can effectively tackle gradient vanishing problem during model updates. 2. The formula encourages sparse activations, which can further suppress overfitting. Batch normalization (BN) is another efficient trick to facilitate deep learning and thus is adopted as a standard process. Furthermore, to avoid the model overfits to training data, at a full connected layer (fc6), we perform dropout on 40% weight of the whole deep model. Other key hyperparameters, such as convolution kernel size, resultant feature map size, and the number of filter banks are also presented in Figure 8.

The neural network training scheme is a vital issue throughout learning system development. We present a general training algorithm pseudo code in Algorithm 1 where $\frac{\partial L}{\partial \theta}$ is the derivative induced by the loss of training data, ϵ is called the learning rate that governs the network update step/speed. From the diagram, we can see the prediction error is iteratively minimized by performing stochastic optimization, and we apply Adaptive Moment Estimation (ADAM) [30] due to the high processing efficiency and low memory usage. It is noteworthy that the data collection is assumed to highly imbalanced that usual observations greatly exceed the number of samples data with debris flow presence. In such scenario, classification rule that predicts the small classes tend to be rare, undiscovered or ignored; consequently, test samples belonging to the small classes, i.e., debris flow-induced patterns, are misclassified more often than those belonging to the prevalent classes. To deal with this issue, we employ penalized classification that imposes an extra cost on the model for making classification mistakes on the minority class (debris flow) during training [32]. The penalization weights for three classes of normal, rain or wind, and debris flow are set to [0.1, 0.2, 0.7], and it will guide the model to make fewer mistakes on debris flow classification throughout the training. In a practical scenario, the pre-trained deep learning system extracts critical information regarding disaster occurrence from the spectrogram of vibration data and then computes the probability of debris flow presence by using softmax function (cl7 layer). Furthermore, to validate the proposed approach, we performed the extensive experimental comparison between the proposed approach with various conventional machine learning algorithms, including Logistic regression (LR), regularized discriminant analysis (RDA) and support vector machine (SVM) [29].

Algorithm 1 Neural Network Training with back propagation

```
procedure TRAIN NEURAL NET(M_t, y_t, W_t)                                  ▷ t=0
    Initialization : W, θ
    for t = 1, 2, ..., T do
        Perform forward propagation : ŷ_t = H(M_t, W_H(θ))
        Compute the class − wise penalized prediction loss : L(y_t, ŷ_t)
        Update weights via back propagation : θ^t ← θ^{t−1} − ε ∂L/∂θ
    return W_H(θ_t)
```

3.2.4. Efficient Sensor Fusion Scheme

Aiming at incorporating multi-channel sensory information for better debris flow monitoring, we adopt an effective computational scheme to exploit varying (relative) contribution concerning the sensor deployment locations. Fundamentally, the idea stems from the fact such as a sensor installed at a higher place is more susceptible to wind noise compare to the ones stay at lower regions; and thus it could be preferable to perform weighted averaging compared to arithmetic averaging. Precisely, we firstly define fusion rule through convex combination as follows:

$$y_{FUSION} = \frac{1}{C}\sum_{1}^{C} \alpha_c \times y_c,\ 0 \leq \alpha_c \leq 1\,, \tag{6}$$

where y_c denote the judgment score estimated at c-th channel data. α_c is the contribution weight with respect to its location (indexed by c). The parameter of α_c can be inferred at validation stage during model training by using linear programming optimizer:

$$\arg\min_{c} \frac{1}{N}\sum_{n=1}^{N}(y_n - (\frac{1}{C}\sum_{1}^{C}\alpha_c \times y_c))^2, \quad \sum_{c=1}^{C}\alpha_c = 1. \tag{7}$$

4. Results

4.1. Data Collection and System Settings

To validate the proposed system for debris flow monitoring, we collected accelerometer data from the 3 June to the 4 July 2017. Onsite rain gauge and anemometer had been used to monitor weather condition. Besides, we applied wire sensors together with a surveillance camera to generate ground truth annotations for debris flow occurrence. Wire sensors can detect the time of debris flow strikes according to the level of the highest wire that has been broken by the flow. In Figure 9, we present three snapshots of monitoring camera showing wire sensor utility. Then, upon an investigation of both wire sensor output (time stamps of wire cut) and video clips, we collect the begin/end time annotations of debris flow occurrence, which are presented in Table 2.

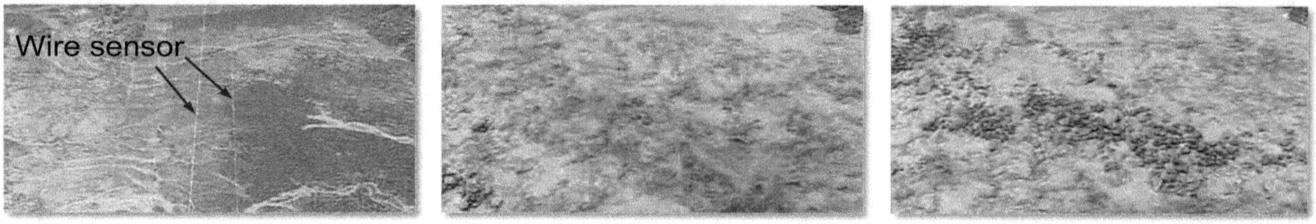

Figure 9. Snapshots of wire sensor utility. The conditions of before/during/after debris flow strikes are shown from left to right, respectively.

Table 2. Details of debris flow occurrence.

Case	Start Time	End Time2
1	7 Jun 2017 16:21:00	7 Jun 2017 17:00:00
2	20 Jun 2017 16:35:00	20 Jun 2017 18:05:00
3	24 Jun 2017 19:03:01	24 Jun 2017 20:05:00

Subsequently, we show the parameter settings of the proposed monitoring system. The sampling frequency of accelerometer data was set to 50 Hz, and streaming data analysis window length is fixed to 5-min. At the anomaly detection stage, the contribution rate η'_K is set as 0.99 to construct the normal vibration pattern subspace. To produce spectrogram of vibration data clips, we set the Fourier analysis window length to 1024 with 3/4 overlapping. To classify the input spectrogram of accelerometer data, we employ a deep convolution neural network (CNN) demonstrated in Section 4.3. During model training, we set the initial learning rate and mini-batch size as 1.0×10^{-3} and 96, respectively. To avoid overfitting, we employ L2 regularization with the regularization parameter set to 0.01; and the maximum updating epoch was set to 40. Besides, as suggested in many previous studies, the momentum parameter is set to 0.9 in stochastic gradient descent optimizer. At the evaluation stage, we performed a particular case of cross-validation called Leave-one-debris flow-out (LODO) scheme. Concretely, at each validation iteration, only one time of debris flow data is used for testing; while the model is trained on all the other debris data collections. As iteration goes, all the debris flow-induced vibration data can be tested. As a result, we obtain the predicted labels to the whole dataset.

4.2. Anomaly Detection Result

This section covers the anomaly detection results for debris flow monitoring. We first present the wind and rain gauge data collected from 3 June to 4 July in Figure 10a,b, respectively. Those climate observations are essential because rainfall event had been deemed to be a significant factor that triggers a debris flow at Sakurajima [5]. Moreover, during the period three debris flow occurred, which were highlighted with pink color. By examining the data, we can see that heavy rain and strong winds often hit together; furthermore, we find that rain falls are necessary for inducing a debris flow, but not sufficient, such as in the case on 10 June. Such fact suggests that rain/wind gauge data cannot provide adequate information for debris flow prediction. Figure 10c shows the anomaly detection result generated by the algorithms demonstrated in Section 3. The whole dataset includes 8835 clips with 5-min length. According to the detection result, 619 segments are marked as suspicious of debris flow occurrence. As a result, the proposed anomaly detection scheme achieved a favorable 93% redundant data reduction without missing any debris flow occurrence. It is noteworthy that there existed much of false alarms which are resort to hostile weather conditions. To suppress such issue, we carry out further content-based classification with machine learning.

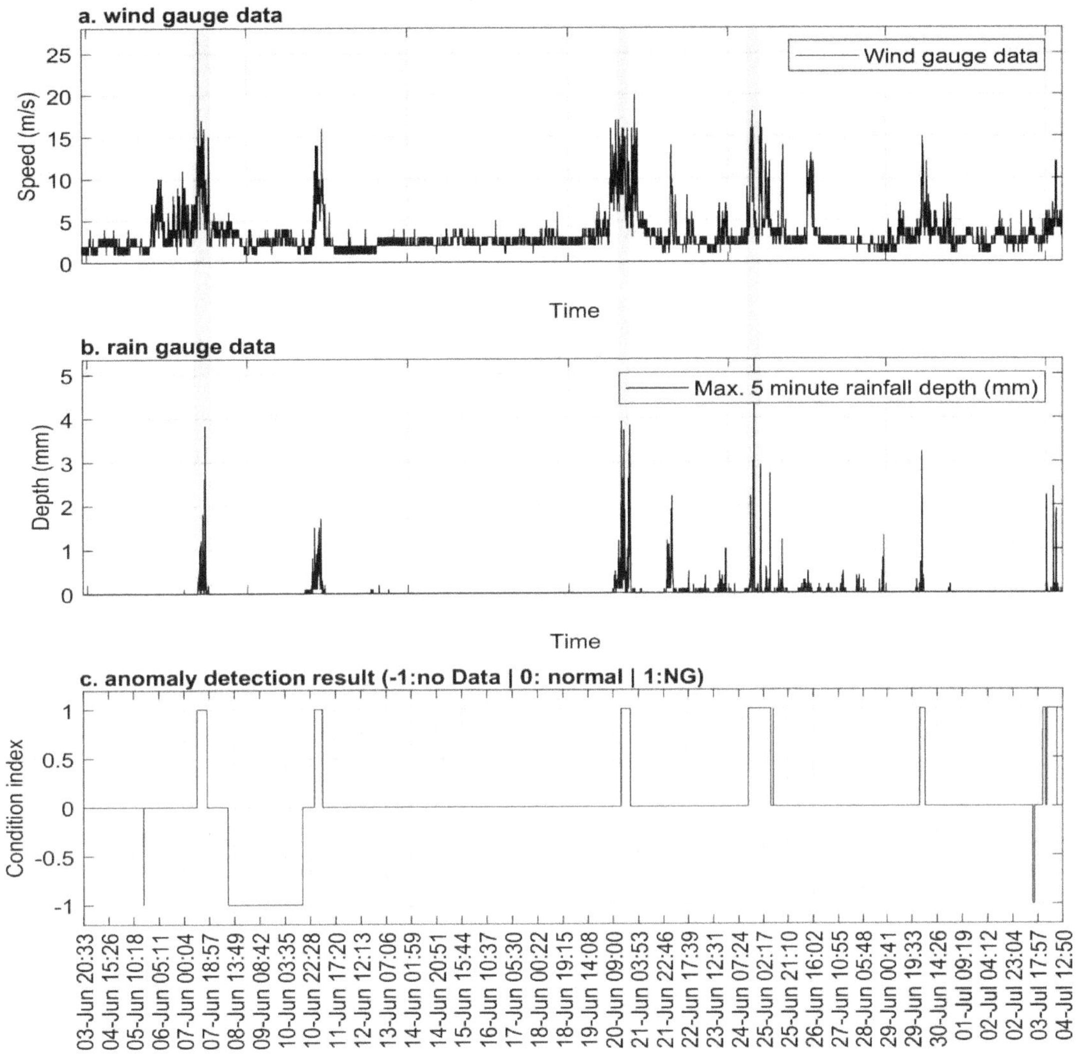

Figure 10. Anomaly detection result for debris flow monitoring, confirming with environmental conditions.

4.3. Debris Flow Identification Performance

Aiming at achieving high precision debris flow-induced vibration data pattern recognition, we conducted an extensive experimental comparison between conventional statistical classifiers and the devised deep learning model. It is noteworthy that those supervised algorithms require both data set and corresponded data labels. To this end, we investigated surveillance video data and assigned 3-valued class labels, in which label 1, 2 and 3 represent regular pattern, raining/wind, and debris flow occurrence on each 5 min data clip, respectively. We applied the convolution neural network demonstrated in Section 3.2.3 to assess the data membership belonging to the three categories. Subsequently, multi-accelerometer score fusion is applied, as explained in Section 3.2.4. We present the prediction results (fusion score) and data annotations on all 619 data clips in Figure 11. It is evident that debris flows can be discerned with a higher fusion scores ranging from 2 to 3. In contrast, non-hazardous data will get fusion scores lower than 2. Upon such observation, we further set the threshold to 2 on fusion score series for debris identification with binary outputs.

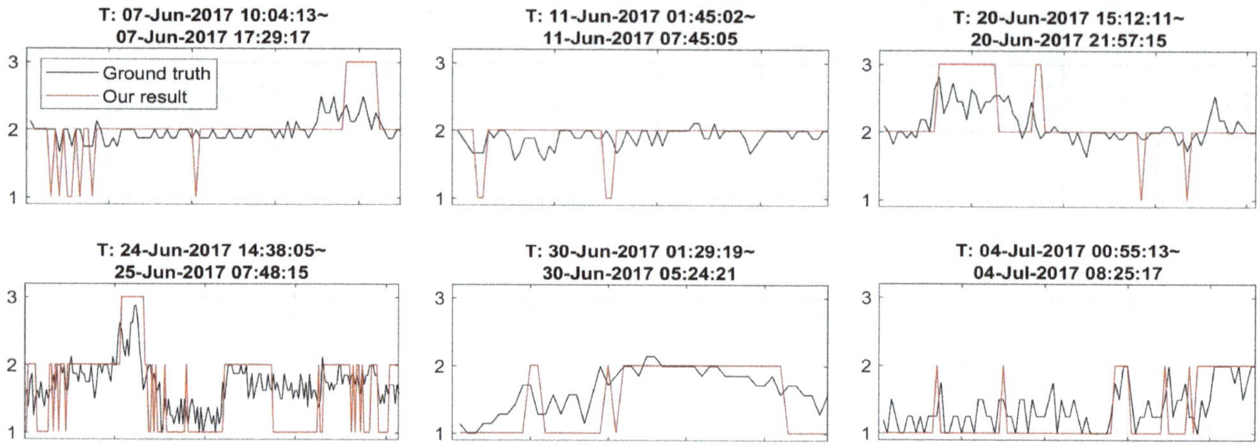

Figure 11. Validation of debris flow identification result, in which the scores above 2 indicate debris flow occurrence.

Moreover, to demonstrate the effectiveness of our approach, we conducted extensive comparison studies by using various feature/classifier combinations for the task. Precisely, in addition to adopting spectrogram of accelerometer data classification by using convolution neural networks, we evaluated the conventional approach by using various summary statistical features shown in Table 1 together with three primary conventional machine learning classifiers, including Logistic regression (LR), regularized discriminant analysis (RDA) and support vector machine (SVM). As for result comparison, receiver operating characteristic (ROC) curve was employed to demonstrate both false alarm and miss detection issues, simultaneously. Figure 12 illustrates all the results under comparison, where *summary stat.* indicates using the conventional features shown in Table 1. The comparison results clarify that using accelerometer spectrogram with CNN model outperformed all other methods with a significant margin, and thus the effectiveness of the proposed approach had been validated.

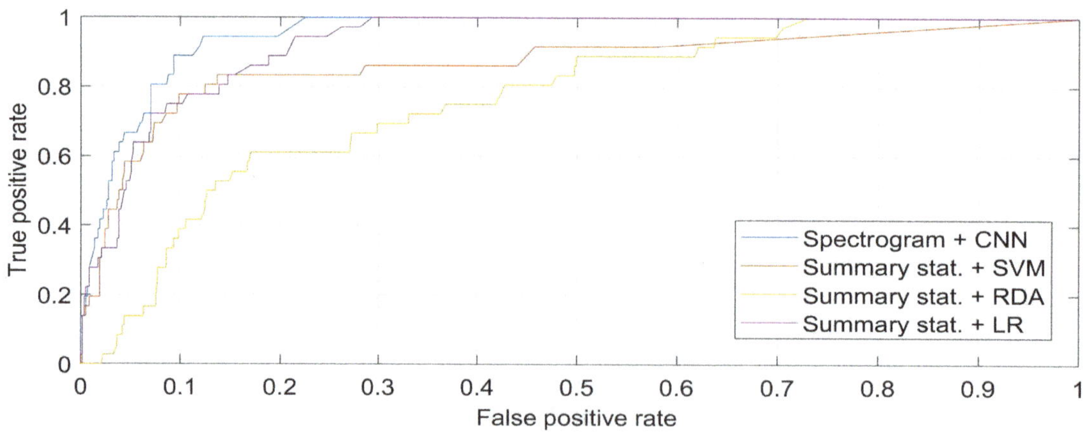

Figure 12. Debris flow identification performance comparison between various data pattern classification methods.

5. Discussion

This study demonstrated that automatic debris flow detection could be achieved by using low-cost wireless accelerometer networks with an advanced machine learning algorithm. According to experimental validations, the proposed approach achieved favorable detection performance; that is, all the three debris flows can be successfully detected. Besides, we investigated the processing efficiency since the provision of timely and effective debris flow progress information are crucial to avoid or reduce the damages. Through evaluation, we confirmed that the developed wireless sensor networks are able to collect vibration data and transmit the data to local analysis workstation in real time even under extreme weather conditions. Subsequently, our debris flow detection algorithm is used to examine the multichannel vibration data clips with a fixed length of 5 min. In other words, the system can take at maximum of 5 min to analyze the latest data while recording the following streaming data. We evaluated the proposed algorithm by using a desktop PC with i9-7900K CPU with 128 GB memory and a laptop PC with i7-7500U CPU with 24 GB memory. Both hardware specifications can complete the data analysis process within a minute. In other words, the system can run for in-situ early detection of debris flow in a *near* real-time manner. Besides, with the increase of massive accelerometer data collection, the data-driven machine learning algorithm, such as deep convolution networks, is anticipated to achieve superior detection precision. Moreover, in our plan, we also expect to expand the current system by adding a new type of sensors for high precision and robust debris flow monitoring.

6. Conclusions

The main objective of this work was to design a computerized debris flow detection system, which consists of wireless accelerometer network hardware and machine learning algorithms to detect debris flows with high accuracy in real time. The proposed system renders a batch of favorable features for debris flow monitoring as follows. First, we devised an efficient wireless triaxial displacement sensor with a self-contained electrical power supply. The sensor is favorably low-cost and thus, allows us to perform (dense) sensor net-based debris flow detection. Secondly, monitoring data is anticipated to be highly redundant since extreme events such as debris flow rarely occur. To get rid of non-hazardous data efficiently, we extracted a list of statistical features from the massive accelerometer signal, and then, performed anomaly detection by using the subspace method. Third, we adopted the latest deep learning scheme to process accelerometer data for precise debris flow identification among other interferences, including strong wind and heavy rain. Furthermore, we integrated all individual sensor's judgment to achieve more accurate and reliable debris flow detection. Experimental results on real data demonstrated the effectiveness of the proposed debris flow monitoring approach. Besides, the developed data pattern investigation framework can be generalized to deal with multi-modal sensory data, such as audio data and geophone signals; and we plan to expand the sensing data source to enhance the monitoring performance further.

Author Contributions: Conceptualization, J.Y. and Y.K.; methodology, J.Y., Y.K., H.T., W.S.; software, J.Y. and Y.K.; validation, J.Y., Y.K. and T.T.; formal analysis, J.Y., Y.K.; investigation, J.Y., Y.K.; resources, H.T., T.K., W.S.; data curation, Y.K., T.K.; writing–original draft preparation, J.Y; writing–review and editing, J.Y., H.T.; visualization, J.Y., Y.K.; supervision, T.K., W.S., H.T.; project administration, T.K., H.T., W.S.

Acknowledgments: We thank the anonymous reviewers for their careful checking of our manuscript and their many insightful comments and suggestions.

References

1. Coussot, P.; Meunier, M. Recognition, classification and mechanical description of debris flows. *Earth Sci. Rev.* **1996**, *40*, 209–227. [CrossRef]
2. Hungr, O.; Evans, S.G.; Hutchinson, I.N. A Review of the Classification of Landslides of the Flow Type. *Environ. Eng. Geosci.* **2001**, *7*, 221–238. [CrossRef]
3. Takahashi, T. A review of Japanese debris flow research. *Int. J. Eros. Control Eng.* **2009**, *2*, 1–4. [CrossRef]
4. Landslide Disaster Cases in Recent Years. Available online: http://www.mlit.go.jp/mizukokudo/sabo/jirei.html (accessed on 15 August 2018).
5. Takeshi, T. Evolution of debris-flow monitoring methods on Sakurajima. *Int. J. Erosion Control Eng.* **2011**, *4*, 21–31. [CrossRef]

6. Marra, F.; Destro, E.; Nikolopoulos, E.I.; Zoccatelli, D.; Creutin, J.D.; Guzzetti, F.; Borga, M. Impact of rainfall spatial aggregation on the identification of debris flow occurrence thresholds. *Hydrol. Earth Syst. Sci.* **2017**, *21*, 4525–4532. [CrossRef]
7. Lee, H.C.; Banerjee, A.; Fang, Y.M.; Lee, B.J.; King, C.T. Design of a Multifunctional Wireless Sensor for In-Situ Monitoring of Debris Flows. *IEEE Trans. Instrum. Meas.* **2010**, *59*, 2958–2967. [CrossRef]
8. Baum, R.L.; Godt, J.W. Early warning of rainfall-induced shallow landslides and debris flows in the USA. *Landslides* **2010**, *7*, 259–272. [CrossRef]
9. Huang, C.J.; Yin, H.Y.; Chen, C.Y.; Yeh, C.H.; Wang, C.L. Ground vibrations produced by rock motions and debris flows. *J. Geophys. Res. Earth Surf.* **2007**, *112*, F02014. [CrossRef]
10. Berti, M.; Genevois, R.; LaHusen, R.; Simoni, A.; Tecca, P.R. Debris flow monitoring in the Acquabona watershed on the Dolomites (Italian Alps). *Phys. Chem. Earth Part B Hydrol. Oceans Atmos.* **2000**, *25*, 707–715. [CrossRef]
11. Arattano, M.; Marchi, L. Systems and sensors for debris-flow monitoring and warning. *Sensors* **2008**, *8*, 2436–2452. [CrossRef]
12. De la Piedra, A.; Benitez-Capistros, F.; Dominguez, F.; Touhafi, A. Wireless sensor networks for environmental research: A survey on limitations and challenges. In Proceedings of the Eurocon 2013, Zagreb, Croatia, 1–4 July 2013; pp. 267–274.
13. Alamdar, F.; Kalantari, M.; Rajabifard, A. Towards multi-agency sensor information integration for disaster management. *Comput. Environ. Urban Syst.* **2016**, *56*, 68–85. [CrossRef]
14. Schimmel, A.; Hübl, J. Automatic detection of debris flows and debris floods based on a combination of infrasound and seismic signals. *Landslides* **2016**, *13*, 1181–1196. [CrossRef]
15. Erdaş, Ç.B.; Atasoy, I.; Açıcı, K.; Oğul, H. Integrating features for accelerometer-based activity recognition. *Procedia Comput. Sci.* **2016**, *98*, 522–527. [CrossRef]
16. Zheng, A.; Amanda, C. *Feature Engineering for Machine Learning: Principles and Techniques for Data Scientists*; O'Reilly Media, Inc.: Newton, MA, USA, 2018.
17. Wan, S.; Lei, T.C. A knowledge-based decision support system to analyze the debris-flow problems at Chen-Yu-Lan River, Taiwan. *Knowl.-Based Syst.* **2009**, *22*, 580–588. [CrossRef]
18. Kern, A.N.; Addison, P.; Oommen, T.; Salazar, S.E.; Coffman, R.A. Machine learning based predictive modeling of debris flow probability following wildfire in the intermountain Western United States. *Math. Geosci.* **2017**, *49*, 717–735. [CrossRef]
19. Dou, J.; Yamagishi, H.; Pourghasemi, H.R.; Yunus, A.P.; Song, X.; Xu, Y.; Zhu, Z. An integrated artificial neural network model for the landslide susceptibility assessment of Osado Island, Japan. *Nat. Hazards* **2015**, *78*, 1749–1776. [CrossRef]
20. Pham, B.T.; Pradhan, B.; Bui, D.T.; Prakash, I.; Dholakia, M.B. A comparative study of different machine learning methods for landslide susceptibility assessment: A case study of Uttarakhand area (India). *Environ. Model. Softw.* **2016**, *84*, 240–250. [CrossRef]
21. Amini, N.; Sarrafzadeh, M.; Vahdatpour, A.; Xu, W. Accelerometer-based on-body sensor localization for health and medical monitoring applications. *Pervasive Mob. Comput.* **2011**, *7*, 746–760. [CrossRef] [PubMed]
22. Lee, W.H.; Lee, R.B. Multi-sensor authentication to improve smartphone security. In Proceedings of the 2015 International Conference on Information Systems Security and Privacy (ICISSP), Angers, France, 9–11 February 2015.
23. Yaseen, Z.M.; Allawi, M.F.; Yousif, A.A.; Jaafar, O.; Hamzah, F.M.; El-Shafie, A. Non-tuned machine learning approach for hydrological time series forecasting. *Neural Comput. Appl.* **2016**, *30*, 1479–1491. [CrossRef]
24. Li, W.; Ni, L.; Li, Z.L.; Duan, S.B.; Wu, H. Evaluation of Machine Learning Algorithms in Spatial Downscaling of MODIS Land Surface Temperature. *IEEE J. Sel. Top. Appl. Earth Obs. Remote Sens.* **2019**. [CrossRef]
25. Murphy, K.P. *Machine Learning: A Probabilistic Perspective*; MIT Press: Cambridge, MA, USA, 2012.
26. LeCun, Y.; Bengio, Y.; Hinton, G. Deep learning. *Nature* **2015**, *521*, 436. [CrossRef] [PubMed]
27. González, S.; Sedano, J.; Villar, J.R.; Corchado, E.; Herrero, Á.; Baruque, B. Features and models for human activity recognition. *Neurocomputing* **2015**, *167*, 52–60. [CrossRef]
28. Machado, I.P.; Gomes, A.L.; Gamboa, H.; Paixão, V.; Costa, R.M. Human activity data discovery from triaxial accelerometer sensor: Non-supervised learning sensitivity to feature extraction parametrization. *Inf. Process. Manag.* **2015**, *51*, 204–214. [CrossRef]
29. Bishop, C.M. *Pattern Recognition and Machine Learning*; Springer: Berlin, Germany, 2006.

30. Goodfellow, I.; Bengio, Y.; Courville, A. *Deep Learning*; MIT Press: Cambridge, MA, USA, 2016.
31. Nweke, H.F.; Teh, Y.W.; Al-Garadi, M.A.; Alo, U.R. Deep learning algorithms for human activity recognition using mobile and wearable sensor networks: State of the art and research challenges. *Expert Syst. Appl.* **2018**, *105*, 233–261. [CrossRef]
32. López, V.; Fernández, A.; Moreno-Torres, J.G.; Herrera, F. Analysis of preprocessing vs. cost-sensitive learning for imbalanced classification. Open problems on intrinsic data characteristics. *Expert Syst. Appl.* **2012**, *39*, 6585–6608. [CrossRef]

9

Satellite-Based Evaluation of the Post-Fire Recovery Process from the Worst Forest Fire Case in South Korea

Jae-Hyun Ryu [1], Kyung-Soo Han [2], Sungwook Hong [3] No-Wook Park [4], Yang-Won Lee [2] and Jaeil Cho [1,*]

1 Department of Applied Plant Science, Chonnam National University, 77 Yongbong-ro, Gwangju 61186, Korea; ryu.jaehyun88@gmail.com

2 Department of Spatial Information Engineering, Pukyong National University, 45 Yongsoro, Namgu, Busan 48513, Korea; kyung-soo.han@pknu.ac.kr (K.-S.H.); modconfi@pknu.ac.kr (Y.-W.L.)

3 Department of Environment, Energy, and Geoinfomatics, Sejong University, 209 Neungdong-ro, Gwangjin-gu, Seoul 05006, Korea; sesttiya@sejong.ac.kr

4 Department of Geoinformatic Engineering, Inha University, 100 Inha-ro, Nam-gu, Incheon 22212, Korea; nwpark@inha.ac.kr

* Correspondence: chojaeil@gmail.com.

Abstract: The worst forest fire in South Korea occurred in April 2000 on the eastern coast. Forest recovery works were conducted until 2005, and the forest has been monitored since the fire. Remote sensing techniques have been used to detect the burned areas and to evaluate the recovery-time point of the post-fire processes during the past 18 years. We used three indices, Normalized Burn Ratio (NBR), Normalized Difference Vegetation Index (NDVI), and Gross Primary Production (GPP), to temporally monitor a burned area in terms of its moisture condition, vegetation biomass, and photosynthetic activity, respectively. The change of those three indices by forest recovery processes was relatively analyzed using an unburned reference area. The selected unburned area had similar characteristics to the burned area prior to the forest fire. The temporal patterns of NBR and NDVI, not only showed the forest recovery process as a result of forest management, but also statistically distinguished the recovery periods at the regions of low, moderate, and high fire severity. The $NBR_{2.1}$ for all areas, calculated using 2.1 μm wavelengths, reached the unburned state in 2008. The NDVI for areas with low and moderate fire severity levels became significantly equal to the unburned state in 2009 ($p > 0.05$), but areas with high severity levels did not reach the unburned state until 2017. This indicated that the surface and vegetation moisture conditions recovered to the unburned state about 8 years after the fire event, while vegetation biomass and health required a longer time to recover, particularly for high severity regions. In the case of GPP, it rapidly recovered after about 3 years. Then, the steady increase in GPP surpassed the GPP of the reference area in 2015 because of the rapid growth and high photosynthetic activity of young forests. Therefore, the concluding scientific message is that, because the recovery-time point for each component of the forest ecosystem is different, using only one satellite-based indicator will not be suitable to understand the post-fire recovery process. NBR, NDVI, and GPP can be combined. Further studies will require more approaches using various terms of indices.

Keywords: forest fire; forest recovery; satellite remote sensing; vegetation index; burn index; gross primary production; South Korea

---

1. Introduction

A forest fire is one of the major disturbances in the ecological diversity, forest succession, the carbon cycle, and hydrological processes of a forest's ecosystem [1–4]. Habitats are altered [5], carbon is released to the atmosphere [6], and runoff and erosion are increased due to the loss of forest from severe forest fires [7,8]. After a forest fire, the evaluation of the damage severity, implications, and spatial patterns is important for forest recovery planning, which plays a critical role in the sustainability of the forest ecosystem and carbon cycle [1,9–11]. It is also necessary to analyze the growth patterns and responses to forest fire disturbance using time series data [12,13]. Thus, the process of forest recovery and the ecological and physiological functions of the burned forest area should be continuously monitored.

The attributes of forest fires, such as fire severity and total area burned, have been conventionally investigated by field observation. The severity of damage to vegetation and soil is generally classified into three levels of low, moderate, and high severity [1,14,15]. However, the field survey has limitations due to frequency of forest fire, assessment difficulty, and the large size of burned areas. For continuous monitoring, field investigators must visit the site multiple times.

Remote sensing techniques, such as satellite imaging, can be useful to regularly observe the burned area and damage severity in real time [16]. The burned area can be measured by the combination of the reflectance values of the visible and infrared channels. Many satellite sensors, such as those of Landsat, Aqua, Terra, Envisat, and SNPP, are capable of identifying forest fires and measuring damage severity in burned areas [17–20]. For example, forest fires can be detected by the brightness temperature of infrared (IR) radiation [21]. In particular, the mid-infrared (MIR) and thermal-infrared (TIR) band are effective to detect forest fire [22]. The burned area and damage severity can be measured using reflectance of near infrared (NIR) and short-wave infrared (SWIR) radiation [23] because the reflectance from living plants and burnt wood are noticeably different. Furthermore, the burned area was extracted from both NIR and the detected forest fire data [19].

The Normalized Burn Ratio (NBR), which uses both NIR and SWIR bands, is widely used to detect burned areas [24]. The Normalized Difference Vegetation Index (NDVI), which uses both red and NIR bands, can identify unhealthy vegetation in burned areas, and has been used to monitor post-fire recovery of forests [25–27]. Thus, these vegetation and burn indices are suitable not only to detect forest fire regions and measure damage severity, but also to evaluate the forest's recovery progress. Van Leeuwen [28] showed that forest fire recovery could be evaluated using moderate-resolution satellite imagery to measure the difference in the NDVI between the burned area and the unburned area every year. Caccamo [29] used the NDVI, Enhanced Vegetation Index (EVI), and Normalized Difference Infrared Index (NDII) to analyze post-fire vegetation recovery. Storey [16] evaluated the sensitivity of vegetation and burn indices to the post-fire recovery of shrubland using seven indices including NBR and NDVI, which were used to evaluate forest recovery.

Previous studies that investigated forest fires using remote sensing indices focused on ecological changes, but the physiological states of plants are also important for understanding the newly established forest ecosystem. Generally, the physiological state of a forest can be obtained by examining the carbon dynamics linked to photosynthesis [11]. For example, gross primary production (GPP) can be useful to interpret the physiological state in a forest post-fire because it is strongly related to photosynthetic activities [30,31].

In this study, we use satellite-based remote sensing data to evaluate forest recovery processes and physiological activity. The worst forest fire in South Korea from 7–15 April 2000 was selected as a case study, and satellite-based vegetation and burn indices and GPP data were used to diagnose both the damage severity and the ecological and physiological recovery levels, depending on the severity level of the forest fire. Further analysis was conducted to minimize the annual variation of meteorological effects. The restoration process was evaluated by comparing the affected areas with an unburned reference area.

2. Materials and Methods

2.1. Satellite Data

Satellite-based NBR and NDVI were used to detect the area affected by forest fire and evaluate post-fire recovery [32]. The NBR was calculated using the reflectance of the NIR and SWIR wavelengths (ρ) as shown in Equation (1):

$$\mathrm{NBR} = \frac{\rho_{\mathrm{NIR}} - \rho_{\mathrm{SWIR}}}{\rho_{\mathrm{NIR}} + \rho_{\mathrm{SWIR}}} \tag{1}$$

The NIR band is an effective spectral band for vegetation monitoring, and the SWIR spectral band effectively represents moisture in soil and vegetation. Sudden changes can occur in the NBR of burned areas because of alterations to the canopy structure and moisture content by forest fires [33], while NBR change is close to zero for unburned areas [32]. Thus, areas affected by the forest fire were identified using the difference between pre-fire and post-fire NBR:

$$\mathrm{dNBR} = \mathrm{NBR}_{\mathrm{pre-fire}} - \mathrm{NBR}_{\mathrm{post-fire}} \tag{2}$$

The NBR was computed using Terra/MODIS (Moderate Resolution Imaging Spectroradiometer) surface reflectance data (MOD09 collection 6). The spatial and temporal resolution for MODIS NBR indices was 500 m and 1 day, respectively. MODIS Band 2 (NIR, 0.86 μm) and Band 7 (SWIR, 2.1 μm) were used to calculate the NBR ($NBR_{2.1}$), and Band 6 (SWIR, 1.6 μm) and Band 2 were also used to calculate another NBR that was defined as $NBR_{1.6}$. In other research fields, $NBR_{1.6}$ is called the Normalized Difference Water Index (NDWI) or NDII [29,34]. $NBR_{2.1}$ is similar to the NBR originally used by Garcia and Caselles [35]. Although 1.6 μm is not commonly used to estimate NBR, this study compared $NBR_{1.6}$ with $NBR_{2.1}$ for monitoring the post-fire forest.

Vegetation health can be represented by the NDVI, a widely used vegetation index [36,37]. The NDVI was computed using NIR and red wavelengths as follows:

$$\mathrm{NDVI} = \frac{\rho_{\mathrm{NIR}} - \rho_{\mathrm{Red}}}{\rho_{\mathrm{NIR}} + \rho_{\mathrm{Red}}} \tag{3}$$

Reflectance of red wavelengths is low when vegetation is healthy and has vital chlorophyll elements [38]. Contrarily, reflectance of NIR wavelengths is higher under such vegetative conditions. The NDVI was calculated from the MOD13A3 data collected by Terra/MODIS from 2000 to 2017. The Terra/MODIS NDVI data's spatial (temporal) resolution was 1 km (1 month). However, there was no Terra/MODIS data available for 1999 to represent the vegetative conditions before the forest fire damage in 2000. Thus, SPOT/Vegetation data from 1999 was used to estimate the NDVI. SPOT/Vegetation NDVI is known be in good agreement with Terra/MODIS NDVI [39]. Weekly SPOT/Vegetation NDVI data was collected in July–August of 1999, with a spatial resolution of 1 km.

To understand the photosynthetic activity of the post-fire forest, the data obtained from Terra/MODIS GPP (MOD17A2H collection 6) was used. Spatial and temporal resolutions were 500 m and 8 days, respectively. MODIS GPP is useful for explaining seasonal vegetation patterns, but it exhibits a slight overestimation when compared with the GPP of flux sites in South Korea [40]. Jung et al. [41] showed that the correlation coefficient of MODIS GPP of three flux sites with forests and croplands was 0.55–0.60.

RGB composite images was calculated using Landsat-5 TM (Thematic Mapper) and Landsat-8/OLI (Operational Land Imager) surface reflectance data. Landsat-5 Band 1 (Blue, 0.49 μm), Band 2 (Green, 0.56 μm), Band 3 (Red, 0.66 μm), and Landsat-8/OLI Band 2 (Blue, 0.48 μm), Band 3 (Green, 0.56 μm), Band 4 (Red, 0.65 μm) were used to compute RGB composite images. The spatial and temporal resolution of the index was 30 m and 8 days.

Forest pixels were extracted from the MODIS Land Cover product. Forest, crop, city, and other land types were divided according to the IGBP (International Geosphere-Biosphere Program) land cover classification scheme (Figure 1c).

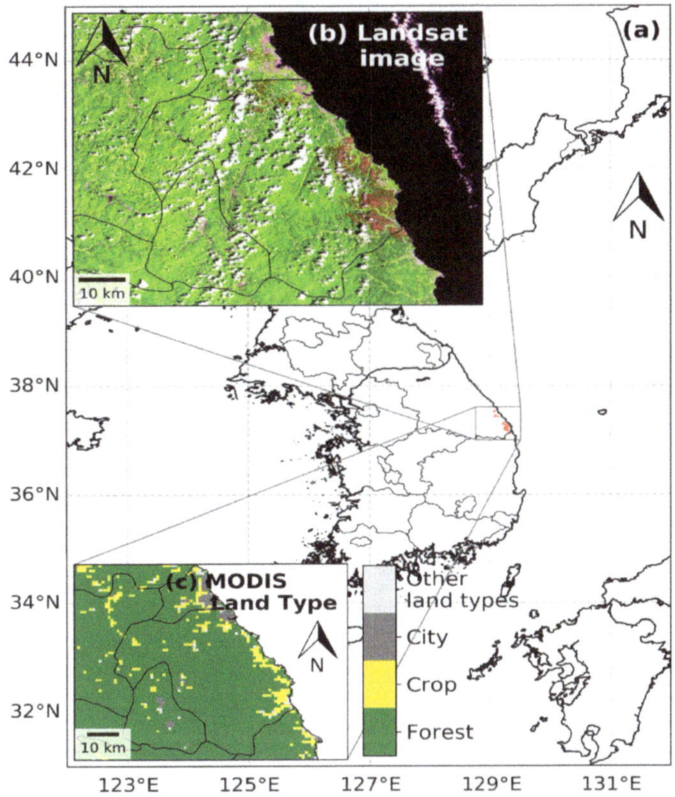

Figure 1. (**a**) Study area; (**b**) Three-band composite image of Landsat-5 consisting of Band 7 (short-wave infrared (SWIR)), Band 4 (near infrared (NIR)), and Band 3 (Red) in 13 August 2000; (**c**) Land type according to the International Geosphere-Biosphere Program (IGBP) land cover classification scheme of MODIS Land Cover.

To match the Terra/MODIS data at a 500 m resolution with the Spot/Vegetation data and MOD13A3, the Terra/MODIS NBR2.1, NBR1.6, and GPP data were unified to 1 km spatial resolution, and a geometric projection was used. The July and August data were averaged for each year because that is the most active vegetation growth period in South Korea (Table 1).

Table 1. Satellite data.

Satellite	Sensor	Band/Product	Spatial Resolution	Period
Terra	MODIS	NIR (Band 2) $SWIR_{1.64\ \mu m}$ (Band 6) $SWIR_{2.13\ \mu m}$ (Band 7) NDVI (MOD13A3) GPP (MOD17A2H)	1 km	May 2000 May 2012 July–August, 2000–2017
SPOT	Vegetation	NDVI	1 km	July–August, 1999
Landsat	TM OLI	Blue (Band 1 or 2) Green (Band 2 or 3) Red (Band 3 or 4)	30 m	August 1999 August 2000 August 2004 August 2007 August 2010 August 2016
Terra & Aqua	MODIS	Land Cover	1 km	2013

2.2. Forest Fires in South Korea and the Study Area

Based on data from 2010, approximately 63% of the land area in South Korea was forest [42]. 59% of forest fires occurred in spring (March–May), and the area damaged by forest fires during spring accounted for approximately 83% of the total area burned that year. Gangwon Province in South Korea often has a dry spring because the air crossing a mountain often changes into dry conditions according to the Föhn phenomenon. Thus, many forest fires have occurred in this area. On 7–15 April 2000, the worst forest fire in South Korean history occurred in this area (the East Coast fire). The forest fire occurred in seven different places and the damaged area covered approximately 23,448 ha. Strong wind speeds (maximum 26.8 m/s) and low relative humidity (minimum 7%) accelerated the spread of the forest fire during this period [15]. After the East Coast fire, the Korea Forest Service (KFS) conducted forest management activities, such as tree planting and artificial regeneration, to aid forest recovery until 2005 [43].

To monitor the post-fire state of the forest, the three severely burned areas (12,697 ha; 4054 ha; and 2244 ha) were investigated. The biggest forest fire happened near the coastline, and the smallest forest fire was in the upper area of our study (Figure 2). The burned areas consisted mainly of forests, but croplands and cities on the coastline were also included. The study area was set from 36.99°N to 37.6°N, and 128.75°E to 129.50°E (Figure 1a,b). Meteorological information in this study is as follows. The average annual air temperature is 12.6 °C, and annual cumulative precipitation is 1278.9 mm, based on 2016 data collected by a meteorological station located in Donghae city. The maximum air temperature is 37.1 °C and the minimum air temperature is −14.0 °C, since May 1992. It rains heavily from July to September.

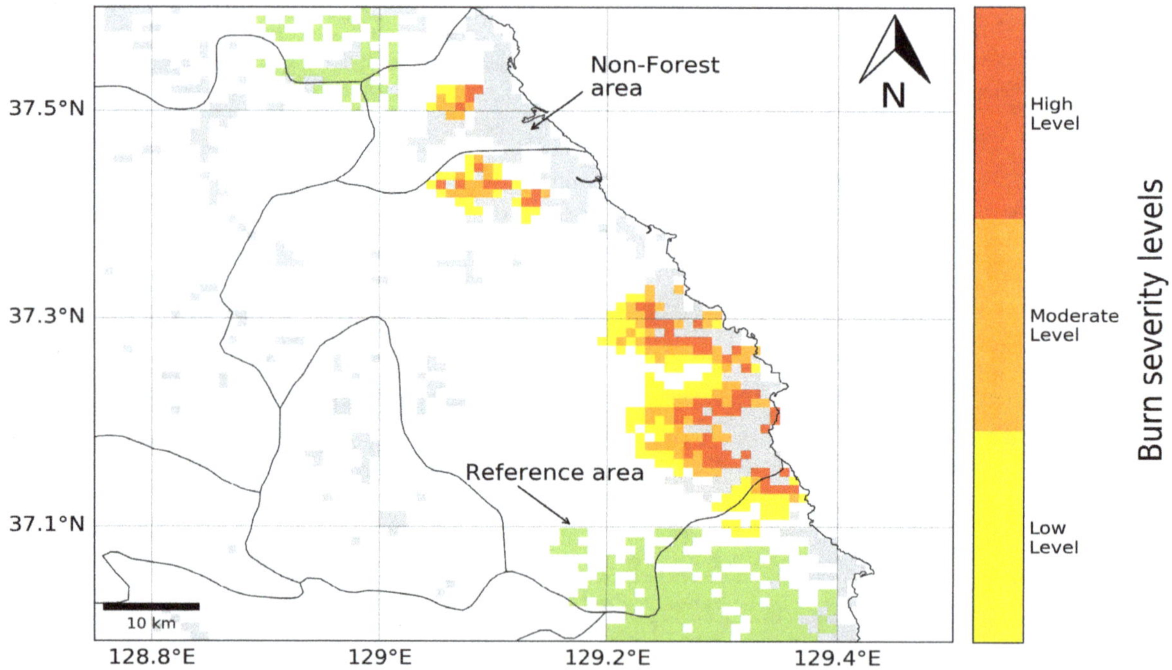

Figure 2. Forest fire areas detected using dNBR of Terra/MODIS. Yellow-Orange-Red pixels indicate forest fire severity divided into three groups using MODIS-based dNBR. Green pixels indicate reference area. Non-forest areas, indicated by grey pixels, were eliminated.

Before the forest fire, the forest was dominated by pine (*Pinus densiflora*) [43,44], which covered approximately 69.5% of the study area. Pine-hardwood and hardwood covered approximately 27.6% and 3% of the study area, respectively [1]. Further, before the forest fire, 20–30 and 30–40 year old trees were spread across 54.44% and 22.62% of the study area, respectively [1].

2.3. Extraction of the Burned Area

The area of forest burned by the East Coast fire was extracted using satellite-based NBR and NDVI. The process of extracting the burned area was as follows. First, Terra/MODIS data from 1999 for the pre-fire forest were unavailable. Through careful study of our preliminary vegetation index analysis output, we assumed that, according to previous studies [16], the moisture condition in the burned area would have mostly recovered after 10 years from the forest fire event and that the vegetative conditions would significantly differ from those immediately after the East Coast fire of 2000. Additionally, we considered the meteorological conditions affecting that area. Mean air temperature was 15.5 (16.3) °C and cumulative precipitation was 63.8 (38.9) mm in May 2000 (2012). Both air temperature and precipitation in 2012 were similar to those in 2000. Thus, the $NBR_{2.1}$ of 2012 was selected as the pre-fire condition for the estimation of dNBR, and MODIS $dNBR_{2.1}$ was calculated in the coastline over Donghae-si, Samcheok-si, and Uljin-gun. The burned area was extracted for a dNBR value greater than 0.10 and excluded for a dNDVI value less than 0.0. The extracted burned area was confirmed using previous studies, and it was consistent with reports in related literature [1,15,45]. Supplementary material shows the availability of this MODIS-derived burned area through comparison with some burned area from Landsat dNBR of more higher-resolution (Figure S1). Second, the burned area identified using MODIS $dNBR_{2.1}$ was classified into three severity levels based on MODIS $dNBR_{2.1}$ values: low, moderate, and high. The ranges of dNBR values for fire severity were flexible and changed according to surface conditions, season, and the interval between data used for calculating the dNBR [33,46–48]. We defined the MODIS dNBR ranges of 0.10–0.15, 0.15–0.20, and 0.20+ as low, moderate, and high levels of severity, respectively. In this study, high level means most of the trees killed in a pixel. Partially or seriously damaged areas in a pixel are defined as moderate level and low level, which might include burned and unburned areas in a pixel. The burned area from Terra/MODIS was expressed by serious damage to the coastline larger rather than in inland, similar to previous studies [45]. Finally, the land types in the burned areas were classified as forest and non-forest using IGBP MODIS land cover data. The non-forest areas were excluded in order to solely monitor changes in forest recovery.

2.4. Definition of the Reference Area

To critically evaluate the forest recovery process, the annually varying meteorological effects on the temporal changes in vegetation and burned indices and GPP were minimized. Meteorological events, such as drought or heavy rainfall, can influence vegetation indices related to forest recovery. Thus, a reference area was necessary to evaluate the recovery of vegetation. The reference area needed to be unburned by the East Coast fire, and the vegetative conditions needed to be similar to the burned area before the fire. In 1999, SPOT/Vegetation NDVI near the burned area was examined. The reference area was selected because the unburned pixels had similar NDVI values to the pixels in the burned area in 1999. The same number of pixels were used for the defined reference area and for the burned area for statistical analysis. An independent two-sample t-test for the burned area and reference area was conducted with IBM SPSS Statistic 23. The P-value was 0.536 at a confidence level of 0.05, indicating that a significant difference did not exist between the two areas. The dominant species in the reference area was similar to that in the burned area. *Pinus densiflora* was dominant in the selected reference area (http://www.forest.go.kr/images/data/down/gispdf_030201_03_5.pdf). The forest age in the reference area was 20~30 years before forest fire. This implied that the two areas had similar surface characteristics in 1999, which was one year before the East Coast fire.

3. Results

3.1. Damage Severity of the Forest Fire

$NBR_{2.1}$, $NBR_{1.6}$, NDVI, and GPP were analyzed to evaluate the damage severity of the East Coast fire. Figure 3 shows the results acquired in May 2000, immediately after the forest fire. The mean of $NBR_{2.1}$ values was 0.480 in the reference area, and $NBR_{2.1}$ values in areas with low, moderate, and high levels of fire severity gradually decreased and were 0.432, 0.381, and 0.321, respectively (Figure 3a). The values in the areas with low, moderate, and high fire severity decreased by 10.03%, 20.52%, and 33.13%, respectively, in comparison with the reference area values. The mean of $NBR_{1.6}$ values was 0.243 in the reference area, which was lower than that of $NBR_{2.1}$ (Figure 3b). $NBR_{1.6}$ values of the burned area decreased when fire severity increased (low: 0.199; moderate: 0.166; high: 0.163); the same trend was observed for $NBR_{2.1}$. $NBR_{1.6}$ values were lower than those of $NBR_{2.1}$, but the percentage decreases in the low, moderate, and high burned area compared to the reference area were 18.14%, 31.72%, and 32.78%, respectively. These results showed that $NBR_{1.6}$ was better at detecting differences between the burned and reference areas than $NBR_{2.1}$, particularly for low and moderate severity levels. However, the difference between low and high damage levels was more distinguished with $NBR_{2.1}$ than with $NBR_{1.6}$.

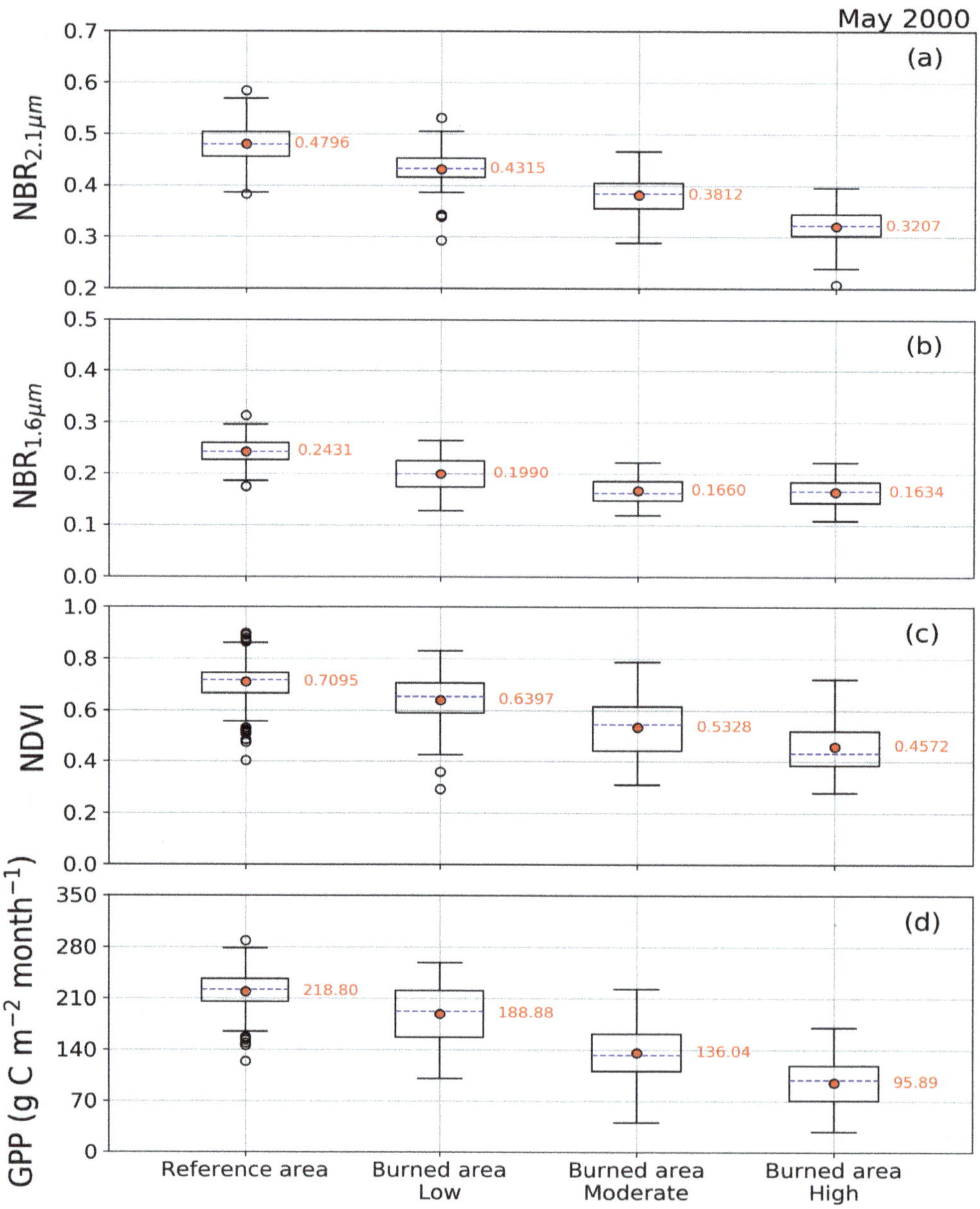

Figure 3. Vegetation indices and gross primary production (GPP) based on forest fire severity. Red circles indicate average values, and blue dash lines indicate median values: (**a**) $NBR_{2.1}$; (**b**) $NBR_{1.6}$; (**c**) NDVI; and (**d**) GPP.

NDVI values were noticeably different depending on the damage severity of the forest fire (Figure 3c). The decreasing pattern of the NDVI with increase in damage severity was similar to that of the NBR. The mean NDVI values in the reference area was 0.710. The percentage decrease of NDVI values in the burned areas was 9.84%, 24.90%, and 35.56% in low, moderate, and high severity areas, respectively. Change in GPP due to the forest fire showed a steeper decline than the $NBR_{2.1}$, $NBR_{1.6}$, and NDVI (Figure 3d). The percentage decrease of GPP in areas with low severity levels (13.67%) was not noticeably different compared with other variables, but the percentage decrease in areas with high severity levels (56.17%) was the largest.

3.2. Temporal Analysis of Forest Recovery

The temporal changes in the study areas were displayed using RGB composite images from Landsat-5/TM and Landsat-8/OLI, which were created from visible red (Band 3 or Band 4), green (Band 2 or Band 3), and blue (Band 1 or Band 2) data (Figure 4). August is the month when the most active plant growth occurs in South Korea. In the 1999 image of the pre-fire forest, the RGB composite image had pixels with a similar green color in the forest areas (Figure 4a). However, brown colored pixels increased immediately after the forest fire (Figure 4b). In images from 2004, four years after the East Coast fire, there were still a large number of brown colored pixels (Figure 4c). Over time, as seen in the 2007 and 2010 images, the brown colored pixels gradually decreased, while the green colored pixels in the burned areas increased. The RGB composite image in 2016 showed that most of the burned area pixels changed from brown to green (Figure 4f).

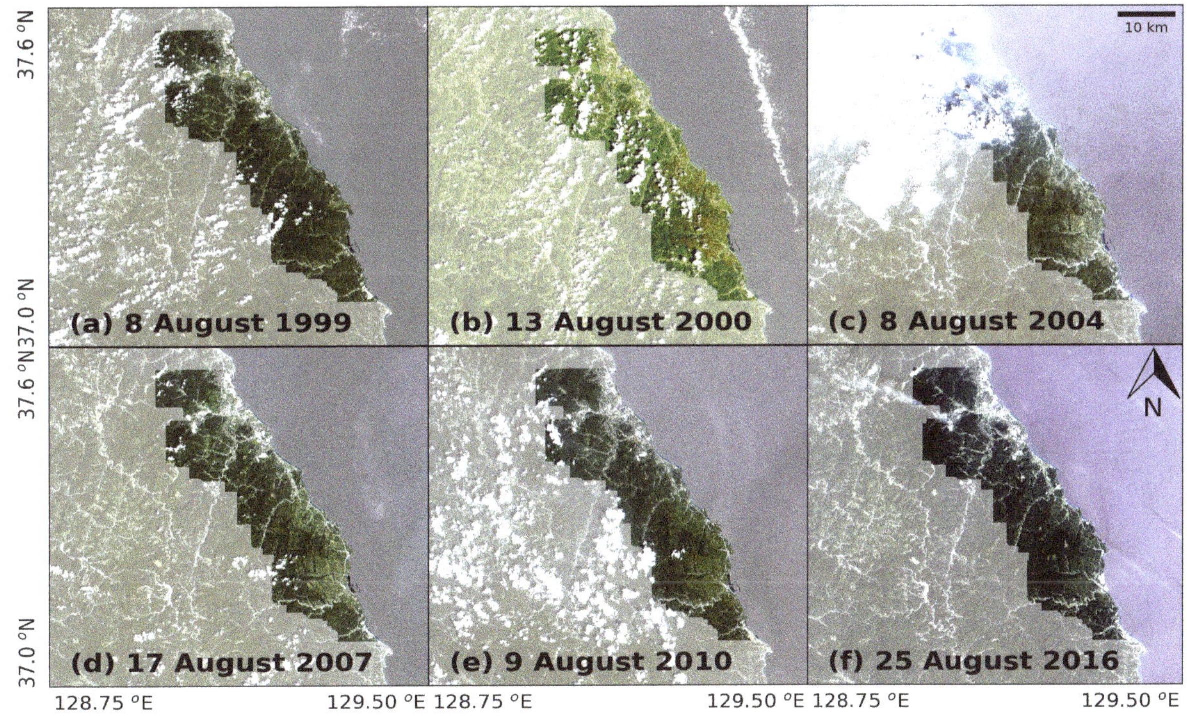

Figure 4. The RGB composite images of Landsat-5 and Landsat-8: (**a**) 8 August 1999; (**b**) 13 August 2000; (**c**) 8 August 2004; (**d**) 17 August 2007; (**e**) 9 August 2010; (**f**) 25 August 2016.

The values of $NBR_{2.1}$, $NBR_{1.6}$, NDVI, and GPP immediately dropped after the forest fire. The extent of the change depended on the severity of the forest fire. However, the initial differences compared with the reference values decreased over time (Figure 5). $NBR_{2.1}$ and $NBR_{1.6}$ showed a similar time series patterns, but the $NBR_{2.1}$ and $NBR_{1.6}$ values in the areas of all damage severity reached the averaged reference value in different years (Figure 5a,b). The recovery-time points were statistically evaluated using an independent two-sample *t*-test (Table 2). In low (high) fire severity areas, $NBR_{2.1}$ was significantly different ($p < 0.05$) compared with the reference area until 2003 (2007), but $NBR_{1.6}$ was significantly different until 2005 (2014). Otherwise, there was significant difference (SD) between $NBR_{2.1}$ and the reference area, and between $NBR_{1.6}$ and the reference area, in areas with low and moderate fire severity after 2014. This was because the values of $NBR_{2.1}$ and $NBR_{1.6}$ in burned areas surpassed those in the reference area.

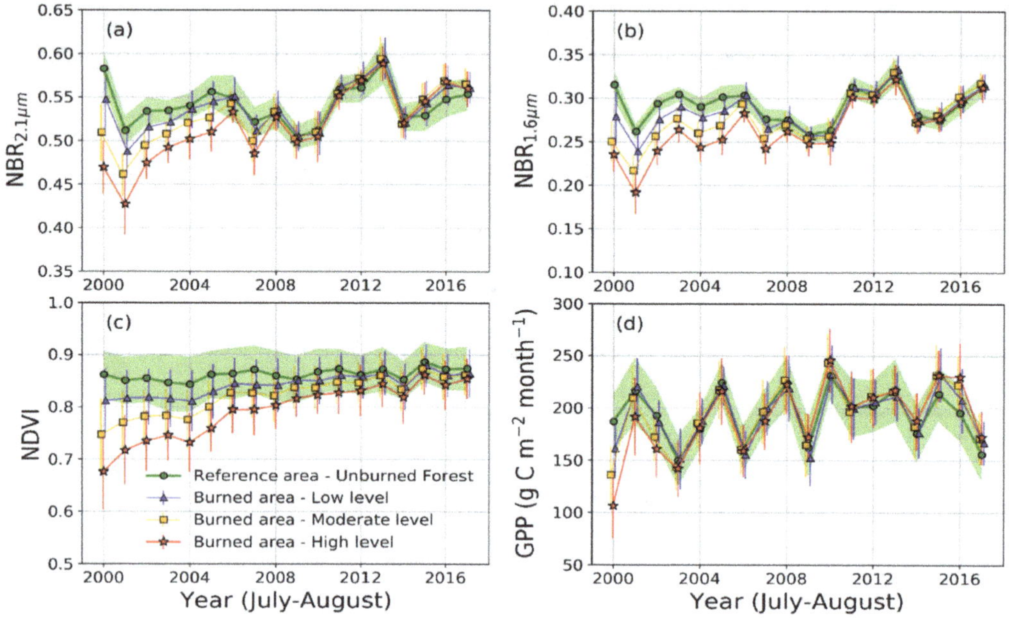

Figure 5. Forest restoration indicated by the time series of the vegetation indices and GPP. Green circles and shadings indicate average and standard deviation values in the unburned reference area, respectively. Blue triangles, yellow squares, and red stars indicate average values in the low, moderate, and high severity burned areas, respectively: (**a**) $NBR_{2.1}$; (**b**) $NBR_{1.6}$; (**c**) NDVI; and (**d**) GPP.

Table 2. Statistically significant differences between the reference area and the burned area calculated using an independent two-sample *t*-test. L, M, and H indicate the low, moderate, and high severity burned areas, respectively. First 'ns' for each variable indicates the year where the reference area values were reached.

Year	$NBR_{2.1}$			$NBR_{1.6}$			NDVI			GPP		
	L	M	H	L	M	H	L	M	H	L	M	H
2000	***	***	***	***	***	***	***	***	***	***	***	***
2001	***	***	***	***	***	***	***	***	***	ns	ns	**
2002	***	***	***	***	***	***	***	***	***	**	***	***
2003	***	***	***	***	***	***	***	***	***	ns	ns	*
2004	ns	***	***	***	***	***	***	***	***	*	ns	ns
2005	***	***	***	***	***	***	***	***	***	ns	ns	ns
2006	ns	ns	***	ns	***	***	**	***	***	*	ns	ns
2007	***	***	***	***	***	***	***	***	***	ns	ns	**
2008	ns	ns	ns	ns	ns	***	**	***	***	*	ns	ns
2009	ns	ns	*	ns	**	***	ns	ns	***	*	ns	ns
2010	ns	ns	ns	*	**	***	**	***	***	ns	*	**
2011	ns	ns	**	ns	**	***	**	**	***	ns	ns	ns
2012	***	***	**	*	*	**	ns	ns	***	ns	*	ns
2013	ns	ns	ns	ns	ns	***	ns	ns	***	*	ns	ns
2014	**	**	*	***	***	***	**	*	***	ns	ns	*
2015	***	***	***	*	**	ns	ns	ns	***	***	**	***
2016	***	***	***	**	***	ns	*	ns	***	**	***	***
2017	***	**	ns	ns	ns	ns	ns	*	**	***	**	*

ns: Not significant at the $p < 0.05$ level. *, **, ***: Significant at the $p < 0.05$, 0.01, and 0.001 levels, respectively.

The NDVI was nearly constant in the reference area (Figure 5c). NDVI trends increased continually for 18 years in the burned area, and the rate of increase was greater in areas with high fire severity levels. SD in the NDVI between areas with low and moderate fire severity and the reference area was shown until 2009. There was also no SD between these values in 2012, 2013, 2015, 2016 (moderate), and 2017 (low). However, in areas with high fire severity levels, SD was shown until 2017.

In the first July–August period after the forest fire, the GPP differences of the areas with low, moderate, and high severity levels existed as $NBR_{2.1}$, $NBR_{1.6}$, and NDVI (Figure 5d). However, the SD between these values ceased earlier than for other indices. In areas with low and moderate fire severity, there was no SD (0.848, 0.134; $p < 0.05$) between GPP in those areas and in the reference area in 2001, which was one year after the forest fire. GPP in areas with high fire severity reached GPP levels in the reference area (0.614; $p < 0.05$) in 2004. Recent GPP in the burned areas was higher than GPP in the reference area, and a significant difference appeared earlier in areas with high forest fire severity.

3.3. *Relative Change in Recovery of the Forest*

To distinguish the effects of the meteorological conditions from the temporal changes in forest recovery, the ratios of variables ($NBR_{2.1}$, $NBR_{1.6}$, NDVI, and GPP) in July–August in the burned areas were compared with the reference area and investigated. These ratios can accurately express the process of forest recovery (Figure 6). Average values of indices in the unburned forest area were ideal criteria for indicating complete forest recovery. The ratios of $NBR_{2.1}$ in burned areas showed an increase after the forest fire (Figure 6a). Immediately after the forest fire, ratios of $NBR_{2.1}$ were 0.940, 0.875, and 0.806 in areas with low, moderate, and high fire severity, respectively. They reached the criteria level (1.0) in 2004, 2006, and 2008 in areas with low, moderate, and high fire severity, respectively (Table 3). These results were consistent with the independent two-sample t-test results. After 2008, the ratios of $NBR_{2.1}$ in all areas exceeded the confidence interval of the criteria level. Ratios of $NBR_{1.6}$ showed similar patterns to those of $NBR_{2.1}$. For example, ratios of $NBR_{2.1}$ and $NBR_{1.6}$ rapidly increased during the first three years after the forest fire, although the initial values and overall temporal changes were different (Figure 6b).

Table 3. Recovery-time point (year) indicated by four indices (i.e., $NBR_{2.1}$, $NBR_{1.6}$, NDVI, and GPP) in regions of three fire severity levels during 18 years. L, M, and H represent the low, moderate, and high severity burned areas, respectively.

Indices	Meaning of Monitoring	Recovery-Time Point (Year)		
		L	M	H
$NBR_{2.1}$	Moisture condition	2004	2006	2008
$NBR_{1.6}$	Moisture condition	2006	2008	2015
NDVI	Vegetation biomass	2009	2009	
GPP	photosynthetic activity	2001	2001	2004

NDVI ratios were 0.942, 0.867, and 0.785 in areas with low, moderate, and high fire severity in July–August 2000, respectively. The ratios increased, particularly during the first three years after the forest fire. These patterns were consistent with the ratios of $NBR_{2.1}$ and $NBR_{1.6}$. The NDVI ratios reached the confidence interval values of the criteria level in 2006 and 2009 for areas with low and moderate fire severity, respectively. After 2009, ratios of NDVI in areas with low and moderate severity levels remained between 0.97 and 1.0. However, in areas with high fire severity, the ratio of NDVI barely met the recovery level in 2017.

Ratios of GPP showed different patterns compared to the NBR and NDVI ratios. The GPP in burned areas reached the confidence interval values of the recovery level in 2001, 2001, and 2003 for areas with all levels of fire severity. Across the whole study period (2001 to 2017), the slope of the GPP ratio in areas with low fire severity was 0.004. However, the slopes of the GPP ratios in areas with moderate and high fire severity were 0.009 and 0.014, respectively. The GPP in these areas exceeded the criteria level (1.0) after 2015, and it, in areas with high severity levels, largely exceeded the GPP in the reference area.

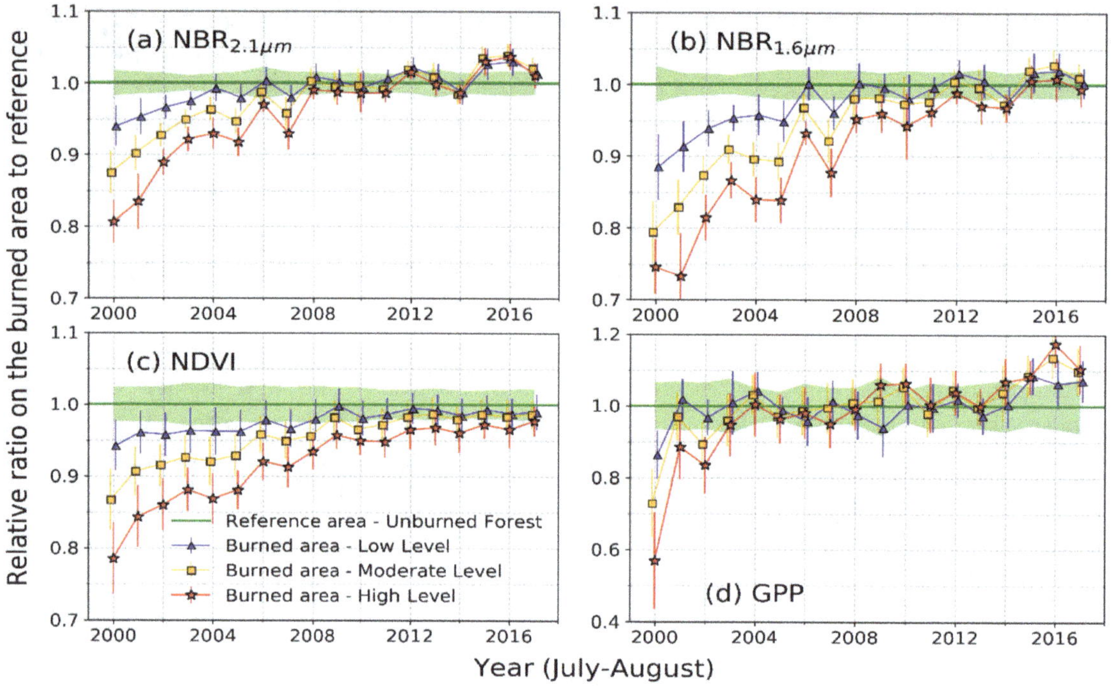

Figure 6. Relative change in vegetation indices and GPP in the burned area compared with the reference area. Green lines indicate average value and green shadings indicate confidence intervals in the unburned forest area. Lines and error bar for blue triangles, yellow squares, and red stars indicate average values and confidence intervals in the low, moderate, and high severity burned areas, respectively: (**a**) $NBR_{2.1}$; (**b**) $NBR_{1.6}$; (**c**) NDVI; and (**d**) GPP.

4. Discussion

Forest biomass recovery began immediately after the forest fire in 2000, and gradually continued through to 2017. Although the July–August period is the most active forest growing season in South Korea, the vegetative activity in July–August 2000 is remarkable because it was so close to the time the forest fire occurred (May 2000). Indeed, in July–August 2000, vegetation increased in all damaged areas. In addition, the rate of recovery of areas with high fire severity was greater than that of areas with low fire severity. According to KFS reports and ground-survey data in the existing literature about the East Coast fire, the natural appearance of vegetation was observed in damaged areas in the first year after the forest fire [1,49]. About 80% of the recovery was caused by the re-growth of surviving sprouts [49]. Also, Lee and Chow [50] showed that there was a rapid recovery for three years after the forest fire. However, the re-growth of surviving sprouts cannot fully explain the faster re-vegetation in areas with high fire severity. Armesto and Pickett [51] concluded that when larger disturbances such as forest fires occur, abundance in certain foliage is observed during the recovery period because of the enhanced re-growth made possible by less competition. The dominant tree type changed from pine to hardwood, such as oak species, after the forest fire [1]. Indeed, pine-hardwood and hardwood trees regenerated relatively rapidly compared to pine trees. The KFS conducted recovery efforts, such as tree planting, until 2005 [42]. For example, about 2-ha pine stand was planted in June 2003 [52]. The $NBR_{2.1}$, $NBR_{1.6}$, NDVI, and GPP data obtained during this study represented the integration of these phenomena. The relative ratio of variables in moderate (high) burned areas increased from about 8.0% (7.3%) to about 11.6% (14.1%) during the first 3–4 months after the fire in 2000. Also, the forest recovery trends for 5 years after the forest fire showed quick changes in four variables (Figure 6). These results mean that the remote sensing based variables effectively indicate the progress of forest recovery.

The NBR can indicate conditions of water and vegetation, and the NDVI can detect the amount of vegetation biomass. GPP is a measure of the rate of photosynthetic activity in vegetation chlorophyll. Therefore, these variables are governed by meteorological conditions. The seasonal variations of vegetation activity and water conditions in a forest are greatly influenced by the annual meteorological variation; this would affect the temporal patterns of forest recovery indicated by NBR, NDVI, and GPP. For example, in the spring drought in 2001, both the $NBR_{2.1}$ and $NBR_{1.6}$ in summer (July–August) 2001 were lower because the water deficit from the spring drought continued into the summer (Figure 5). Oppositely, the NDVI and GPP in summer 2001 were higher, possibly because of the higher photosynthetic activity caused by the reduction of clouds during the drought period [53].

To minimize the annual variation of the meteorological effects, a comparison between the burned and reference areas was carried out (Figure 6). All values of the indicators of forest recovery (i.e., $NBR_{2.1}$, $NBR_{1.6}$, NDVI, and GPP), which represented the conditions of the burned area, gradually became closer to the values of the reference forest. However, the time that it took for the values of the burned area to match the reference area values was different. This means that the complete recovery-time point can be evaluated differently depending on the monitoring indicators chosen. The $NBR_{2.1}$ and $NBR_{1.6}$ values exceeded the levels of the reference area in 2017, but $NBR_{2.1}$ reached the reference level earlier than $NBR_{1.6}$. Given that the 2.1 μm wavelength is more sensitive to water at low moisture levels than the 1.6 μm wavelength [54], $NBR_{2.1}$ might be able to detect the recovery of moisture conditions in a forest earlier.

On the other hand, the NDVI almost reached the reference level in 2017, but it did not exceed the level during our analysis period. Therefore, the complete recovery-time point indicated by the NBR was faster than that of the NDVI. This result is consistent with the results of previous studies [16]. Ahn [44] compared two camera images taken immediately after and six years after the East Coast fire, and a lot of the area had been recovered by vegetation, but canopy height was still low. Although surface moisture conditions are generally preserved under vegetation cover, the complete recovery of vegetation biomass and health conditions will be achieved after the recovery of the moisture conditions in the forest. Polychronaki [55] showed that the complete recovery of vegetation after a severe forest fire required more than 20 years.

Immediately after the East Coast fire, GPP was reduced to an extent dependent on the severity of the fire. This was consistent with the result of a previous study [56]. The GPP in the burned areas reached reference levels in 2004. This was the fastest recovery among $NBR_{2.1}$, $NBR_{1.6}$, and NDVI. Further, after 2015, the GPP in burned areas was much higher than the GPP in the reference area, particularly in areas with high fire severity. This result could have been caused by technical and ecological factors. First, the fraction of absorbed photosynthetically active radiation (fAPAR), estimated based on the relationship with NDVI, is an important parameter for calculating MODIS GPP. However, the recovery pattern shown by the NDVI was different than that of the GPP. We did not test the GPP algorithm in this study, but Bolton [11] argued that satellite-based GPP can be influenced by canopy structure. Second, the GPP of young forests after disturbances, such as forest fire, shows dramatic growth, while GPP in older forests slightly declines [57,58].

5. Conclusions

NBRs, NDVI, and GPP in terms of properties of moisture condition, vegetation biomass, and photosynthetic activity were applied to monitor the temporal patterns of forest recovery and identify the recovery-time point after the worst forest fire in South Korean history, according to three levels of fire severity. Further, the change of moisture condition was separately evaluated by two NBR types of 2.1 and 1.6 μm bands, and those NBRs were also used to detect the burned areas. These four remote-sensing variables on the forest recovery progress had similar temporal patterns representing ecological functions. However, their recovery rate was different in the region of each fire severity level.

In NBR2.1 (NBR1.6), the burned area became close to the unburned reference area after 4, 6, and 8 (6, 8, and 15) years in low, moderate, and high levels (Table 3). In NDVI, 9 years were required to recover in the low and moderate levels, but the burned area in the higher level needed more time. On the other hand, the GPP in the burned area continuously increased during our whole study period, and excessed the value in the reference area after 15 years. Thus, we concluded that the application using those indices of different properties could be suitable for evaluating the progress of forest recovery from a variety of perspectives. A single remote-sensing indictor should be not necessary for decision-making in forest management.

GPP is the outcome of complex biogeochemical processes of a forest ecosystem. Thus, it is commonly considered as a useful indicator of the ecological condition of a forest. However, the ground-based GPP in the post-fire region is not well appropriately interpreted because spatio-temporal data are not rarely produced. Satellite-based GPP is useful to monitor photosynthetic activity over a large area, but it is relatively difficult to estimate, unlike the simple calculations for NBR and NDVI from the satellite sensors. A process-based biogeochemical model for GPP estimation might be one of the effective ways to understand the cause and effect of the forest recovery processes in a time series. In addition, the satellite derived NBR and NDVI observational data should contribute to the model performance. In this study, we identified the possibility of using $NBR_{2.1}$, $NBR_{1.6}$, NDVI, and GPP to evaluate the recovery of burned areas. In future work, an integrated approach of satellite observation and biogeochemical modeling will be necessary and further long-term monitoring will be required.

Author Contributions: All authors contributed this paper. J.-H.R., the main author, designed the research, analyzed the remote sensing data, and wrote the manuscript; K.-S.H., S.H., N.-W.P. and Y.W.L., the co-authors, reviewed the paper and contributed to the discussion; J.C., corresponding author, designed the research and wrote the manuscript.

Acknowledgments: We are also grateful to the three anonymous reviewers for their constructive comments. This work was supported by "Development of Hydrology, wildfire, and statistical Applications" project, funded by ETRI, which is a subproject of "Development of Geostationary Meteorological Satellite Ground Segment (NMSC-2017-01)" program funded by NMSC (National Meteorological Satellite Center) of KMA (Korea Meteorological Administration).

References

1. Choung, Y.; Lee, B.-C.; Cho, J.-H.; Lee, K.-S.; Jang, I.-S.; Kim, S.-H.; Hong, S.-K.; Jung, H.-C.; Choung, H.-L. Forest responses to the large-scale east coast fires in Korea. *Ecol. Res.* **2004**, *19*, 43–54. [CrossRef]
2. Seedre, M.; Taylor, A.R.; Brassard, B.W.; Chen, H.Y.H.; Jõgiste, K. Recovery of Ecosystem Carbon Stocks in Young Boreal Forests: A Comparison of Harvesting and Wildfire Disturbance. *Ecosystems* **2014**, *17*, 851–863. [CrossRef]
3. Nicholson, Á.; Prior, L.D.; Perry, G.L.W.; Bowman, D.M.J.S. High post-fire mortality of resprouting woody plants in Tasmanian Mediterranean-type vegetation. *Int. J. Wildland Fire* **2017**, *26*, 532–537. [CrossRef]
4. Houle, G.P.; Kane, E.S.; Kasischke, E.S.; Gibson, C.M.; Turetsky, M.R. Recovery of carbon pools a decade after wildfire in black spruce forests of interior Alaska: Effects of soil texture and landscape position. *Can. J. For. Res.* **2018**, *48*, 1–10. [CrossRef]
5. Dale, V.H.; Joyce, L.A.; McNulty, S.; Neilson, R.P.; Ayres, M.P.; Flannigan, M.D.; Hanson, P.J.; Irland, L.C.; Lugo, A.E.; Peterson, C.J.; et al. Climate change and forest disturbances: Climate change can affect forests by altering the frequency, intensity, duration, and timing of fire, drought, introduced species, insect and pathogen outbreaks, hurricanes, windstorms, ice storms, or landslides. *BioScience* **2001**, *51*, 723–734. [CrossRef]
6. Isaev, A.S.; Korovin, G.N.; Bartalev, S.A.; Ershov, D.V.; Janetos, A.; Kasischke, E.S.; Shugart, H.H.; French, N.H.F.; Orlick, B.E.; Murphy, T.L. Using remote sensing to assess Russian forest fire carbon emissions. *Clim. Chang.* **2002**, *55*, 235–249. [CrossRef]
7. Meyer, V.F.; Redente, E.F.; Barbarick, K.A.; Brobst, R.B.; Paschke, M.W.; Miller, A.L. Plant and Soil Responses to Biosolids Application following Forest Fire. *J. Environ. Qual.* **2004**, *33*, 873–881. [CrossRef] [PubMed]
8. Cerdá, A.; Doerr, S.H. Influence of vegetation recovery on soil hydrology and erodibility following fire: An 11-year investigation. *Int. J. Wildland Fire* **2005**, *14*, 423–437. [CrossRef]

9. y Silva, F.R.; González-Cabán, A. 'SINAMI': A tool for the economic evaluation of forest fire management programs in Mediterranean ecosystems. *Int. J. Wildland Fire* **2010**, *19*, 927–936. [CrossRef]
10. Arnold, K.T.; Murphy, N.P.; Gibb, H. Post-fire recovery of litter detritivores is limited by distance from burn edge. *Austral Ecol.* **2016**, *42*, 94–102. [CrossRef]
11. Bolton, D.K.; Coops, N.C.; Hermosilla, T.; Wulder, M.A.; White, J.C. Assessing variability in post-fire forest structure along gradients of productivity in the Canadian boreal using multi-source remote sensing. *J. Biogeogr.* **2017**, *44*, 1294–1305. [CrossRef]
12. Amiro, B.D.; Chen, J.M.; Liu, J. Net primary productivity following forest fire for Canadian ecoregions. *Can. J. For. Res.* **2000**, *30*, 939–947. [CrossRef]
13. Goetz, S.J.; Fiske, G.J.; Bunn, A.G. Using satellite time-series data sets to analyze fire disturbance and forest recovery across Canada. *Remote Sens. Environ.* **2006**, *101*, 352–365. [CrossRef]
14. Keeley, J.E. Fire intensity, fire severity and burn severity: A brief review and suggested usage. *Int. J. Wildland Fire* **2009**, *18*, 116–126. [CrossRef]
15. Lee, J.-M.; Lee, S.-W.; Lim, J.-H.; Won, M.-S.; Lee, H.-S. Effects of heterogeneity of pre-fire forests and vegetation burn severity on short-term post-fire vegetation density and regeneration in Samcheok, Korea. *Landsc. Ecol. Eng.* **2013**, *10*, 215–228. [CrossRef]
16. Storey, E.A.; Stow, D.A.; O'Leary, J.F. Assessing postfire recovery of chamise chaparral using multi-temporal spectral vegetation index trajectories derived from Landsat imagery. *Remote Sens. Environ.* **2016**, *183*, 53–64. [CrossRef]
17. Miller, J.D.; Thode, A.E. Quantifying burn severity in a heterogeneous landscape with a relative version of the delta Normalized Burn Ratio (dNBR). *Remote Sens. Environ.* **2007**, *109*, 66–80. [CrossRef]
18. Lanorte, A.; Danese, M.; Lasaponara, R.; Murgante, B. Multiscale mapping of burn area and severity using multisensor satellite data and spatial autocorrelation analysis. *Int. J. Appl. Earth Obs.* **2013**, *20*, 42–51. [CrossRef]
19. Alonso-Canas, I.; Chuvieco, E. Global burned area mapping from ENVISAT-MERIS and MODIS active fire data. *Remote Sens. Environ.* **2015**, *163*, 140–152. [CrossRef]
20. Nioti, F.; Xystrakis, F.; Koutsias, N.; Dimopoulos, P. A Remote Sensing and GIS Approach to Study the Long-Term Vegetation Recovery of a Fire-Affected Pine Forest in Southern Greece. *Remote Sens.* **2015**, *7*, 7712–7731. [CrossRef]
21. Giglio, L.; Descloitres, J.; Justice, C.O.; Kaufman, Y.J. An Enhanced Contextual Fire Detection Algorithm for MODIS. *Remote Sens. Environ.* **2003**, *87*, 273–282. [CrossRef]
22. He, L.; Li, Z. Enhancement of a fire-detection algorithm by eliminating solar contamination effects and atmospheric path radiance: Application to MODIS data. *Int. J. Remote Sens.* **2011**, *32*, 6273–6293. [CrossRef]
23. Roy, D.P.; Boschetti, L. Southern Africa Validation of the MODIS, L3JRC, and GlobCarbon Burned-Area Products. *IEEE Trans. Geosci. Remote Sens.* **2009**, *47*, 1032–1044. [CrossRef]
24. Key, C.H.; Benson, N.C. Measuring and remote sensing of burn severity: The CBI and NBR. In Proceedings of the Joint Fire Science Conference and Workshop, Boise, ID, USA, 15–17 June 1999.
25. Lentile, L.B.; Holden, Z.A.; Smith, A.M.S.; Falkowski, M.J.; Hudak, A.T.; Morgan, P.; Lewis, S.A.; Gessler, P.E.; Benson, N.C. Remote sensing techniques to assess active fire characteristics and post-fire effects. *Int. J. Wildland Fire* **2006**, *15*, 319–345. [CrossRef]
26. Hope, A.; Tague, C.; Clark, R. Characterizing post-fire vegetation recovery of California chaparral using TM/ETM+ time-series data. *Int. J. Remote Sens.* **2007**, *28*, 1339–1354. [CrossRef]
27. Meng, R.; Dennison, P.E.; Huang, C.; Moritz, M.A.; D'Antonio, C. Effects of fire severity and post-fire climate on short-term vegetation recovery of mixed-conifer and red fir forests in the Sierra Nevada Mountains of California. *Remote Sens. Environ.* **2015**, *171*, 311–325. [CrossRef]
28. Van Leeuwen, W.J.D.; Casady, G.M.; Neary, D.G.; Bautista, S.; Alloza, J.A.; Carmel, Y.; Wittenberg, L.; Malkinson, D.; Orr, B.J. Monitoring post-wildfire vegetation response with remotely sensed time-series data in Spain, USA and Israel. *Int. J. Wildland Fire* **2010**, *19*, 75–93. [CrossRef]
29. Caccamo, G.; Bradstock, R.; Collins, L.; Penman, T.; Watson, P. Using MODIS data to analyse post-fire vegetation recovery in Australian eucalypt forests. *J. Spat. Sci.* **2014**, *60*, 341–352. [CrossRef]
30. Padfield, D.; Lowe, C.; Buckling, A.; Ffrench-Constant, R.; Jennings, S.; Shelley, F.; Olafsson, J.S.; Yvon-Durocher, G. Metabolic compensation constrains the temperature dependence of gross primary production. *Ecol. Lett.* **2017**, *20*, 1250–1260. [CrossRef] [PubMed]

31. Zhou, S.; Zhang, Y.; Ciais, P.; Xiao, X.; Luo, Y.; Caylor, K.K.; Huang, Y.; Wang, G. Dominant role of plant physiology in trend and variability of gross primary productivity in North America. *Sci. Rep.* **2017**, *7*, 41366. [CrossRef] [PubMed]
32. Laneve, G.; Fusilli, L.; Marzialetti, P.; De Bonis, R.; Bernini, G.; Tampellini, L. Development and Validation of Fire Damage-Severity Indices in the Framework of the PREFER Project. *IEEE J. Sel. Top. Appl. Earth Obs. Remote Sens.* **2016**, *9*, 2806–2817. [CrossRef]
33. Miller, J.D.; Knapp, E.E.; Key, C.H.; Skinner, C.N.; Isbell, C.J.; Creasy, R.M.; Sherlock, J.W. Calibration and validation of the relative differenced Normalized Burn Ratio (RdNBR) to three measures of fire severity in the Sierra Nevada and Klamath Mountains, California, USA. *Remote Sens. Environ.* **2009**, *113*, 645–656. [CrossRef]
34. Chen, D.; Huang, J.; Jackson, T.J. Vegetation water content estimation for corn and soybeans using spectral indices derived from MODIS near- and short-wave infrared bands. *Remote Sens. Environ.* **2005**, *98*, 225–236. [CrossRef]
35. García, M.J.L.; Caselles, V. Mapping burns and natural reforestation using thematic Mapper data. *Geocarto Int.* **1991**, *6*, 31–37. [CrossRef]
36. Huete, A.; Didan, K.; Miura, T.; Rodriguez, E.P.; Gao, X.; Ferreira, L.G. Overview of the radiometric and biophysical performance of the MODIS vegetation indices. *Remote Sens. Environ.* **2002**, *83*, 195–213. [CrossRef]
37. Kogan, F.; Gitelson, A.; Zakarin, E.; Spivak, L.; Lebed, L. AVHRR-Based Spectral Vegetation Index for Quantitative Assessment of Vegetation State and Productivity. *Photogramm. Eng. Remote Sens.* **2003**, *69*, 899–906. [CrossRef]
38. Serrano, L.; Filella, I.; Penuelas, J. Remote Sensing of Biomass and Yield of Winter Wheat under Different Nitrogen Supplies. *Crop Sci.* **2000**, *40*, 723–731. [CrossRef]
39. Yin, H.; Udelhoven, T.; Fensholt, R.; Pflugmacher, D.; Hostert, P. How Normalized Difference Vegetation Index (NDVI) Trendsfrom Advanced Very High Resolution Radiometer (AVHRR) and Système Probatoire d'Observation de la Terre VEGETATION (SPOT VGT) Time Series Differ in Agricultural Areas: An Inner Mongolian Case Study. *Remote Sens.* **2012**, *4*, 3364–3389. [CrossRef]
40. Shim, C.; Hong, J.; Hong, J.; Kim, Y.; Kang, M.; Malla Thakuri, B.; Kim, Y.; Chun, J. Evaluation of MODIS GPP over a complex ecosystem in East Asia: A case study at Gwangneung flux tower in Korea. *Adv. Space Res.* **2014**, *54*, 2296–2308. [CrossRef]
41. Jung, C.G.; Lee, Y.G.; Kim, S.J.; Jang, C.H. Quantitative Study of CO_2 based on Satellite Image for Carbon Budget on Flux Tower Watersheds. *J. Korean Soc. Agric. Eng.* **2015**, *57*, 109–120, (In Korean with English abstract). [CrossRef]
42. Park, M.; Lee, H. Forest Policy and Law for Sustainability within the Korean Peninsula. *Sustainability* **2014**, *6*, 5162–5186. [CrossRef]
43. Choi, C.-Y.; Lee, E.-J.; Nam, H.-Y.; Lee, W.-S.; Lim, J.-H. Temporal changes in the breeding bird community caused by post-fire treatments after the Samcheok forest fire in Korea. *Landsc. Ecol. Eng.* **2013**, *10*, 203–214. [CrossRef]
44. Ahn, Y.S.; Ryu, S.-R.; Lim, J.; Lee, C.H.; Shin, J.H.; Choi, W.I.; Lee, B.; Jeong, J.H.; An, K.W.; Seo, J.I. Effects of forest fires on forest ecosystems in eastern coastal areas of Korea and an overview of restoration projects. *Landsc. Ecol. Eng.* **2013**, *10*, 229–237. [CrossRef]
45. Lee, B.; Kim, S.Y.; Chung, J.; Park, P.S. Estimation of fire severity by use of Landsat TM images and its relevance to vegetation and topography in the 2000 Samcheok forest fire. *J. For. Res.* **2008**, *13*, 197–204. [CrossRef]
46. Key, C.H.; Benson, N.C. *Landscape Assessment Sampling and Analysis Methods*; General Technical Report RMRS-GRT-164-CD; USDA Forest Service, Rocky Mountain Research Station: Ogden, UT, USA, 2006.
47. Soverel, N.O.; Perrakis, D.D.B.; Coops, N.C. Estimating burn severity from Landsat dNBR and RdNBR indices across western Canada. *Remote Sens. Environ.* **2010**, *114*, 1896–1909. [CrossRef]
48. Mallinis, G.; Mitsopoulos, I.; Chrysafi, I. Evaluating and comparing Sentinel 2A and Landsat-8 Operational Land Imager (OLI) spectral indices for estimating fire severity in a Mediterranean pine ecosystem of Greece. *GISci. Remote Sens.* **2017**, *55*, 1–18. [CrossRef]
49. Kim, C.-G.; Shin, K.; Joo, K.Y.; Lee, K.S.; Shin, S.S.; Choung, Y. Effects of soil conservation measures in a partially vegetated area after forest fires. *Sci. Total Environ.* **2008**, *399*, 158–164. [CrossRef] [PubMed]

50. Lee, R.J.; Chow, T.E. Post-wildfire assessment of vegetation regeneration in Bastrop, Texas, using Landsat imagery. *GISci. Remote Sens.* **2015**, *52*, 609–626. [CrossRef]
51. Armesto, J.J.; Pickett, S.T.A. Experiments on disturbance in old-field plant communities: Impact on species richness and abundance. *Ecology* **1985**, *66*, 230–240. [CrossRef]
52. Kim, Y.S.; Byun, J.K.; Kim, C.; Park, B.B.; Kim, Y.K.; Bae, S.W. Growth response of Pinus densiflora seedlings to different fertilizer compound ratios in a recently burned area in the eastern coast of Korea. *Landsc. Ecol. Eng.* **2012**, *10*, 241–247. [CrossRef]
53. Korea Meteorological Administration. Available online: https://data.kma.go.kr (accessed on 13 May 2018).
54. Wang, L.; Qu, J.J.; Hao, X.; Zhu, Q. Sensitivity studies of the moisture effects on MODIS SWIR reflectance and vegetation water indices. *Int. J. Remote Sens.* **2008**, *29*, 7065–7075. [CrossRef]
55. Polychronaki, A.; Gitas, I.Z.; Minchella, A. Monitoring post-fire vegetation recovery in the Mediterranean using SPOT and ERS imagery. *Int. J. Wildland Fire* **2014**, *23*, 631–642. [CrossRef]
56. Dore, S.; Kolb, T.E.; Montes-Helu, M.; Sullivan, B.W.; Winslow, W.D.; Hart, S.C.; Kaye, J.P.; Koch, G.W.; Hungate, B.A. Long-term impact of a stand-replacing fire on ecosystem CO_2 exchange of a ponderosa pine forest. *Glob. Chang. Biol.* **2008**, *14*, 1801–1820. [CrossRef]
57. Brown, S.; Lugo, A.E.; Chapman, J. Biomass of tropical tree plantations and its implications for the global carbon budget. *Can. J. For. Res.* **1986**, *16*, 390–394. [CrossRef]
58. Quintero-Méndez, M.; Jerez-Rico, M. Heuristic forest planning model for optimizing timber production and carbon sequestration in teak plantations. *iForest* **2017**, *10*, 430–439. [CrossRef]

10

Retrieval of Total Precipitable Water from Himawari-8 AHI Data: A Comparison of Random Forest, Extreme Gradient Boosting and Deep Neural Network

Yeonjin Lee [1], Daehyeon Han [2], Myoung-Hwan Ahn [1,*], Jungho Im [2] and Su Jeong Lee [1]

[1] Department of Atmospheric Sciences and Engineering, Ewha Womans University, 52 Ewhayeodae-gil, Seodaemun-gu, Seoul 03760, Korea

[2] School of Urban and Environmental Engineering, Ulsan National Institute of Science and Technology, Ulsan 44919, Korea

* Correspondence: terryahn65@ewha.ac.kr.

Abstract: Total precipitable water (TPW), a column of water vapor content in the atmosphere, provides information on the spatial distribution of moisture. The high-resolution TPW, together with atmospheric stability indices such as convective available potential energy (CAPE), is an effective indicator of severe weather phenomena in the pre-convective atmospheric condition. With the advent of high performing imaging instrument onboard geostationary satellites such as Advanced Himawari Imager (AHI) onboard Himawari-8 of Japan and Advanced Meteorological Imager (AMI) onboard GeoKompsat-2A of Korea, it is expected that unprecedented spatiotemporal resolution data (e.g., AMI plans to provide 2 km resolution data at every 2 min over the northeast part of East Asia) will be provided. To derive TPW from such high-resolution data in a timely fashion, an efficient algorithm is highly required. Here, machine learning approaches—random forest (RF), extreme gradient boosting (XGB), and deep neural network (DNN)—are assessed for the TPW retrieved from AHI over the clear sky in Northeast Asia area. For the training dataset, the nine infrared brightness temperatures (BT) of AHI (BT8 to 16 centered at 6.2, 6.9, 7.3, 8.6, 9.6, 10.4, 11.2, 12.4, and 13.3 μm, respectively), six dual channel differences and observation conditions such as time, latitude, longitude, and satellite zenith angle for two years (September 2016 to August 2018) are used. The corresponding TPW is prepared by integrating the water vapor profiles from InterimEuropean Centre for Medium-Range Weather Forecasts Re-Analysis data (ERA-Interim). The algorithm performances are assessed using the ERA-Interim and radiosonde observations (RAOB) as the reference data. The results show that the DNN model performs better than RF and XGB with a correlation coefficient of 0.96, a mean bias of 0.90 mm, and a root mean square error (RMSE) of 4.65 mm when compared to the ERA-Interim. Similarly, DNN results in a correlation coefficient of 0.95, a mean bias of 1.25 mm, and an RMSE of 5.03 mm when compared to RAOB. Contributing variables to retrieve the TPW in each model and the spatial and temporal analysis of the retrieved TPW are carefully examined and discussed.

Keywords: total precipitable water; Himawari-8 AHI; machine learning; random forest; deep neural network; XGBoost

---

1. Introduction

Water vapor, one of the most influential constituents of the atmosphere, is responsible for determining the amount of precipitation that a region can receive [1]. Total precipitable water (TPW) is a meteorological factor that shows the amount of water vapor contained in the column of air per unit area of the atmosphere in terms of the depth of liquid [2]. Although TPW does not represent the vertical structure of moisture, it describes horizontal gradients of integrated water vapor content.

Furthermore, the amount of water vapor contained in the troposphere has significant implications for determining the strength and severity of a severe weather event [3]. Therefore, TPW is one of the critical variables used by forecasters when severe weather conditions are expected [4].

Himawari-8, Japan's geostationary (GEO) meteorological satellite launched in October 2014, includes a primary instrument called Advanced Himawari Imager (AHI). AHI has 16 channels—four visible (VIS), two near-infrared (NIR), and 10 infrared (IR) channels—with a temporal resolution of 10 min for the full disk (less than 10 min for limited areas) and a spatial resolution of 2 km at nadir for the infrared channels (less than 2 km for the visible channels) [5]. Unlike low earth orbit (LEO) satellites, which have limited temporal resolution, GEO satellites can produce data more timely and frequently. The retrieved high temporal resolution TPW from GEO satellite sensor data can be utilized to monitor pre-convective environments and predict heavy rainfall, convective storms, and clouds that may cause serious damage to human life and infrastructure [6–8]. For example, Lee et al. [8] showed that the 10-min interval measurements from the AHI sensor successfully provided information about the pre-landfall environment for typhoon Nangka that occurred in 2015.

In the remote sensing field, there are several approaches for the retrieval of TPW from IR channels of GEO satellite observations, including (1) a physical method using the one-dimensional variational system, (2) a split-window algorithm, and (3) machine learning algorithms. The physical modeling based on a nonlinear optimal estimation method has been traditionally used for vertical profiles of temperature and humidity (T-q profile) [9]. While the TPW derived from the T-q profile retrieved with a physical method usually has high accuracy, it does not fully use the original resolution information from satellite observations due to the high computing load. The split-window method is based on a different absorptive response to the water vapor content at two channels (near 11.0 and 12.0 μm) [10–12]. This method can be classified into the linear approach, look-up table approach, physical approach, and covariance-variance ratio approach [13–16]. These approaches have limitations which are easily affected by uncertainties in the surface emissivity, errors in first guess field, and the instrument noise [16,17]. On the other hand, machine learning approaches such as random forest (RF), extreme gradient boosting (XGB), and deep neural network (DNN) are capable of predicting the nonlinear relationship between the parameters [18] when compared to the other methods. Since these approaches have been often adopted as ways for the fast and reliable calculation of a target variable, various machine learning techniques have been used to derive TPW from satellite sensor data. For instance, Wang et al. [19] used a multi-layer perceptron neural network with NIR radiances from Moderate-resolution Imaging Spectro-radiometer (MODIS) as input to retrieve TPW. Zhang et al. [20] also employed a radial basis function neural network algorithm using infrared data from the hyperspectral sounder, Atmospheric Infrared Sounder, to retrieve TPW. Although these machine learning based models have shown good performance, there has been little exploration in comparison of multiple machine learning techniques especially for GEO satellite data-based TPW retrievals.

In this study, machine learning based approaches for the retrieval of TPW from GEO satellite data over clear sky pixels are proposed. The objectives of this study are to (1) develop the TPW retrieval algorithms from Himawari-8 AHI data using RF, XGB, and DNN over Northeast Asia, (2) quantify and examine relative variable importance and contribution by model, and (3) analyze the temporal and spatial distribution of the errors compared to ERA-Interim in Northeast Asia for 2 years. Section 2 describes the study area and data used for the algorithm development and Section 3 explains the retrieval methods (RF, XGB, and DNN) based on the machine learning approaches. Variable importance, validation results, and discussion are described in Section 4. The conclusions are presented in Section 5.

2. Data

Himawari-8 launched on October 7, 2014 carries out a meteorological mission on a GEO orbit (centered on 140.68°E). The considerably improved AHI sensor in temporal, spatial, and spectral resolutions over its predecessors is now better suited for nowcasting and has improved the accuracy of numerical forecasts [21]. AHI has a total of 16 spectral channels (0.47–13.3 μm) consisting of VIS, NIR, and IR—short-wavelength infrared (SWIR), mid-wavelength infrared (MWIR) and thermal infrared (TIR) channels (Table 1). AHI collects data every 10 min from the full disk and every 2.5 min from the north-eastern and south-western areas of Japan. The brightness temperature (BT) data from Himawari-8 AHI infrared channels are used as input variables to develop machine learning-based models to retrieve TPW.

Table 1. Himawari-8 Advanced Himawari Imager (AHI) specifications.

Channel Number	Central Wavelength (μm)	Band Width (μm)	Spatial Resolution at Sub Satellite Point (km)
1	0.47	0.05	1
2	0.51	0.02	1
3	0.64	0.03	0.5
4	0.86	0.02	1
5	1.6	0.02	2
6	2.3	0.02	2
7	3.9	0.22	2
8	6.2	0.37	2
9	6.9	0.12	2
10	7.3	0.17	2
11	8.6	0.32	2
12	9.6	0.18	2
13	10.4	0.30	2
14	11.2	0.20	2
15	12.4	0.30	2
16	13.3	0.20	2

ERA-Interim, the reanalysis of the global atmospheric dataset, has been released by the European Centre for Medium-range Weather Forecasts (ECMWF). The data have been continuously updated and provided since 1979 with one month to two months delay. The data consist of analysis fields provided four times (00, 06, 12, and 18 UTC) a day and forecasts fields provided with 3, 6, 9, and 12 h steps at 00 and 12 UTC [22]. ERA-Interim TPW covering both ocean and land with a spatial resolution of 0.125° × 0.125° (Table 2) are selected as a target variable corresponding to the input variables [23,24]. TPW of ERA-Interim can be downloaded from the ECMWF website (https://apps.ecmwf.int/datasets/data/interim-full-daily/levtype=sfc/).

Table 2. Description of the reference data and usage.

Reference Data	Temporal and Spatial Resolution	Period and Usage
ERA-Interim	6 h/ 0.125° × 0.125°	September 2016–August 2018 Four days per month (5th, 10th, 15th, 20th) Training data (90%)/test data (10%) September 2016–August 2018 Two days per month (1st, 25th) Validation data
RAOB	12 h [1]/ 13 sites over Northeast Asia	September 2016–August 2018 Two days per month (1st, 25th) Validation data

[1] The nominal time of a radiosonde launch is 0000 or 1200 UTC. Occasionally, this launches at 0600 and 1800 UTC.

Radiosondes measure the state of the atmosphere at each altitude level and thus the radiosonde-derived water vapor content can be considered as true reference data. For the quantitative validation of the retrieved TPW, radiosonde observations (RAOB) from the University of Wyoming were used (Table 2). However, it should be noted that there might be uncertainty in RAOB, especially humidity profiles, depending on the type of the sensor used. Most sensors used in the study are located in Northeast Asia and have high biases compared to the humidity profiles from ECMWF [25]. Thus, only the data from 13 observation stations (Figure 1) located in Northeast Asia (mainly in Japan) having the same sensor types (i.e., Vaisala RS92 and Meisei RS-11G) with generally good accuracy are used to assess the retrieved TPW. These radiosonde data are available on the website from the University of Wyoming (http://weather.uwyo.edu/upperair/sounding.html).

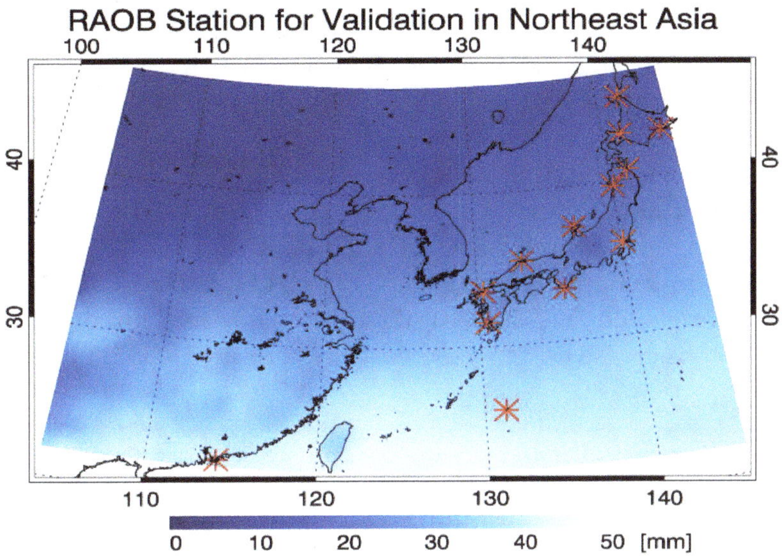

Figure 1. The red asterisks indicate 13 radiosonde observations (RAOB) sites with high accuracy for validation within the study area. The averaged ERA-Interim total precipitable water (TPW) from October 2016 to August 2018 was used as a background image.

For a comparison between two reference data (i.e., RAOB and ERA-interim TPW), ERA-Interim TPW was collocated at the grid-point near the 13 radiosonde stations in Northeast Asia during the validation period. The comparison result is shown in Figure 2. The error metrics were calculated from 130 collocated matches over clear sky regions. Specifically, the two reference data agree with each other with mean bias (ERA-Interim TPW—RAOB TPW) of −0.56 mm and root mean square error (RMSE) of 2.78 mm.

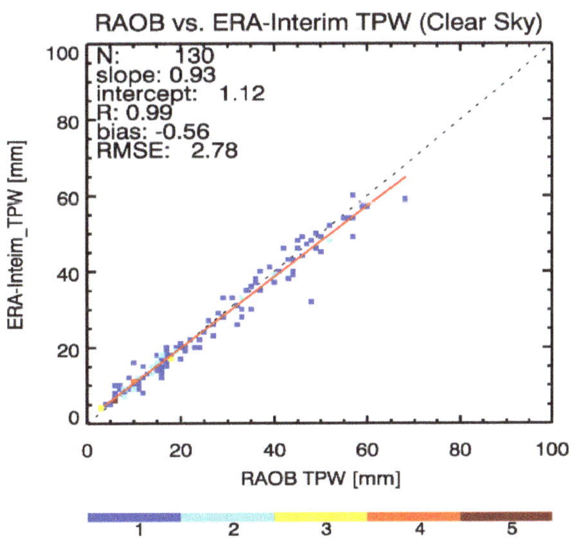

Figure 2. Comparison between RAOB TPW and ERA-Interim TPW in Northeast Asia during the validation period (1st, 25th per month from September 2016 to August 2018). ERA-Interim TPW is collocated with the 13 RAOB stations over a clear sky region. Scatter plots are colored by density: the x-axis is RAOB TPW and the y-axis is ERA-Interim TPW. The red line represents a regression line.

3. Methods

3.1. Preparation of Training Dataset

Himawari-8 AHI infrared BTs and ERA-Interim TPW were used as the input variables and the target variable of machine learning models, respectively. Table 3 shows the input variables used in model training. The channel centered at 3.9 μm among the AHI IR channels was excluded from the input variables due to the contamination problem by solar radiation during the daytime [26]. The BT at each channel used as the training data contributes to the models at a different level. For example, the window channels (i.e., BT11, BT13, BT14, and BT15) are related to the temperature of land and sea surfaces, while the water vapor channels (i.e., BT8, BT9, and BT10) have characteristics related with the distribution of water vapor in three different vertical layers. BT12 and BT16 channels, which correspond to O_3 and CO_2 absorption bands respectively, are related to the information of atmospheric air mass [26]. In addition, dual channel differences (DCD), including the difference between the window channel and water vapor channel (i.e., DCD BT14−BT8, DCD BT14−BT9, DCD BT14−BT10), the difference between window channels (i.e., DCD BT14−BT11, DCD BT14−BT15), and the difference between water vapor channels (i.e., DCD BT10−BT8) carry information on the column water vapor amount. The DCDs tend to increase as TPW increases except for DCD BT14−BT15, which shows an opposite tendency.

Table 3. Summary of the physical characteristics of input variables in the retrieval models of TPW. Here, each brightness temperature (BT) of channels 8 to 16 are named BT8 to BT16, respectively.

Input Variable	Physical Characteristics
Cyclic_day	Temporal characteristics
Latitude	Spatial characteristics
Longitude	Spatial characteristics
Satellite zenith angle	Optical depth
BT8 (IR 6.2 μm)	Upper tropospheric water vapor
BT9 (IR 6.9 μm)	Mid and upper tropospheric water vapor
BT10 (IR 7.3 μm)	Mid tropospheric water vapor
BT11 (IR 8.6 μm)	Total water for stability, dust, SO_2, rainfall
BT12 (IR 9.63 μm)	Total ozone
BT13 (IR 10.4 μm)	Surface
BT14 (IR 11.2 μm)	Sea surface temperature and rainfall
BT15 (IR 12.4 μm)	Total water and SST
BT16 (IR 13.3 μm)	Air temperature
DCD BT14−BT8	Upper tropospheric moisture
DCD BT14−BT9	Mid and upper tropospheric moisture
DCD BT14−BT10	Mid tropospheric moisture
DCD BT14−BT11	Amount of water vapor
DCD BT14−BT15	Split-window channels (amount of water vapor)
DCD BT10−BT8	Difference between water vapor channels

Training data were collected in Northeast Asia region, four days a month (5th, 10th, 15th, and 20th of each month) and four times a day (00, 06, 12, and 18 UTC for each day) from September 2016 to August 2018. The study area over Northeast Asia centered at 37.588 N, 124.044 E (4300 km in E-W direction and 2900 km in the N-S direction) is covering South Korea, North Korea, Japan, Taiwan, and parts of China and Russia (Figure 1). To prevent cloud contamination, cloudy pixels were removed based on the empirical cloud mask algorithm proposed by Lee et al. (2019) [27]. When 5×5 neighboring pixels of a grid of target data (ERA-Interim TPW) are all clear, the AHI data are averaged to consider different spatial scales. The construction of the representative training data is crucial to develop successful retrieval models using machine learning [28–30]. Since the raw data have a skewed distribution toward low TPW (Figure 3a), it is necessary to adjust them to have a balanced distribution to avoid biased training [30]. Thus, the original dataset was randomly divided into the training dataset

(80%, 698,033 samples) and the testing dataset (20%, 174,510 samples) with the same number of data for each bin (i.e., 1 mm in TPW) as shown in Figure 3b,c. Here, each subset has the same distributions of the original dataset for the balanced samples. For the validation, data that are not used for the training are selected from two days a month (1st and 25th of each month), four times a day (00, 06, 12, and 18 UTC for each day) from September 2016 to August 2018.

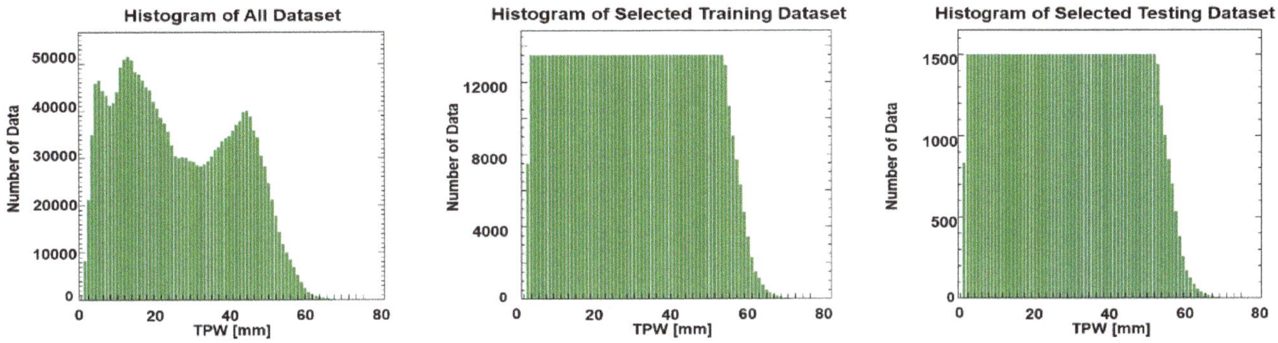

(a) (b) (c)

Figure 3. Histograms of training data for machine learning in Northeast Asia for two years (i.e., September 2016 to August 2018). The x-axis is the TPW range (0–80 mm) and the y-axis is histogram density. The bin size is 1 mm. (**a**) Histogram of all TPW during the period, (**b**) histogram of the selected TPW (13,500 datasets per bin size) for training, and (**c**) histogram of the selected TPW (1500 datasets per bin size) for testing.

3.2. Machine Learning Approaches

Machine learning-based approaches have been widely applied in the field of remote sensing for both classification and regression [4,6,19,20,23,24,31–47]. Here, three machine learning models (i.e., RF, XGB, and DNN) for TPW retrieval using AHI data are assessed. The process flow of the TPW retrieval based on machine learning algorithms is illustrated in Figure 4. For the cloud screening, the radiative transfer model (RTM) BT simulation was conducted and here, the Radiative Transfer for TIROS Operational Vertical Sounder (RTTOV) version 11.2 [48] was used to simulate BT from the unified model analysis data. Through an empirical thresholding test [27] using the simulated BT and the observed AHI BT, the clear-sky data were selected. If data were cloud free, TPW was retrieved by the machine learning models.

RF is an ensemble of rule-based algorithm based on classification and regression trees (CART) [49]. RF has been widely applied to various fields such as remote sensing and geographic information science [31–38]. Each independent tree in RF is created by the randomly selected subsets of training samples and input variables. The results from a multitude of independent trees are averaged to produce the final output of RF. For a fast implementation of RF in the R environment, a method of "ranger" is used to take advantage of the parallel processing of RF especially suited for the large and high dimensional dataset [50]. The RF has two basic model parameters to tune: the number of trees (num.trees) and the number of variables sampled in random when splitting at each node (mtry). The other parameters except for the two most representative parameters are used by applying default values. To find two optimum parameters, mtry is tested with 2, 4, 8, 10, and 19 and num.trees is tested with 100, 250, 500, and 1000. The parameter optimization is conducted based on the mean square error (MSE). The mtry and num.trees were determined at the 10 and 1000, respectively.

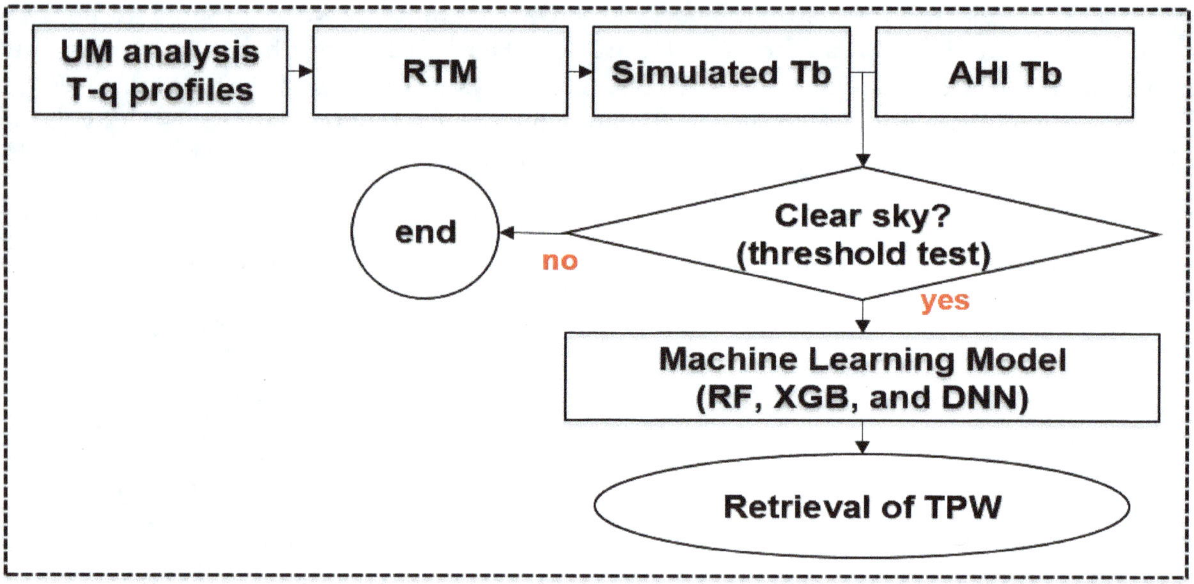

Figure 4. Process flow of the proposed machine learning (random forest (RF), extreme gradient boosting (XGB), and deep neural network (DNN))-based TPW retrieval models from AHI observations under clear-sky conditions.

Extreme gradient boosting (XGB) is a kind of tree-based boosting method, which is a sequential ensemble learning model of a decision tree. XGB weighs data with unexpected error from training to make the model better predict such data iteratively. The XGB based algorithm has been widely used in the latest studies [38,51,52]. The algorithm is highly effective in reducing the computing time and provides optimal use of memory resources because of parallel and distributed computing [51]. XGB has produced higher classification accuracy and faster execution time when compared to other models in several studies [41–43]. The XGB model is developed in the Python environment using XGBoost packages, here. For parameter optimization of the XGB model, Bayesian optimization is used as an effective tool for optimizing parameters in machine learning models [52]. Four parameters—number of trees used (n_trees), maximum depth of a tree (max_depth), minimum loss reduction required to make a further partition on a leaf node of the tree (gamma), and the subsample ratio of columns when constructing each tree (colsample_bytree)—were empirically tuned based on root mean square error (RMSE). The optimum n_trees, gamma, colsample_bytree, and max_depth were 4553, 0.7, 1.0, and 10, respectively.

Artificial neural network (ANN) is a biologically inspired machine learning approach that consists of neurons showing particularly good performance at modeling the nonlinear relationships between input and target data using a backpropagation [53]. Backpropagation adjusts the connection strength (weight) between neurons by minimizing the prediction error iteratively. ANN has been used for various purposes in remote sensing fields [42,43]. DNN is a subset of ANN with multiple hidden layers. Here, the DNN is developed with "Keras" which is a high-level Python library for deep learning. The DNN model has parameters including the number of hidden layers and neurons, optimizer, and activation function. The DNN parameters were adjusted to achieve high performance for the prepared training dataset. The activation functions, i.e., linear, sigmoid, tanh, and rectified linear unit (ReLU), hidden layers with 1 to 4 and neurons with 5 to 30 in five intervals in each hidden layer were tested. Besides, several widely used optimizers (i.e., stochastic gradient descent, rmsprop, and Adam) that update the weights in the direction that error decreases were tested by comparing the calculated results with the target value per iteration. The other parameters of the DNN are set as follows: batch_size = 256, dropout_rate = 0.5, stop_steps = 100 (if there is no improvement within n steps, training will be terminated), and learning rate = 0.001. Through the empirical testing, ReLU was set as the activation function with the number of the hidden layers = 4, and the number of the hidden neuron = 60 using Adam optimizer.

3.3. Accuracy Metrics

For the selection of model configuration, K-fold cross-validation is used [54]. This method has been widely used to estimate overall performance. When a specific value for k is chosen (here, k is 10), datasets are randomly and equally distributed into k groups. One group is the test fold and the (k−1) groups are the training folds. In total k-times validation, performance is calculated by using the different test folds for each validation. Finally, the averaged validation results are used to tune the hyperparameters of each model. Typical accuracy metrics including correlation coefficient (R), mean bias, and RMSE are used (Equations (1) to (3), respectively). These statistical error metrics are calculated between the target TPW value (or reference data) and averaged TPW using the retrieved value based on the collocation criteria (Table 4).

$$\mathrm{R} = \frac{\sum_{i=1}^{n}\left(A_i - \overline{A}\right)\left(B_i - \overline{B}\right)}{\sqrt{\sum_{i=1}^{n}\left(A_i - \overline{A}\right)^2 \sum_{i=1}^{n}\left(B_i - \overline{B}\right)^2}}, \tag{1}$$

$$\mathrm{bias} = \frac{1}{n}\sum_{i=1}^{n}\left(A_i - B_i\right)^2, \tag{2}$$

$$\mathrm{RMSE} = \sqrt{\frac{\sum_{i=1}^{n}\left(A_i - B_i\right)^2}{n}}. \tag{3}$$

Table 4. Description of reference data and collocation criteria for the validation of the proposed models.

Reference Data	Temporal Resolution	Spatial Resolution	Collocation Criteria
ERA-Interim	6 h	0.125° × 0.125°	Averaging one or more retrieved TPW within a 0.1-degree radius
RAOB	12 h [1]	13 sites over Northeast Asia	Averaging one or more retrieved TPW within a 0.1-degree radius

[1] The nominal time of a radiosonde launch is 0000 or 1200 UTC. Occasionally, this launches at 0600 and 1800 UTC.

Additionally, to determine the contribution of each input variable to the performance of the three models, relative variable importance indicating how much a given model uses that variable to make accurate predictions is analyzed. To measure variable importance in a tree-based model for regression, the impurity reductions are summed over all split nodes in the tree [43]. In contrast to tree models, there is no clear way to assess the variable importance in the DNN model. For the same criterion for relative importance, the RMSE difference between the results obtained using all variables and the results calculated through the 'leave-one-variable-out' method is utilized. This method starts with all variables and removes each variable iteratively to determine the performance and robustness of the model [55].

4. Results and Discussion

4.1. Model Performance

Figure 5 illustrates the averaged model performance using 10-fold cross-validation with the accuracy metrics (Table 2). The XGB model shows the highest accuracy with the RMSE of 2.46 mm (Figure 5b), followed by RF with 2.63 mm (Figure 5a) while the DNN model results in relatively less accurate performance with the RMSE of 2.69 mm (Figure 5c). The mean biases are under the absolute value of 0.15 mm for all models (0.03 mm for RF, 0.00 mm for XGB, and 0.13 mm for DNN) and the correlation coefficients are 0.99 in all models. In all statistical results, the RF and XGB models are very similar and show higher performance than the DNN model. These model performances using the test dataset imply that the rule-based algorithms (i.e., RF and XGB) are more suitable for the retrieval of TPW. Overall, the XGB model outperforms the RF and DNN models. This might be because the XGB models highly weigh the weak learner.

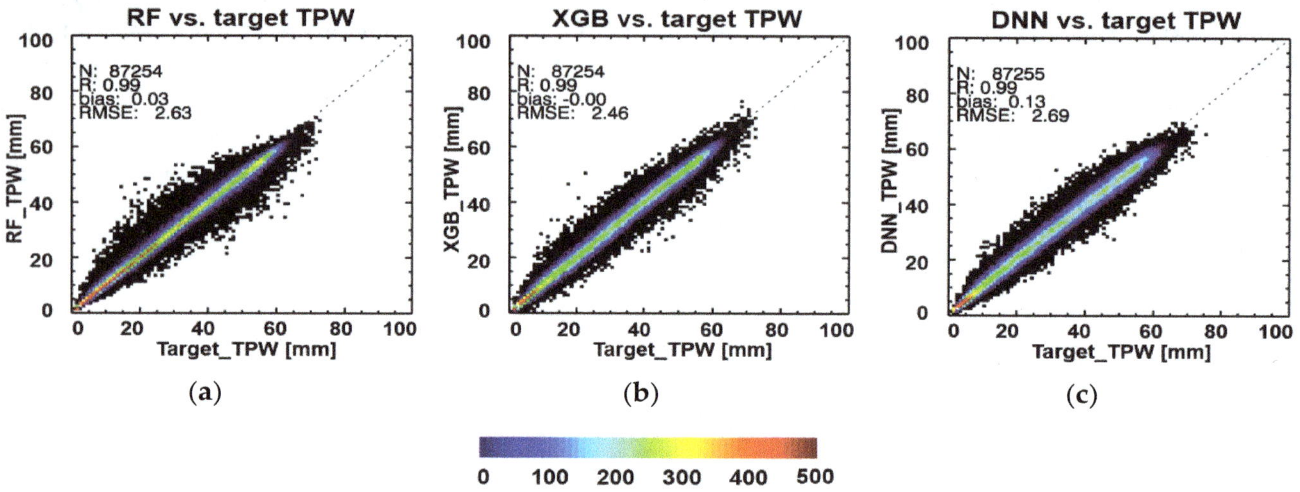

Figure 5. Model performance of machine learning models (RF, XGB, and DNN) using the test dataset. Scatter plots are colored by density: the x-axis is the target TPW (ERA-Interim TPW) and the y-axis is the retrieved TPW using each machine learning model (**a**) RF, (**b**) XGB, and (**c**) DNN.

4.2. Variable Importance

Figure 6a–c shows the summary of calculated variable importance results of the RF, XGB, and DNN. BT16 is identified as the most contributing variable to the retrieval of TPW in all models while the cyc_day and longitude are considered next significant in the RF and XGB and BT12, cyc_day, and latitude are considered next significant in the DNN. The BT16, which is a sensitive channel to carbon dioxide, is used for the estimation of mean tropospheric air temperature. The cyc_day reflects natural variations of TPW considering seasonal characteristics [56]. The longitude and latitude represent the spatial distribution of TPW. The BT12 is a channel sensitive to water vapor as well as ozone. Unlike the variable importance results identified in the DNN, some variables in the RF and XGB show improved accuracy when they are excluded (i.e., negative RMSE difference in Figure 6a,b). This implies that the variables might have redundant information and not be necessary for the RF and XGB to produce the best performance [38].

The ensemble tree models (RF and XGB) provide variable importance during model training, while DNN does not provide such information. Figure 6d,e shows the variable importance identified by using the provided library from RF and XGB. While the BT12 and the difference of window channels (i.e., BT14−BT11 and BT14−BT15) are identified to be very significant for the RF model, the cyc_day and longitude are used as the most important variables in the XGB model. The BT14−BT11 and BT14−BT15, differences between two window channels, imply that the amount of atmospheric water vapor calculated by the difference in absorption coefficients is directly related to TPW. While the provided XGB variable importance (Figure 6e) has a similar pattern to the variable importance in terms of RMSE difference (Figure 6b), in RF, the rank of variable importance shows different patterns among the variables. This might be because the optimized RF model learns using randomly collected variables (here, mtry is 10) at each split time unlike XGB (1.0 for colsample_bytree) and DNN (use all variables). To identify the linearity of the input variables to TPW, the correlation coefficients between TPW and each input variable (Figure 6f) are examined. If a high correlation variable has a high value in the variable importance identified by the model, it might indicate that the linearity between the input variables and TPW is relatively strong. In contrast, the nonlinearity is strong in the opposite case. Interestingly, despite the lowest correlation of cyc_day and longitude with TPW, they have the highest variable importance in the XGB model. These results imply that the XGB model well utilizes the nonlinear characteristics of the variables when compared to the other two models.

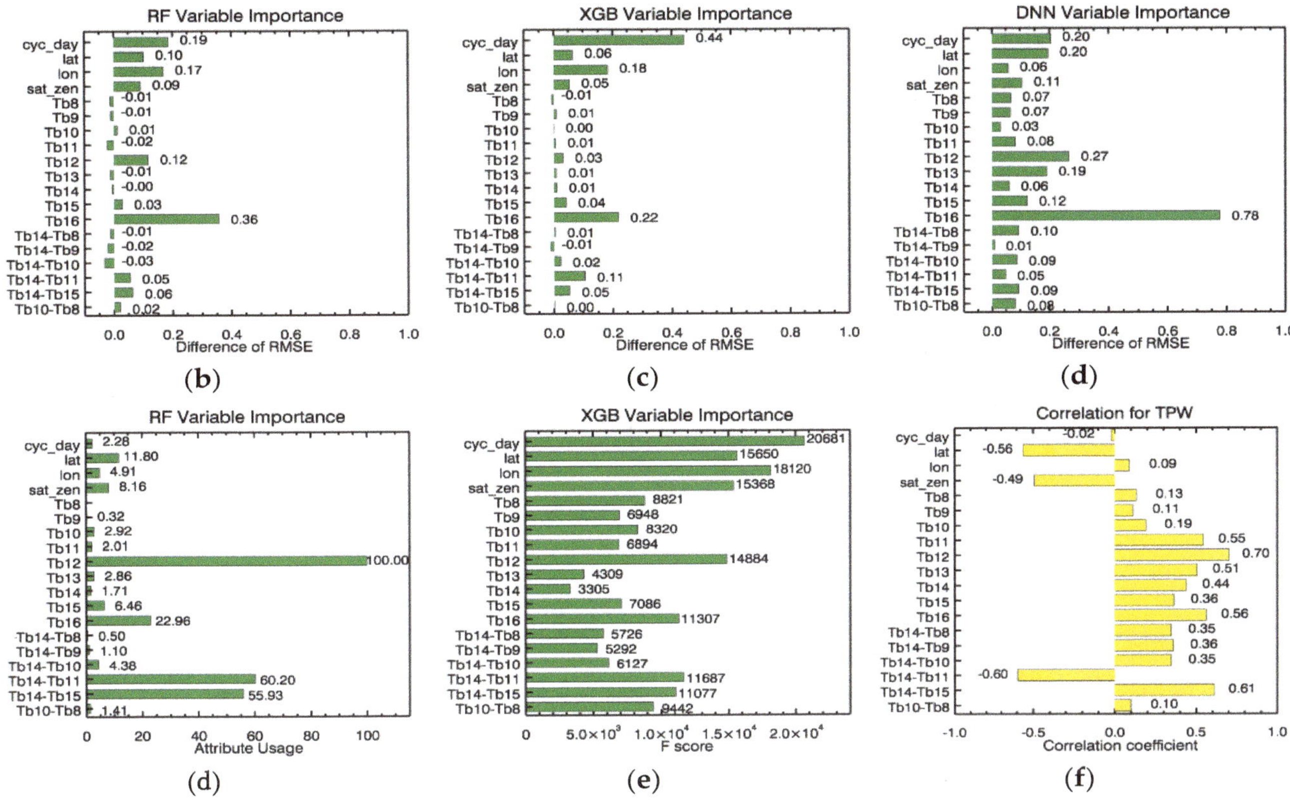

Figure 6. Relative variable importance identified by (**a**) RF, (**b**) XGB, and (**c**) DNN. Bar plots using input variables of the training dataset: the x-axis is the difference of root mean square error (RMSE) and the y-axis is the corresponding name of input variables. The difference in RMSE (the original RMSE − new RMSE) is calculated using the leave-one-variable-out method. The original RMSE of RF, XGB, and DNN are 2.63, 2.46, 2.69 mm, respectively. The variable importance is calculated after the construction of ensemble trees for the rule-based models (**d**) RF and (**e**) XGB. The longer the length of the bar is, the more important the variable is. (**f**) summarizes correlation coefficients between the input variables and TPW.

4.3. Validation Results

To validate the developed RF, XGB, and DNN models, the observed Himawari-8 AHI data that were not used for both the training and testing were utilized (validation dataset in Section 3.1). Table 5 shows the quantitative validation results between ERA-Interim TPW and the retrieved TPW from each model over Northeast Asia region during the validation period (Table 2). Approximately 500,000 collocated data are used for each model over the 'all', 'land', 'sea', and 'coast' regions as categorized in the table. In all areas, the DNN model yields the highest performance in terms of RMSE, followed by the XGB and RF models. The algorithms tend to overestimate TPW. To analyze the spatial distribution of the averaged errors (i.e., bias and RMSE), the retrieved TPW and the ERA-Interim TPW for two years (September 2016 to August 2018) are mapped over Northeast Asia (Figure 7). The mean biases of the RF, XGB, and DNN are about 1.60, 1.30, and 1.47 mm, respectively. The mean RMSE value is 5.02 mm for RF, 4.79 mm for XGB, and 4.56 mm for DNN. The spatial distributions of the averaged bias and RMSE show the characteristics of the discontinuity between land and sea. The lowest performance along the coastal regions (Table 5) can be attributed to these characteristics in the ERA-Interim grid. Additionally, the averaged errors of the RF, XGB, and DNN models appear relatively high in the black circled regions as shown in Figure 7g. These regions have relatively lower surface pressures when compared to the surrounding areas and the models tend to overestimate TPW over these regions with low surface pressure. It is clear that the relatively lower surface pressure is related to the relatively higher terrain elevation and this is the main cause for the overestimated TPW in the region. Thus, further study will need to consider including elevation as one of the predictors. In the meantime, the XGB model has a lower bias in the region with lower surface pressure compared to the RF and DNN model since the XGB model learns repetitively to generate the weighted mean of weak learners [44].

Table 5. Accuracy assessment based on ERA-Interim TPW for the RF, XGB, and DNN in Northeast Asia (1st, 25th per month from September 2016 to August 2018). Validation metrics (i.e., bias and RMSE) are calculated over the 'all', 'land', 'sea', and 'coast' regions, respectively.

	All		**Land**		**Sea**		**Coast**	
	bias	**RMSE**	**bias**	**RMSE**	**bias**	**RMSE**	**bias**	**RMSE**
RF	0.62	5.09	0.66	4.94	0.55	5.29	0.88	5.98
XGB	0.70	4.85	0.67	4.63	0.72	5.14	1.04	5.65
DNN	0.90	4.65	0.94	4.50	0.83	4.86	1.20	5.22

In the validation results (Table 5), the overall accuracies of three models decrease compared to the results of the model performance as shown in Section 4.1. For example, the retrieved TPW using test datasets from all models agrees well with the ERA-Interim TPW overall both on land and sea within 0.15 mm in terms of bias while they have positive bias ranging from 0.62 to 0.90 in the validation results. The datasets randomly selected for the evaluation of the model performance can be different from those used for the validation of retrieval accuracy, and this discrepancy might be caused by overfitting since the optimal model is not chosen based on the final validation result. The model performance is used as an index to internally validate each model and to optimally tune the model that is sufficiently learned. The model accuracy needs to be evaluated with a dataset that is not used for the training or testing.

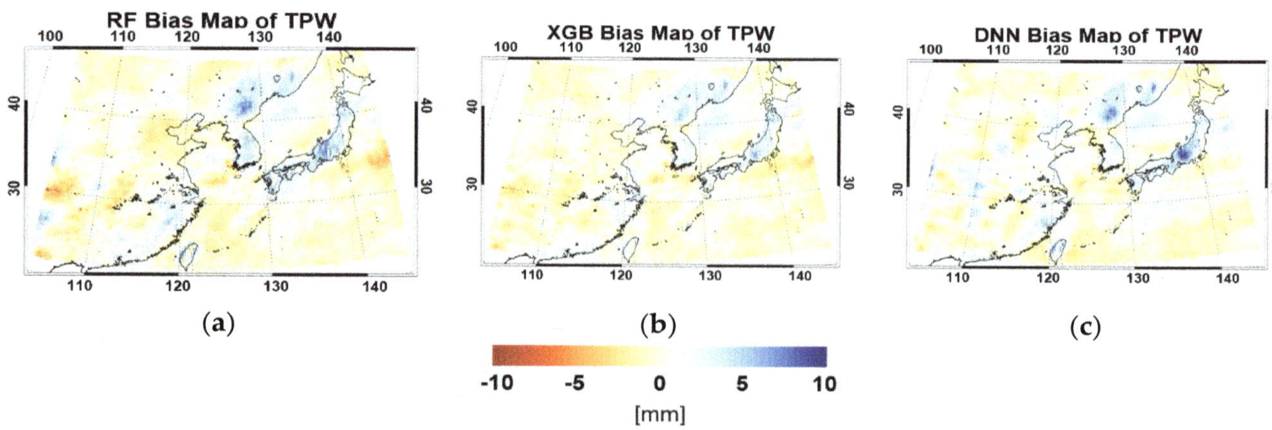

Figure 7. *Cont.*

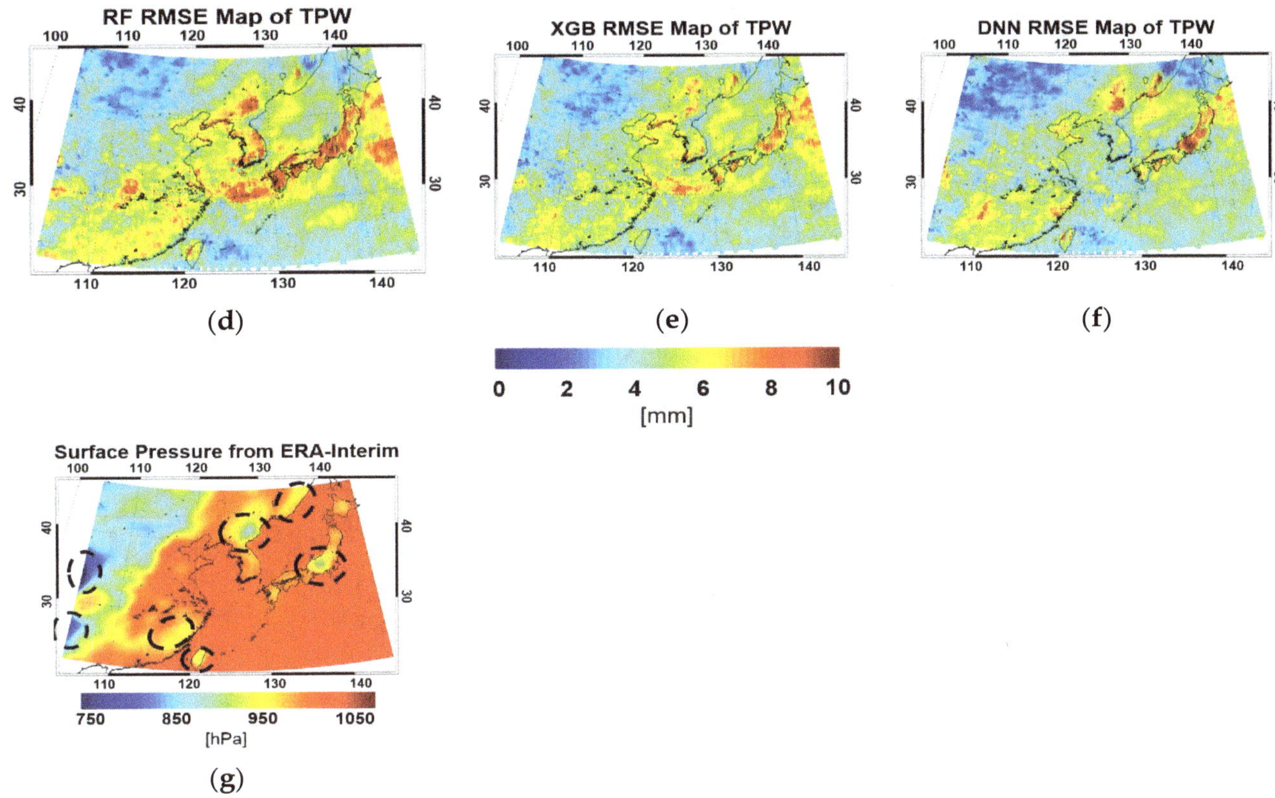

Figure 7. Mean error maps of TPW in Northeast Asia for two years (September 2016 to August 2018). The upper figures represent mean bias (ERA-Interim TPW – retrieved TPW) maps (**a–c**) and the lower figures represent mean RMSE maps (**d–f**). The left, middle, and right figures are the retrieved results from the RF, XGB, and DNN, respectively. (**g**) is the averaged surface pressures using ERA-Interim data from October 2016 to August 2018 in the study region. The black dotted circles indicate relatively lower surface pressures when compared to the surrounding areas.

To verify the performance of the models, a quantitative validation was carried out using RAOB from the University of Wyoming with an untrained dataset (Table 2). In addition to the ERA-interim data, RAOB data were also utilized for quantitative validation of model performance using an untrained dataset (Figure 8). Since only RAOB stations with high accuracy sensors were used, the number of in situ measurements used for the validation is relatively small (130 collocation data over the clear sky). The comparisons of TPW from the DNN, XGB, and RF models with RAOB and ERA-Interim show good performance in the order of RMSE during the validation period. The results from all models yield positive bias (2.39, 1.76, and 1.25 mm, respectively). This implies that all models tend to overestimate the TPW values concerning RAOB, which is similar to the validation results using ERA-Interim. The biases (retrieved TPW – reference TPW) using RAOB are smaller than the bias using ERA-Interim. The averaged difference value (about 0.56 mm) coincides with the difference between the two reference data (Figure 2). Additionally, the RMSE over the coastal region is larger compared to other regions (Table 5) since RAOB is collected mainly over land and coastal locations and also over islands as shown in Figure 1. The DNN is identified as the most optimal model for the retrieval of TPW through both validation results using ERA-Interim and RAOB. The RMSE value is comparable to or even better than the performance of TPW retrieval based on optical sensor data. For example, the neural network model developed from the MODIS near-infrared data over Western Europe and western Africa shows a validation RMSE of 6.4 mm when compared to RAOB [19].

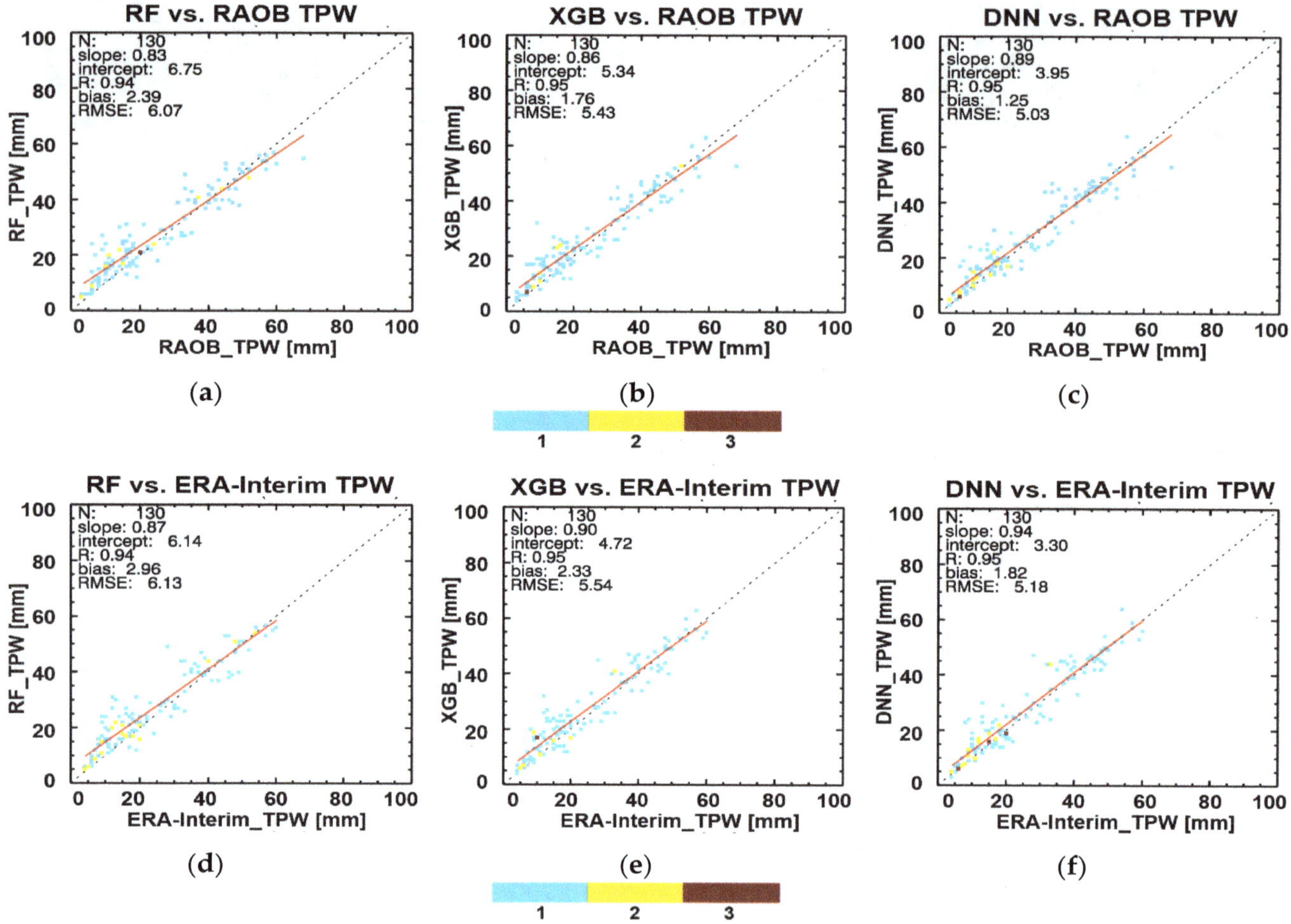

Figure 8. Accuracy assessment results of the retrieved TPW from the DNN, RF, and XGB using RAOB (**a**–**c**) and ERA-Interim (**d**–**f**) in Northeast Asia (1st, 25th per month from September 2016 to August 2018). The retrieved TPW from each model and ERA-Interim TPW is collocated with the 13 RAOB stations over clear sky regions. Scatter plots are colored by density: the x-axis is the reference data (RAOB and ERA-Interim) and the y-axis is the retrieved TPW from each model. The red lines are the regression lines.

Figure 9 displays the time series of the averaged mean bias and RMSE between the retrieved TPW from the DNN model and ERA-Interim TPW per scene between October 2016 and August 2018 in the Northeast Asia area. The variabilities of the mean bias and RMSE are relatively high during the humid summer season, followed by fall and spring. On the other hand, both errors are relatively low and consistently stable during the dry winter season. This is because humidity errors are larger under moist conditions [57]. As described earlier, the study utilizes the criteria based on the single channel or dual-channel differences to detect clouds and they may not effectively discriminate snow from clouds, causing a low retrieval rate in the Northern Hemisphere winter [27]. The decreased retrieval rate can lead to relatively poor performance of the retrieval algorithm in winter.

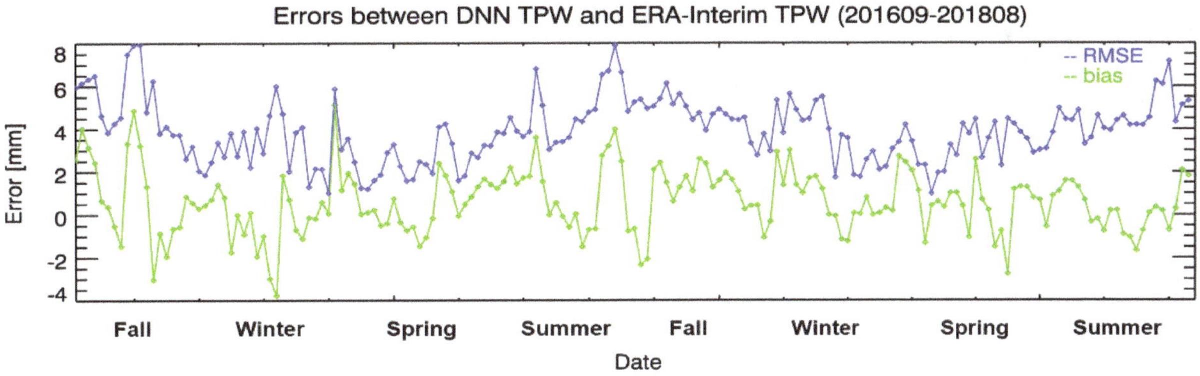

Figure 9. Time series (1st, 25th per month from September 2016 to August 2018) of bias and RMSE of the retrieved TPW from DNN for ERA-Interim TPW. The x-axis shows time and the y-axis shows the value of errors. The blue line is RMSE and the green line is bias.

4.4. Novelty and Limitations

The machine learning approaches (RF, XGB, and DNN) for the TPW retrieval from Himawari-8 AHI data in Northeast Asia were compared and analyzed in terms of their spatial and temporal characteristics of performances. The DNN model shows an overall good agreement with both types of reference data (ERA-Interim and RAOB data) and the RF model showed the lowest performance. The variable importance of each variable was calculated using the 'leave-one-variable-out' method. The retrieved TPW, provided every 10 min with 2 km resolution at nadir, together with atmospheric stability indices such as the Lifted index or CAPE, plays a good predictor of severe weather phenomena in the pre-convective atmospheric condition [4]. In addition to 10 min of AHI observations, the rapid scanning mode (about 2 min) data are also available, which can be used to monitor details of the temporal changes of the atmosphere. This near-real-time monitoring can give a promising result for the severe weather forecast for a more smooth transition of the atmosphere and information between cloudy scenes [8].

Aside from the novelties of this study, there are several limitations. One of the main limitations is the relatively low accuracy of the cloud mask used. This problem can cause uncertainty in the retrieval of TPW and validation results. As a result, there are little data during the wintertime especially over land, leading to relatively high RMSE of the retrieved TPW. Another limitation is the robustness of the model depending on the dataset. This is typically caused by overfitting [58] and makes the model difficult to directly apply to other cases. One possible solution to mitigate the difference between the accuracy of model tuning and validation is 'online learning' keeping the up-to-date dataset by constantly updating new data to the model [59]. This can help the model generalization since the difference between the training, testing, and validation dataset is decreased. Another solution is to combine different models. This is called 'ensemble method' to create a new model with a various model combination. This method has the advantage to complement the weaknesses of each other. It is important to choose the model considering the given problem.

5. Conclusions

In this study, TPW retrieval models based on the machine learning approaches (RF, XGB, and DNN) were developed for Himawari-8 AHI data over Northeast Asia. Nine AHI BTs (BT8 to BT16 centered at 6.2, 6.9, 7.3, 8.6, 9.6, 10.4, 11.2, 12.4, and 13.3 μm), six kinds of dual channel differences, time, location information (latitude and longitude), and satellite zenith angles were used as the input variables while the TPW calculated from the atmospheric temperature and humidity profiles from ERA-Interim were used as a target variable for the models. The parameters of each model were optimized through 10-fold cross-validation with the testing dataset (10% of the training dataset). The BT16, temperature sounding channel, is identified as the most contributing variable to the TPW retrievals in all models. The ERA-Interim and in-situ data (i.e., RAOB) are used for the model validation and characterization of each model. The DNN model yields the highest accuracy metrics (R, mean bias, and RMSE) regardless of the reference data used (i.e., R is 0.96, mean bias is 0.90 mm, and RMSE is 4.65 mm when using ERA-Interim TPW and R is 0.95 mean bias is 1.25 mm, and RMSE is 5.03 mm with respect to RAOB) and the RF model showed the lowest performance. The distribution of spatially averaged errors of each model reveals that the TPW is overestimated particularly in the regions with relatively lower surface pressures, mainly due to the relatively high elevations of the regions. This suggests the importance of considering the elevation as a predictor in a TPW retrieval study using machine learning techniques. The validation results also show a decreased accuracy compared to the accuracy obtained during the model training, implying a tendency for the overfitting. Nevertheless, the retrieved TPW with finer spatiotemporal resolution (2-min intervals with about 2 km spatial resolution) together with atmospheric instability is expected to provide quite useful information for analyzing and predicting possible severe weather phenomena such as convective storms and heavy rainfall in pre-convective environments. In the future, the proposed models are going to be used for the TPW retrievals from Advanced Meteorological Imager loaded in the Geostationary Korea Multi-Purpose Satellite (GeoKompsat)-2A, a South Korea's second geostationary meteorological satellite, which has similar specifications to Himawari-8 AHI.

Author Contributions: Y.L. led manuscript writing and contributed to data analysis and research design. D.H. contributed to data analysis and model design. M.-H.A. supervised this study, contributed to the research design and manuscript writing, and served as the corresponding author. J.I. contributed to research design and manuscript writing. S.J.L. contributed to the discussion of the results.

References

1. Trenberth, K.E.; Dai, A.; Rasmussen, R.M.; Parsons, D.B. The changing character of precipitation. *Bull. Am. Meteorol. Soc.* **2003**, *84*, 1205–1218. [CrossRef]
2. Viswanadham, Y. The relationship between total precipitable water and surface dew point. *J. Appl. Meteorol.* **1981**, *20*, 3–8. [CrossRef]
3. Manning, T.; Zhang, K.; Rohm, W.; Choy, S.; Hurter, F. Detecting severe weather using GPS tomography: An Australian case study. *J. Glob. Position. Syst.* **2012**, *11*, 58–70. [CrossRef]
4. Lee, S.J.; Ahn, M.H.; Lee, Y. Application of an artificial neural network for a direct estimation of atmospheric instability from a next-generation imager. *Adv. Atmos. Sci.* **2016**, *33*, 221–232. [CrossRef]
5. Bessho, K.; Date, K.; Hayashi, M.; Ikeda, A.; Imai, T.; Inoue, H.; Inoue, H.; Kumagai, Y.; Miyakwa, T.; Murata, H.; et al. An introduction to Himawari-8/9—Japan's new-generation geostationary meteorological satellites. *J. Meteorol. Soc. Jpn.* **2016**, *94*, 151–183. [CrossRef]
6. Martinez, M.A.; Velazquez, M.; Manso, M.; Mas, I. Application of LPW and SAI SAFNWC/MSG satellite products in pre-convective environments. *Atmos. Res.* **2007**, *83*, 366–379. [CrossRef]
7. Liu, Z.; Min, M.; Li, J.; Sun, F.; Di, D.; Ai, Y.; Li, Z.; Qin, D.; Li, G.; Lin, Y.; et al. Local Severe Storm Tracking and Warning in Pre-Convection Stage from the New Generation Geostationary Weather Satellite Measurements. *Remote Sens.* **2019**, *11*, 383. [CrossRef]
8. Lee, Y.K.; Li, J.; Li, Z.; Schmit, T. Atmospheric temporal variations in the pre-landfall environment of typhoon Nangka (2015) observed by the Himawari-8 AHI. *Asia-Pac. J. Atmos. Sci.* **2017**, *53*, 431–443. [CrossRef]
9. Lee, S.J.; Ahn, M.-H.; Chung, S.-R. Atmospheric Profile Retrieval Algorithm for Next Generation Geostationary Satellite of Korea and Its Application to the Advanced Himawari Imager. *Remote Sens.* **2017**, *9*, 1294. [CrossRef]
10. Wan, Z.; Dozier, J.A. Generalized split-window algorithm for retrieving land-surface temperature from space. *IEEE Trans. Geosci. Remote Sens.* **1996**, *34*, 892–905. [CrossRef]
11. Tang, B.; Bi, Y.; Li, Z.L.; Xia, J. Generalized split-window algorithm for estimate of land surface temperature from Chinese geostationary FengYun meteorological satellite (FY-2C) data. *Sensors* **2008**, *8*, 933–951. [CrossRef] [PubMed]
12. Chesters, D.; Robinson, W.D.; Uccellini, L.W. Optimized retrievals of precipitable water from the VAS "Split Window". *J. Appl. Meteorol. Climatol.* **1987**, *26*, 1059–1066. [CrossRef]
13. Dalu, G. Satellite remote sensing of atmospheric water vapour. *Int. J. Remote Sens.* **1986**, *7*, 1089–1097. [CrossRef]
14. Sobrino, J.A.; Jimenez, J.C.; Raissouni, N.; Soria, G. A simplified method for estimating the total water vapor

content over sea surfaces using NOAA-AVHRR channels 4 and 5. *IEEE Trans. Geosci. Remote Sens.* **2002**, *40*, 357–361. [CrossRef]

15. Schroedter-Homscheidt, M.; Drews, A.; Heise, S. Total water vapor column retrieval from MSG-SEVIRI split window measurements exploiting the daily cycle of land surface temperatures. *Remote Sens. Environ.* **2008**, *112*, 249–258. [CrossRef]
16. Barton, I.J.; Prata, A.J. Difficulties associated with the application of covariance–variance techniques to retrieval of atmospheric water vapor from satellite imagery. *Remote Sens. Environ.* **1999**, *69*, 76–83. [CrossRef]
17. Knabb, R.D.; Fuelberg, H.E. A comparison of the first-guess dependence of precipitable water estimates from three techniques using GOES data. *J. Appl. Meteorol.* **1997**, *36*, 417–427. [CrossRef]
18. Nielsen, M.A. *Neural Networks and Deep Learning*; Determination Press: San Francisco, CA, USA, 2015; Available online: http://neuralnetworksanddeeplearning.com/ (accessed on 29 December 2017).
19. Wang, W.; Sun, X.; Zhang, R.; Li, Z.; Zhu, Z.; Su, H. Multi-layer perceptron neural network based algorithm for estimating precipitable water vapour from MODIS NIR data. *Int. J. Remote Sens.* **2006**, *27*, 617–621. [CrossRef]
20. Zhang, S.L.; Xu, L.S.; Ding, J.L.; Liu, H.L.; Deng, X.B. Precipitable Water Vapor Retrieval Using Neural Network from Infrared Hyperspectral Soundings. *Key Eng. Mater.* **2012**, *500*, 390–396. [CrossRef]
21. Lee, Y.-K.; Li, Z.; Li, J. Evaluation of the GOES-R ABI LAP Retrieval Algorithm Using the GOES-13 Sounder. *J. Atmos. Ocean. Technol.* **2014**, *31*, 3–19. [CrossRef]
22. Dee, D.P.; Uppala, S.M.; Simmons, A.J.; Berrisford, P.; Poli, P.; Kobayashi, S.; Andrae, U.; Balmaseda, M.A.; Balsamo, G.; Bauer, P.; et al. The ERA-Interim reanalysis: Configuration and performance of the data assimilation system. *Q. J. R. Meteorol. Soc.* **2011**, *137*, 553–597. [CrossRef]
23. Basili, P.; Bonafoni, S.; Mattioli, V.; Pelliccia, F.; Ciotti, P.; Carlesimo, G.; Pierdicca, N.; Venuti, G.; Mazzoni, A. Neural-network retrieval of integrated precipitable water vapor over land from satellite microwave radiometer. In Proceedings of the 2010 11th Specialist Meeting on Microwave Radiometry and Remote Sensing of the Environment IEEE, Washington, DC, USA, 1–4 March 2010; pp. 161–166. [CrossRef]
24. Bonafoni, S.; Mattioli, V.; Basili, P.; Ciotti, P.; Pierdicca, N. Satellite-based retrieval of precipitable water vapor over land by using a neural network approach. *IEEE Trans. Geosci. Remote Sens.* **2011**, *49*, 3236–3248. [CrossRef]
25. Ingleby, B. On the Accuracy of Different Radiosonde Types–ECMWF. TECO-2016 Madrid, 30 September 2016. Available online: https://www.wmo.int/pages/prog/www/IMOP/publications/IOM-125_TECO_2016/Session_4/O4(8)_pres_Ingleby_TECO_types_4_8.pdf (accessed on 7 May 2019).
26. Ebell, K.; Orlandi, E.; Hünerbein, A.; Löhnert, U.; Crewell, S. Combining ground-based with satellite-based measurements in the atmospheric state retrieval: Assessment of the information content. *J. Geophys. Res. Atmos.* **2013**, *118*, 6940–6956. [CrossRef]
27. Lee, S.J.; Ahn, M.H.; Ha, S. Total Column Ozone Retrieval From the Infrared Measurements of a Geostationary Imager. *IEEE Trans. Geosci. Remote Sens.* **2019**, 1–9. [CrossRef]
28. Zhou, L.; Pan, S.; Wang, J.; Vasilakos, A.V. Machine learning on big data: Opportunities and challenges. *Neurocomputing* **2017**, *237*, 350–361. [CrossRef]
29. Karahoca, A. *Advances in Data Mining Knowledge Discovery and Applications*; InTech: Rijeka, Croatia, 2012; ISBN 978-953-51-0748-4. [CrossRef]
30. Yu, L.; Wang, S.; Lai, K.K. An integrated data preparation scheme for neural network data analysis. *IEEE Trans. Knowl. Data Eng.* **2006**, *18*, 217–230. [CrossRef]
31. Amani, M.; Salehi, B.; Mahdavi, S.; Granger, J.; Brisco, B. Wetland classification in Newfoundland and Labrador using multi-source SAR and optical data integration. *GISci. Remote Sens.* **2017**, *54*, 779–796. [CrossRef]
32. Liu, T.; Abd-Elrahman, A.; Morton, J.; Wilhelm, V.L. Comparing fully convolutional networks, random forest, support vector machine, and patch-based deep convolutional neural networks for object-based wetland mapping using images from small unmanned aircraft system. *GISci. Remote Sens.* **2018**, *55*, 243–264. [CrossRef]
33. Guo, Z.; Du, S. Mining parameter information for building extraction and change detection with very high-resolution imagery and GIS data. *GISci. Remote Sens.* **2017**, *54*, 38–63. [CrossRef]
34. Richardson, H.J.; Hill, D.J.; Denesiuk, D.R.; Fraser, L.H. A comparison of geographic datasets and field measurements to model soil carbon using random forests and stepwise regressions (British Columbia, Canada). *GISci. Remote Sens.* **2017**, *54*, 573–591. [CrossRef]

35. Forkuor, G.; Dimobe, K.; Serme, I.; Tondoh, J.E. Landsat-8 vs. Sentinel-2: Examining the added value of sentinel-2's red-edge bands to land-use and land-cover mapping in Burkina Faso. *GISci. Remote Sens.* **2018**, *55*, 331–354. [CrossRef]
36. Georganos, S.; Grippa, T.; Vanhuysse, S.; Lennert, M.; Shimoni, M.; Kalogirou, S.; Wolff, E. Less is more: Optimizing classification performance through feature selection in a very-high-resolution remote sensing object-based urban application. *GISci. Remote Sens.* **2018**, *55*, 221–242. [CrossRef]
37. Zhang, C.; Smith, M.; Fang, C. Evaluation of Goddard's LiDAR, hyperspectral, and thermal data products for mapping urban land-cover types. *GISci. Remote Sens.* **2018**, *55*, 90–109. [CrossRef]
38. Sonobe, R.; Yamaya, Y.; Tani, H.; Wang, X.; Kobayashi, N.; Mochizuki, K.I. Assessing the suitability of data from Sentinel-1A and 2A for crop classification. *GISci. Remote Sens.* **2017**, *54*, 918–938. [CrossRef]
39. Santos, L.D. GPU Accelerated Classifier Benchmarking for Wildfire Related Tasks. Ph.D. Thesis, NOVA University of Lisbon, Lisbon, Portugal, 2018. Available online: https://run.unl.pt/handle/10362/61547 (accessed on 11 May 2019).
40. Nisa, I.; Siegel, C.; Rajam, A.S.; Vishnu, A.; Sadayappan, P. Effective Machine Learning Based Format Selection and Performance Modeling for SpMV on GPUs. In Proceedings of the 2018 IEEE International Parallel and Distributed Processing Symposium Workshops (IPDPSW), Vancouver, BC, Canada, 21–25 May 2018; pp. 1056–1065.
41. Babajide Mustapha, I.; Saeed, F. Bioactive molecule prediction using extreme gradient boosting. *Molecules* **2016**, *21*, 983. [CrossRef] [PubMed]
42. Yang, J.; Guo, A.; Li, Y.; Zhang, Y.; Li, X. Simulation of landscape spatial layout evolution in rural-urban fringe areas: A case study of Ganjingzi District. *GISci. Remote Sens.* **2019**, *56*, 388–405. [CrossRef]
43. Omrani, H.; Tayyebi, A.; Pijanowski, B. Integrating the multi-label land-use concept and cellular automata with the artificial neural network-based Land Transformation Model: An integrated ML-CA-LTM modeling framework. *GISci. Remote Sens.* **2017**, *54*, 283–304. [CrossRef]
44. Antonanzas, J.; Urraca, R.; Aldama, A.; Fernández-Jiménez, L.A.; Martínez-de-Pisón, F.J. *Single and Blended Models for Day-Ahead Photovoltaic Power Forecasting*; Springer International Publishing: La Rioja, Spain, 2017; pp. 427–434, ISBN 978-3-319-59649-5.
45. Pan, B. Application of XGBoost algorithm in hourly PM2.5 concentration prediction. In Proceedings of the IOP Conference Series: Earth and Environmental Science, Harbin, China, 8–10 February 2018.
46. Just, A.; De Carli, M.; Shtein, A.; Dorman, M.; Lyapustin, A.; Kloog, I. Correcting Measurement Error in Satellite Aerosol Optical Depth with Machine Learning for Modeling PM2.5 in the Northeastern USA. *Remote Sens.* **2018**, *10*, 803. [CrossRef]
47. EUMETSAT. Product Tutorial on Total Precipitable Water Content Products. 2014. Available online: http://www.eumetrain.org/data/3/359/print_3.htm#page_1.0.0 (accessed on 5 May 2019).
48. Hocking, J.; Rayer, P.; Saunders, R.; Madricardi, M.; Geer, A.; Brunel, P.; Vidot, J. RTTOV v11 Users Guide. NWP SAF, Version 1.4. September 2015. Available online: https://www.nwpsaf.eu/site/download/documentation/rtm/docs_rttov11/users_guide_11_v1.4.pdf (accessed on 5 July 2019).
49. Breiman, L. Random forests. *Mach. Learn.* **2001**, *45*, 5–32. [CrossRef]
50. Wright, M.N.; Ziegler, A. Ranger: A fast implementation of random forests for high dimensional data in C++ and R. *J. Stat. Softw.* **2017**, 77. [CrossRef]
51. Chen, T.; Guestrin, C. XGBoost: A Scalable Tree Boosting System. In Proceedings of the 22nd ACM Sigkdd International Conference on Knowledge Discovery and Data Mining, San Francisco, CA, USA, 13–17 August 2016; pp. 785–794.
52. Dewancker, I.; McCourt, M.; Clark, S. Bayesian Optimization Primer. 2015. Available online: https://app.sigopt.com/static/pdf/SigOpt_Bayesian_Optimization_Primer.pdf (accessed on 11 May 2019).
53. Blackwell, W.J.; Chen, F.W. *Neural Networks in Atmospheric Remote Sensing*; Artech House: Norwood, MA, USA, 2009; p. 234, ISBN 978-1-59693-372-9.
54. Bengio, Y.; Grandvalet, Y. No unbiased estimator of the variance of k-fold cross-validation. *J. Mach. Learn. Res.* **2004**, *5*, 1089–1105.
55. Khalid, S.; Khalil, T.; Nasreen, S. A survey of feature selection and feature extraction techniques in machine learning. In Proceedings of the Science and Information Conference (SAI), London, UK, 27–29 August 2014.
56. Mears, C.A.; Santer, B.D.; Wentz, F.J.; Taylor, K.E.; Wehner, M.F. Relationship between temperature and precipitable water changes over tropical oceans. *Geophys. Res. Lett.* **2007**, *34*. [CrossRef]

57. Noh, Y.-C.; Sohn, B.-J.; Kim, Y.; Joo, S.; Bell, W. Evaluation of Temperature and Humidity Profiles of Unified Model and ECMWF Analyses Using GRUAN Radiosonde Observations. *Atmosphere* **2016**, *7*, 94. [CrossRef]
58. Bastani, O.; Ioannou, Y.; Lampropoulos, L.; Vytiniotis, D.; Nori, A.; Criminisi, A. *Measuring Neural Net Robustness with Constraints*; Curran Associates Inc.: Red Hook, NY, USA, 2016; pp. 2613–2621, ISBN 978-1-5108-3881-9.
59. Shalev-Shwartz, S. Online learning and online convex optimization. *Found. Trends® Mach. Learn.* **2012**, *4*, 107–194. [CrossRef]

11

Flash Flood Risk Analysis Based on Machine Learning Techniques in the Yunnan Province, China

Meihong Ma [1,2], Changjun Liu [1,*], Gang Zhao [3,*], Hongjie Xie [4], Pengfei Jia [5], Dacheng Wang [6], Huixiao Wang [2] and Yang Hong [7]

1 China Institute of Water Resources and Hydropower Research, Beijing 100038, China; mmhkl2007@163.com
2 College of water sciences, Beijing Normal University, Beijing 100875, China; Mmhjpf2016@163.com
3 School of Geographical Sciences, University of Bristol, Bristol BS8 1SS, UK
4 Department of Geological Sciences University of Texas at San Antonio, San Antonio, TX 78249, USA; Hongjie.Xie@utsa.edu
5 CITIC Construction Co., Ltd., Beijing 100027, China; jiapf2011@163.com
6 Lab of Spatial Information Integration, Institute of Remote Sensing and Digital Earth, Chinese Academy of Sciences, Beijing 100101, China; wangdc@radi.ac.cn
7 School of Earth and Space Sciences, Peking University, Beijing 100871, China; yanghong588@pku.edu.cn
* Correspondence: lcj2005@iwhr.com (C.L.); gang.zhao@bristol.ac.uk (G.Z.).

Abstract: Flash flood, one of the most devastating weather-related hazards in the world, has become more and more frequent in past decades. For the purpose of flood mitigation, it is necessary to understand the distribution of flash flood risk. In this study, artificial intelligence (Least squares support vector machine: LSSVM) and classical canonical method (Logistic regression: LR) are used to assess the flash flood risk in the Yunnan Province based on historical flash flood records and 13 meteorological, topographical, hydrological and anthropological factors. Results indicate that: (1) the LSSVM with Radial basis function (RBF) Kernel works the best (Accuracy = 0.79) and the LR is the worst (Accuracy = 0.75) in testing; (2) flash flood risk distribution identified by the LSSVM in Yunnan province is near normal distribution; (3) the high-risk areas are mainly concentrated in the central and southeastern regions, where with a large curve number; and (4) the impact factors contributing the flash flood risk map from higher to low are: Curve number > Digital elevation > Slope > River density > Flash Flood preventions > Topographic Wetness Index > annual maximum 24 h precipitation > annual maximum 3 h precipitation.

Keywords: flash flood; risk; LSSVM; China

---

1. Introduction

Flash flood is one of the most devastating natural disasters with characteristics of high-velocity runoff, short lead-time and fast-rising water [1]. Economic losses caused by flash flood increase year by year with the increase of population and infrastructure in flood-prone areas [2]. For instance, a total of 28,826 flash flood events happened in the United States between 2007 and 2015 and 10% of flash flood resulted in damages exceeding $100,000 [3]. According to the China Floods and Droughts Disasters Bulletin of 2015, an average of 935 people dies each year by flash flood disasters from 2000 to 2015. Owing to the impact of climate change, the flash flood risk is predicted to increase with the frequent extreme precipitation and sea level rise [4]. Therefore, an accurate risk assessment is critical for flash flood prevention.

Flash floods risk is a combination of flood hazard and vulnerability of an area [5,6]. Flood risk is widely assessed by hydrological models or data-driven model based on historical flood inventories. The hydrological model has a clear physical mechanism that reflects the process of flood generation and transportation. One of the most widely used models is 1–2-dimension routing model such as MIKE 21, which can truly reflect the flooding scope and water depth during flooding. The flood risk is assessed by combining water depth and local vulnerability [7,8]. However, since the simulation of the actual hydrological process is affected by many factors (e.g., model's parameter, structure, input data), the model accuracy and uncertainty need to be further explored [9]. Meanwhile, different regions require different types of hydrological models, resulting in high data requirements and time-consuming on model development [10,11].

In terms of this, data-driven models were proposed for flood risk assessment. Data-driven models adopt black-box models and uses various intelligent algorithms to establish optimal mathematical relationships between disaster and explanatory factors, such as analytic hierarchy process (AHP), set pair analysis method (SPAM) and so forth. AHP is a simple and effective multi-criteria decision-making method, which effectively solves the lack of quantitative data in flood risk assessment and the complex relationship involving multiple risk factors [12]. SPAM is a method for systematic analysis of uncertain problems, effectively dealing with the incompleteness of information for flood risk prediction [13]. However, AHP and SPAM are all based on expert opinions in choosing the indicator weighting that introducing uncertainty and subjectivity in assessment [14]. With the development of artificial intelligence, machine learning (ML) models, including support vector machine (SVM), Random Forest (RF) and Decision Tree (DT), has been proposed and applied in flood risk assessment. Machine learning models avoid the subjective determination of weights by learning the relationship between flood risk and explanatory factors. Among them, SVM is a popular ML model that can solve linear and nonlinear regression problems and has gained extensive applications in pattern recognition, data mining and speech recognition [15]. Least Squares Support Vector Machine (LSSVM) is a simple SVM that uses least squares and linear equations to improve model efficiency [16]. Flash flood data is often complex and incomplete and the relationships between variables can be strongly nonlinear and involve high-order interactions. Therefore, it is of great value to explore the flash flood risk assessment by LSSVM method.

Nowadays, with the in-depth application of 3S technologies (Remote Sensing, Geography Information Systems and Global Positioning Systems) in hydrology, the acquisition of spatial information on the underlying surface of the basin have been significantly improved [17]. Meanwhile, a series of intelligent algorithms based on big data have been proposed that are valuable to use in hydrology. In this study, we developed a flash flood assessment framework based on machine learning models. We utilize the LSSVM method with three kernel functions (linear: LN; radial basis function: RBF; polynomial: PL) and classical logistic regression (LR) method to assess flash flood risk based on the official statistics of flash flood events. The performances of our proposed method are evaluated with five indices and ROC curve in Section 3.1. The distribution of flash flood risk in the study area and the relationship between flood risk and flood trigger factors are discussed in Section 3.2.

2. Materials and Methods

2.1. Study Area

Yunnan Province (20°8′–29°16′N, 97°31′–106°12′E) is located in southwestern China with an area of 383,210 km^2. It is one of the most flooded provinces in China and the economy relies mainly on natural resources. In 2016, Yunnan province had a population of 47.7 million, a gross domestic product (GDP) of 1.49 billion yuan. Yunnan province is located in the low latitude plateau and the terrain is dominated by mountains, with a canyon in the west, a plateau in the east and a major river running through the deep valley. From the southeastern mountainous area to the northwest Hengduan Mountains, the altitude ranges from less than 100 m to more than 6000 m, with an average elevation of 1980 m. The mountainous area, plateau area and watershed area account for 80%, 12.5% and 7.5% of the total area respectively. About 39% of slopes exceed 25° in mountainous areas and the slopes of the northeast and northwest mountainous areas even reach 60–90%. The soil texture is loose, of which more than 50% is krasnozem. The climate is mainly affected by atmospheric circulation, which is a low mountain monsoon climate. The annual average precipitation is 1102 mm, with significant spatial-temporal differences [18]. Meanwhile, extreme weather events occur frequently, especially during the summer flood season (June to September), with rainfall accounting for 85–95% between May and October.

China has implemented the construction of non-structural measures for flash flood prevention since 2011. In Yunnan Province, there are 206 flash floods events from 2011 to 2015, causing 237 deaths. Especially in 2014 and 2015, the number of deaths accounted for 22.2% and 8.1% of the national total, respectively, which were the most affected by the flash floods. In order to defend against flash flood, Yunnan has launched the construction of non-structural flood prevention measures covering 129 counties since 2010. The average construction fund is $0.87 million for each county. The preventive measures implemented include: encrypting automatic rainfall stations to improve the quality of monitoring data, installing simple rainfall equipment with alarms, building an alarm system consisting of radio broadcasts and simple alarm devices. Obviously, although Yunnan Province already has a certain defense base, it still suffers from severe flash flood disasters. Therefore, it is of great significance to study the flash flood risk in Yunnan Province. Figure 1 shows the historical flash floods in Yunnan Province from 2011 to 2015. Obviously, flash floods mainly occur on lower slopes, mainly because the air rises on the windward slope and the water vapor condenses easily to form precipitation, which causes runoff to accumulate in the valley and triggers flash floods. The leeward slope is not easy to form precipitation due to the air sinking and the temperature moving downward [19].

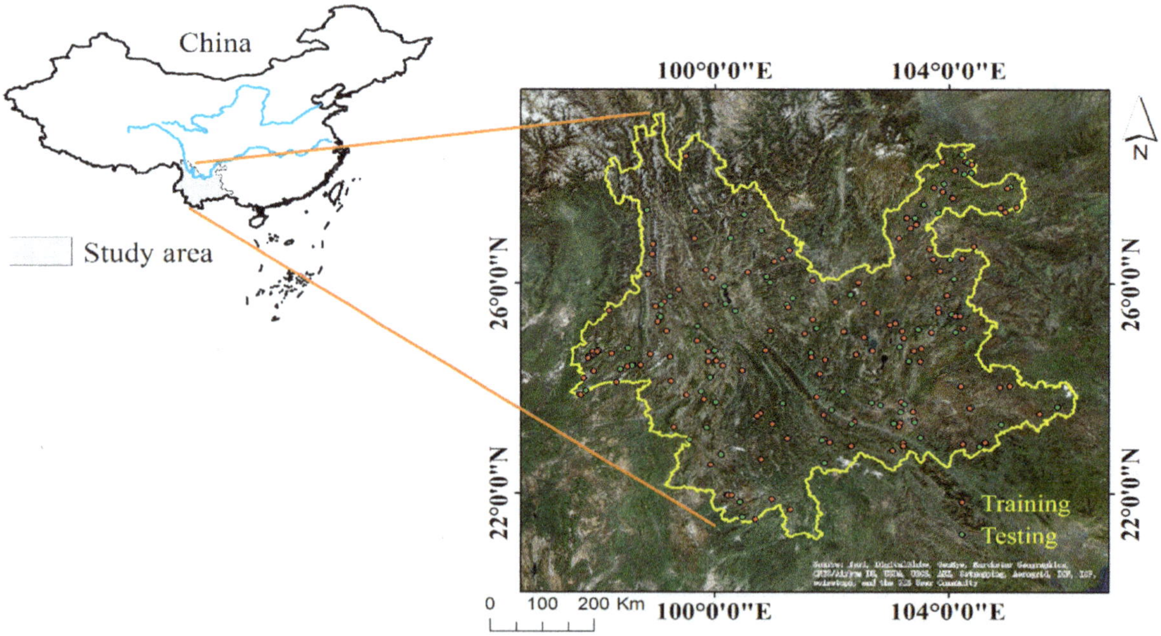

Figure 1. Location of the study area and the distribution of flash flood inventories (red for training and green for testing) from 2011 to 2015 in Yunnan Province, China.

2.2. Data

The flash flood records are mainly from official authoritative departments, such as the Ministry of Water Resources (MWR), the Ministry of Land and Resources and some local government agencies in Yunnan province. These data are divided into training and testing datasets, 70% of which are randomly selected for training and the remaining 30% data for testing. The principle of the distribution ratio is that the samples are evenly distributed and have certain representativeness (Figure 1). It is important to emphasize that all the flash floods studied in this paper involve death or missing; regardless of incidents that do not cause casualties. The remote sensing data and other data covered in this paper are shown in Table 1.

Table 1. Factors, flood inventories and data sources.

Name		Source	Time
Abbreviation	**Meaning**		
3-H-P	Annual maximum 3 h precipitation	China Meteorological Forcing Dataset	2011–2015
24-H-P	Annual maximum 24 h precipitation	China Meteorological Forcing Dataset	2011–2015
AP	Annual precipitation	China Meteorological Forcing Dataset	2011–2015
DEM	Digital elevation model	Shuttle Radar Topography Mission (SRTM)	2000
SL	Slope	Shuttle Radar Topography Mission (SRTM)	2000
RD	River density	Basic vector format dataset of China	-
VC	Vegetation coverage	MODIS products	2011–2015
CN	Curve number	NRCS CN global dataset	2011–2015
TWI	Topographic wetness index	Shuttle Radar Topography Mission (SRTM)	2000
SM	Soil moisture	ESA's SMOS dataset	2011–2015
Pop	Population	Data Center for Resources and Environmental Sciences Chinese Academy of Sciences (RESDC)	2010
GDP	Gross domestic product	Data Center for Resources and Environmental Sciences Chinese Academy of Sciences (RESDC)	2010
FFP	Flash flood preventions	Statistic bulletin from the Ministry of Water Resources and local governments	2012–2015

2.3. Flash Flood Triggering Factors

Flash flood disasters are mainly affected by meteorological, topographical hydrological, anthropological factors. The related factors affecting flash flood risk are shown in Figure 2 and are described as followed:

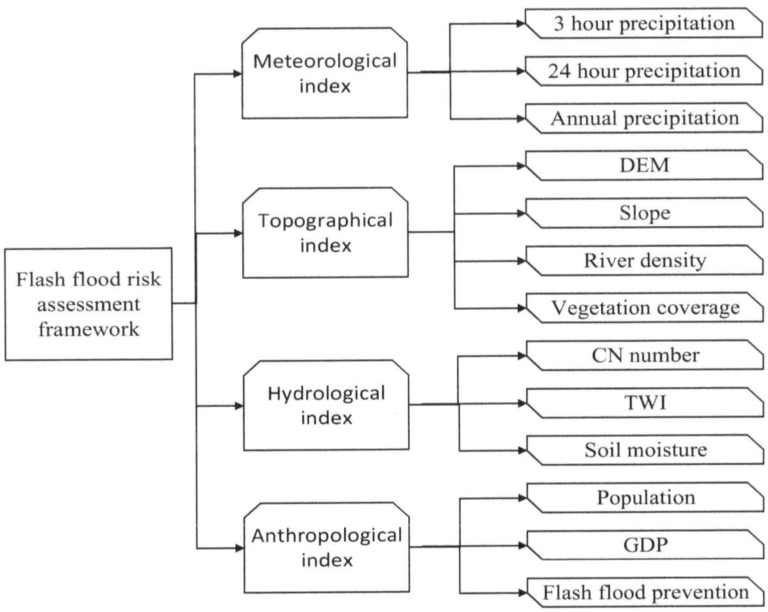

Figure 2. Explanatory factors affecting flash flood risk in this study.

(1) Meteorological factors

Three meteorological factors including 3-H-P, 24-H-P and AP are the main factors leading to flash floods, with 3-H-P and 24-H-P reflecting the frequency and characteristics of short-term rainfall and AP reflecting the characteristics of long-term rainfall. The precipitation data comes from the China Meteorological Forcing Dataset (CMFD), produced by the Institute of Tibetan Plateau Research, Chinese Academy of Sciences (hereafter ITPCAS). The dataset is based primarily on the existing Princeton reanalysis data, Global Land Data Assimilation System (GLDAS) data, Global Energy and Water cycle Experiment—Surface Radiation Budget (GEWEX-SRB) radiation data and Tropical Rainfall

Measuring Mission (TRMM) precipitation data in the world, combined conventional CMA weather observations were produced with temporal and spatial resolutions of 3 h and 0.1° × 0.1°, respectively.

(2) Topographical factors

Digital elevation model (DEM) retrieved from NASA SRTM, a 90-m raster in 2000. DEM resolution mainly affects the watershed topography, which in turn affects the accuracy of runoff generation and convergence. The higher the DEM resolution, the higher the accuracy of the extracted watershed features. However, high-resolution DEM over-emphasizes the computational burden of the model, greatly restricting the runtime of the model [20]. Slope (SL) refers to the ratio of the vertical height of the slope to the horizontal direction, which is suitable for the sensitivity analysis of floods. Generally, the SL is calculated from the DEM data using the ArcGIS tool [17]. River density (RD) utilizes China's basic vector format dataset, which is related to the area of the grid and the length of the river in the grid [21]. Vegetation coverage (VC) is calculated by an average multi-year normalized difference vegetation index (NDVI) based on MODIS images. It represents vegetation distribution and biomass levels from 2011 to 2015 [22].

(3) Hydrological factors

The Curve Number (CN) derived from the soil conservation service curve number (SCS-CN) model is a comprehensive indicator calculated according to the National Engineering Handbook of US, which primarily reflects the potential capacity of runoff generation in different grids. It is a non-dimensional index with a theoretical value between 0 (no runoff) and 100 (no infiltration). For details of CN, please refer to Zeng et al. (2017) [23]. The topographic wetness index (TWI), combined with the local uphill contribution area and the entire slope, is widely used to quantify the topographical control of flood concentration processes and can be calculated from DEM [24]. Soil moisture (SM) data is from the European Space Agency (ESA) with a spatial accuracy of 50 km. It can estimate moisture in the soil surface (down to 5 cm) which is important for hydrological modeling. SM indicates the non-linear partitioning of the precipitation into infiltration and runoff, affecting runoff by affecting infiltration [25].

(4) Anthropological factors

The effects of flood risks are often related to anthropology, manifested as loss of economic property and casualties. The losses generally increase with the population growth in flood-prone areas, especially in economically developed and densely populated areas. Therefore, Gross Domestic Product (GDP) and population (Pop) are selected as anthropological factors for flash flood assessment. DDP is defined as "an aggregate measure of production equal to the sum of the gross values added of all resident and institutional units engaged in production (plus any taxes and minus any subsidies, on products not included in the value of their outputs), mainly reflecting the economic situation of the study area. Moreover, GDP is a total indicator, which basically organizes indicators describing various aspects of the national economy through a series of scientific principles and methods. Therefore, GDP contained contributing indicators such as over-exploitation [26]. The 1-km gridded GDP and population of Yunnan Province are collected from the Data Center for Resources and Environmental Sciences Chinese Academy of Sciences (RESDC). In 2010, the Chinese government initiated the construction of national-level non-structural measures for flash flood prevention. This investment is the largest non-structural project in China, involving a total area of 3.86 million km^2 in 29 provinces (autonomous regions and municipalities). The preventive measures include the national flash flood investigation and evaluation, the establishment of construction monitoring and early warning platforms, automatic rainfall stations and water level stations, mass observations and mass prevention and so forth. The FFP data is mainly from the MWR and local governments and utilizing the investment funds to comprehensively reflect the flash flood prevention situation [27,28]. The related factors affecting flash flood risk in the LSSVM method are shown in Figure 3.

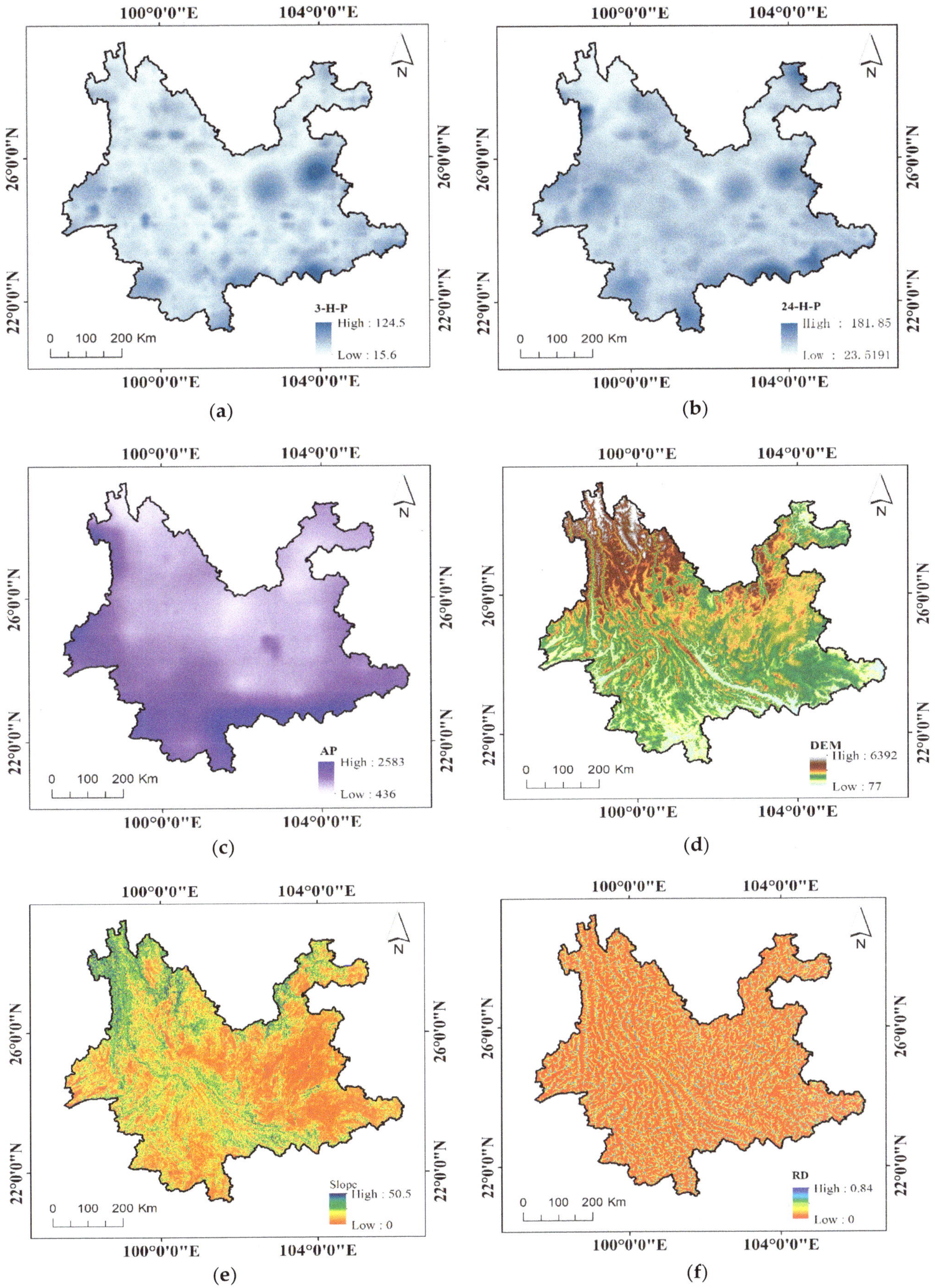

Figure 3. *Cont.*

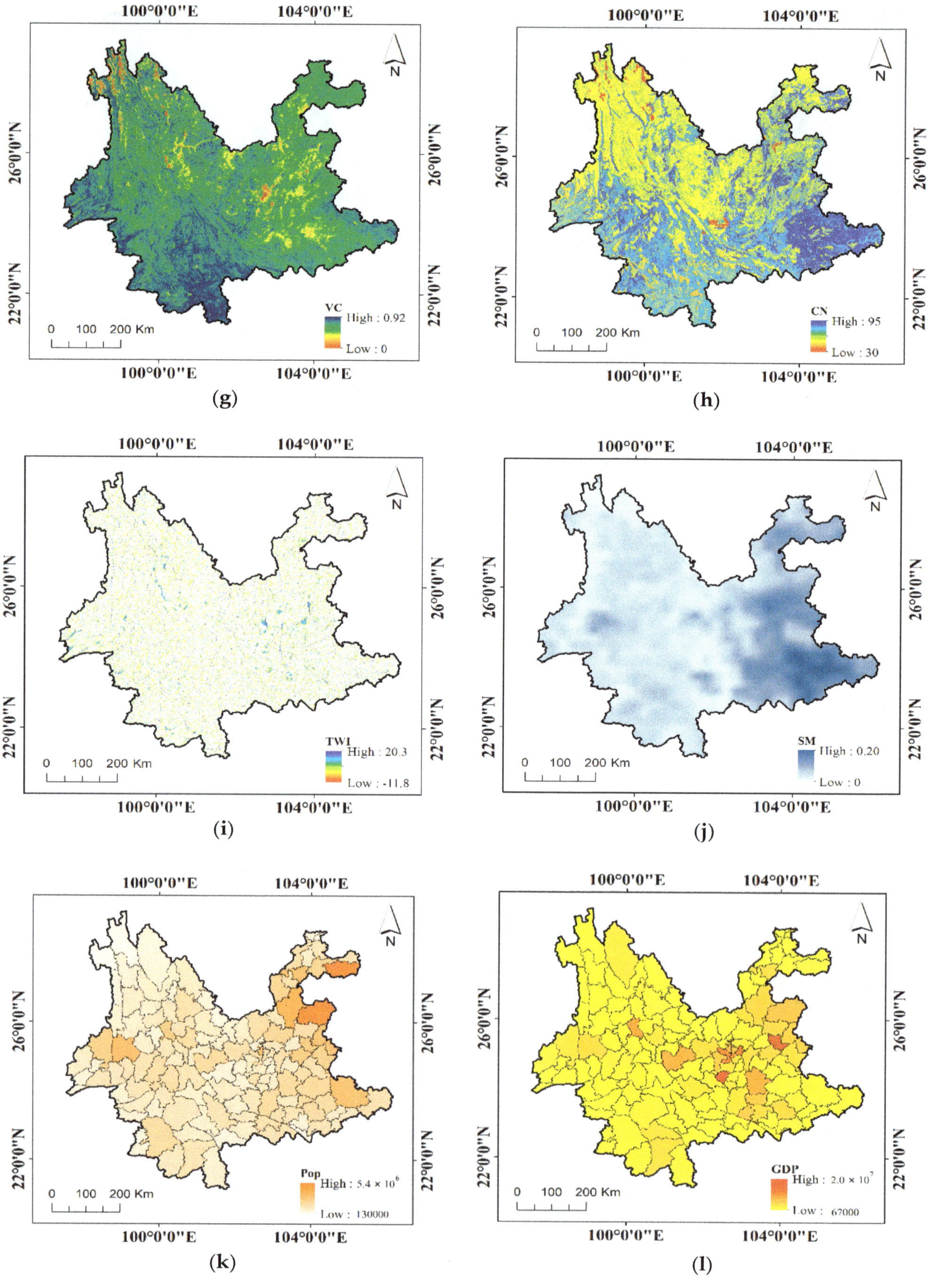

Figure 3. *Cont.*

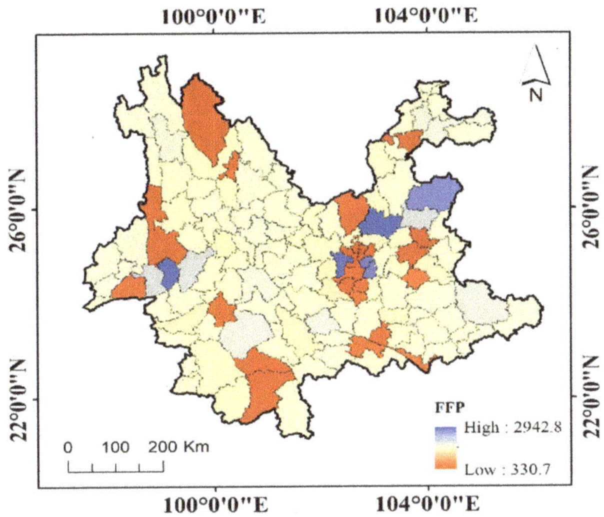

(m)

Figure 3. Explanatory factors of flash flood risk. (**a**) Annual Maximum 3 h Precipitation (**b**) Annual Maximum 24 h Precipitation; (**c**) Annual Precipitation (**d**) Digital Elevation Model; (**e**) Slope (**f**) River Density; (**g**) Vegetation Coverage (**h**) Curve Number; (**i**) Topographic Wetness Index (**j**) Soil Moisture; (**k**) Population (**l**) Gross Domestic Product; (**m**) Flash Flood Preventions.

2.4. Methodology

(1) LSSVM

LSSVM utilizes a set of linear equations to minimize the complexity of the optimization process. The constraint optimization problems can be solved using Lagrange multipliers. Consider a given training set x_i, y_i, $i = 1,2,\ldots,f$ with input data x_i and output data y_i, the LSSVM equation can be indicated as follows:

$$\min W(m,n) = \frac{1}{2}M^H M + \frac{1}{2}\beta\sum_{i=1}^{f} n_i{}^2 \tag{1}$$

Subject to

$$y_i = m^T\Phi(x_i) + b + n_i, i = 1,2,\ldots,f \tag{2}$$

where m is the weight vector, β is the penalty parameter, n_i is the approximation error, f is the number of autoregressive terms in the LR model, $\Phi(x_i)$ is the nonlinear mapping function and b is the bias term. The corresponding Lagrange function can be obtained by Equation (3):

$$W(m,n,\ \alpha,b) \ = \ J(m,n) - \sum_{i=1}^{f}\alpha_i m^T\phi(x_i) + b + n_i - y_i \tag{3}$$

where α_i is the Lagrange multiplier. Using the Karush-Kuhn-Tucker (KKT) conditions, the solutions can be obtained by partially differentiating with respect to m, b, n_i and α_i:

$$\begin{cases} \frac{\partial W}{\partial m} = 0 \to m = \sum\limits_{i=1}^{f} \alpha_i\Phi(x_i) \\ \frac{\partial W}{\partial b} = 0 \to \sum\limits_{i=1}^{f} \alpha_i = 0 \\ \frac{\partial W}{\partial n_i} = 0 \to \alpha_i = \beta n_i \\ \frac{\partial W}{\partial \alpha_i} = 0 \to w^T\phi(x_i) + b + n_i - y_i = 0 \end{cases} \tag{1}$$

By elimination w and n_i, the equations can be changed into

$$\begin{bmatrix} b \\ \alpha \end{bmatrix} = \begin{bmatrix} 0 & I_v{}^T \\ I_v & \psi + \beta^{-1} I \end{bmatrix}^{-1} \begin{bmatrix} 0 \\ y \end{bmatrix} \quad (2)$$

where $y = \left[y_1, y_2, \ldots, y_f\right]^T$, $I_v = [1, 1, \ldots 1]^T$, $\alpha = [\alpha_1, \alpha_2, \ldots, \alpha_f]$ and the Mercer condition has been applied to the matrix $\Omega_{km} = \phi(x_k)^T \Phi(x_o), k, m = 1, 2, \ldots, f$. Therefore, the LSSVM for regression can be obtained from Equation (6):

$$y(x) = \sum_{i=1}^{f} \alpha_i K(x_i, x) + b \quad (3)$$

where $K\ (x, x_i)$ is the kernel function. For LSSVM, there are many kernel functions including linear (Equation (7)), polynomial (ploy) (Equation (8)), radial basis function (RBF) (Equation (9)), sigmoid and so forth. However, most widely used kernel functions are RBF and polynomial Kernel.

$$\text{Linear (LN) Kernel: } K\ (x_i, x) = \langle x_i, x \rangle \quad (4)$$

$$\text{Polynomial (PL) Kernel: } K\ (x_i, x) = (\gamma\ \langle x_i, x \rangle\ +\ \tau)^d\ \gamma > 0 \quad (5)$$

$$\text{Radial basis function (RBF) Kernel: } K\ (x_i, x) = \exp\ (-\gamma\ \|x_i - x\|^2),\ \gamma > 0 \quad (6)$$

where γ, τ and d are Kernel parameters.

The Matlab toolbox named LSSVMLab is used to implement LSSVM in this study. The parameters of LSSVM are automatically calibrated during training with 10-fold cross-validation method. More details regarding the principles and application of LSSVM can be found in the LSSVMLab Toolbox User's Guide [29,30].

(2) LR

LR is a probabilistic statistical classification procedure used to predict the dependent variable based on one or more independent variables. The advantage is that the dependent variable has only two cases, that is, occurrence and non-occurrence. In contrast, the stochastic gradient ascent algorithm is generally used to reduce the periodic fluctuations and the computational complexity of the iterative algorithm to further optimize the LR model, which can be calculated by the following equation [31]:

$$\log it(y) = \beta_0 + \beta_1 x_1 + \cdots + \beta_i x_i + e \quad (7)$$

where y is the dependent variable, x_i is the i-th explanatory variable, β_0 is a constant, β_i is the i-th regression coefficient and e is the error. The probability (p) of the occurrence of y is

$$p = \frac{e^{\beta_0 + \beta_1 x_1 + \cdots \beta_i x_i}}{1 + e^{\beta_0 + \beta_1 x_1 + \cdots \beta_i x_i}} \quad (8)$$

If the estimated probability is greater than 0.5 (or other user-defined thresholds), the object is classified as a successful group; otherwise, the object belongs to the failed group. In addition, we train 1 for flash flood, 0 for no flash flood, the values scale from 0 to 1 corresponding to the flash flood sensitivity of the basin from minimum to maximum. The result is the probability that each point is assigned as 0 to 1 training set. Similarly, equal interval classification is used to categorize the probability index of the flash flood into five risk zones of lowest (0–0.2), low (0.2–0.4), moderate (0.4–0.6), high (0.6–0.8) and the highest (0.8–1).

(3) Evaluation index

In the study, five indices including Precision(P), Recall(R), Accuracy (ACC), Kappa(K) and F-score(F) are used to evaluate the results from four models. ACC is the proportion of correctly classified cases to all cases in the set but there is no way to better deviate from the test data to evaluate the model. P is the fraction of recognized instances that are relevant, while R is the fraction of relevant instances retrieved. A better choice is the F-score, which can be interpreted as a weighted average of recalls and precision. Equations (12)–(15) shows how each index calculated, to measure the accuracy of model prediction.

$$\text{Precision}: \; P = \frac{TP}{TP + FP} \tag{9}$$

$$\text{Recall}: \; R = \frac{TP}{TP + FN} \tag{13}$$

$$\text{Accuracy}: \; A = \frac{TP + TN}{TP + FP + TN + FN} \tag{14}$$

$$\text{F} - \text{score}: \; F = \frac{(2 * P * R)}{(P + R)} \tag{15}$$

where *TP*, *FN*, *TN* and *FP* denote the number of true positive, false negative, true negative and false positive, respectively.

Cohen's kappa measures the observer's consistency. It is used to assess the consistency between two or more raters when categorizing a measurement scale. The values are between 1 and 0, corresponding to a perfect agreement and no agreement, respectively. Equation (18) is calculated the Kappa score:

$$\text{Kappa}: \; K = \frac{p_p - p_{\text{exp}}}{1 - p_{\text{exp}}} \tag{16}$$

where P_p is the relatively observed consistency among evaluators and P_{exp} is a hypothetical probability of coincidence, using the observed data to calculate the probability that each observer randomly sees each category. If the raters are in complete agreement, then $k = 1$. If, except by chance, no agreement is reached among the raters (as given by P_{exp}), $k \leq 0$.

3. Results and Discussion

3.1. Comparison of Results Obtained by Four Models

Table 2 shows model performances in the testing period. The accuracy, precision, recall, F-score and kappa range are 0.75 to 0.79, 0.76 to 0.82, 0.74 to 0.77, 0.75 to 0.79 and 0.5 to 0.59, respectively. Obviously, all models have relatively high precision. Although there is no significant difference between the three different kernel functions of the LSSVM model. They are all better than the LR method and the model 2 (LSSVM with RBF kernel) simulates the best.

Table 2. Result of models in testing period.

Index	Model 1	Model 2	Model 3	Model 4
Accuracy	0.78	0.79	0.76	0.75
Precision	0.81	0.82	0.79	0.76
Recall	0.74	0.77	0.74	0.74
F-score	0.78	0.79	0.76	0.75
Kappa	0.56	0.59	0.53	0.50

Model 1: LSSVM + LN, model 2: LSSVM + RBF, model 3: LSSVM + PL, model 4: LR.

Receiver Operating Characteristics (ROC) curves, created by plotting the TP Rate against the FP Rate, are graphical tools applied to the analysis of classification effects over the entire class distribution. Area Under Curve (AUC) is the area under the ROC curve and usually in the range of 0.5 and 1. The AUC equal 0.5 and 1 are accidental classification and perfect classification, respectively. Figure 4 shows the good AUC results obtained by four models but the LSSVM with the RBF kernel has the highest AUC (0.81), followed by LSSVM + LN (0.80) and LSSVM + PL (0.80), the classic LR model (0.78) is relatively poor.

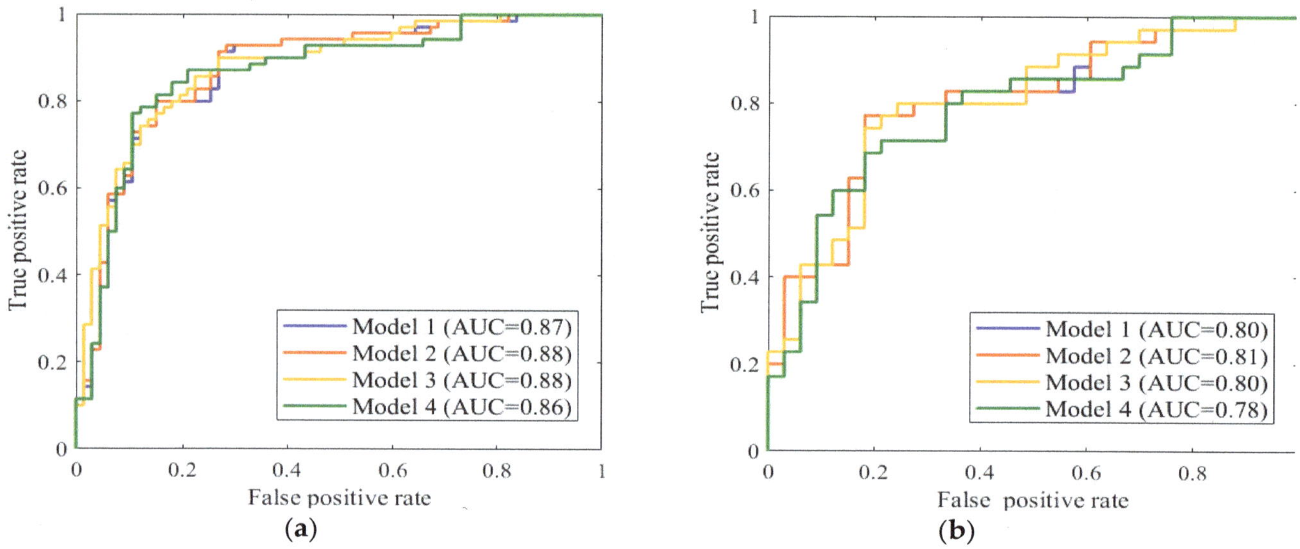

Figure 4. ROC of four models in training (**left**) and testing (**right**). (Model 1: LSSVM + LN, model 2: LSSVM + RBF, model 3: LSSVM + PL, model 4: LR). (**a**) training (**b**) testing.

3.2. Flash Flood Risk Map Comparison

Based on the LR model and the LSSVM model with three kernels of LN, RBF and PL, the flood risk maps of Yunnan Province are generated in the GIS environment. As shown in Figure 5, the high-risk areas are mainly concentrated in the south-central region, accounting for 32% of the total area. Although LSSVM is not significantly better than LR in the training and testing, the risk distribution is significantly different. Figure 6 shows that the flash flood risk obtained by LSSVM is approximately a normal distribution, which is consistent with the previous study in Yunnan Province, China [32,33]. While the risk obtained by LR is a uniform distribution. Therefore, the flood risk maps obtained by LSSVM are more reliable than LR.

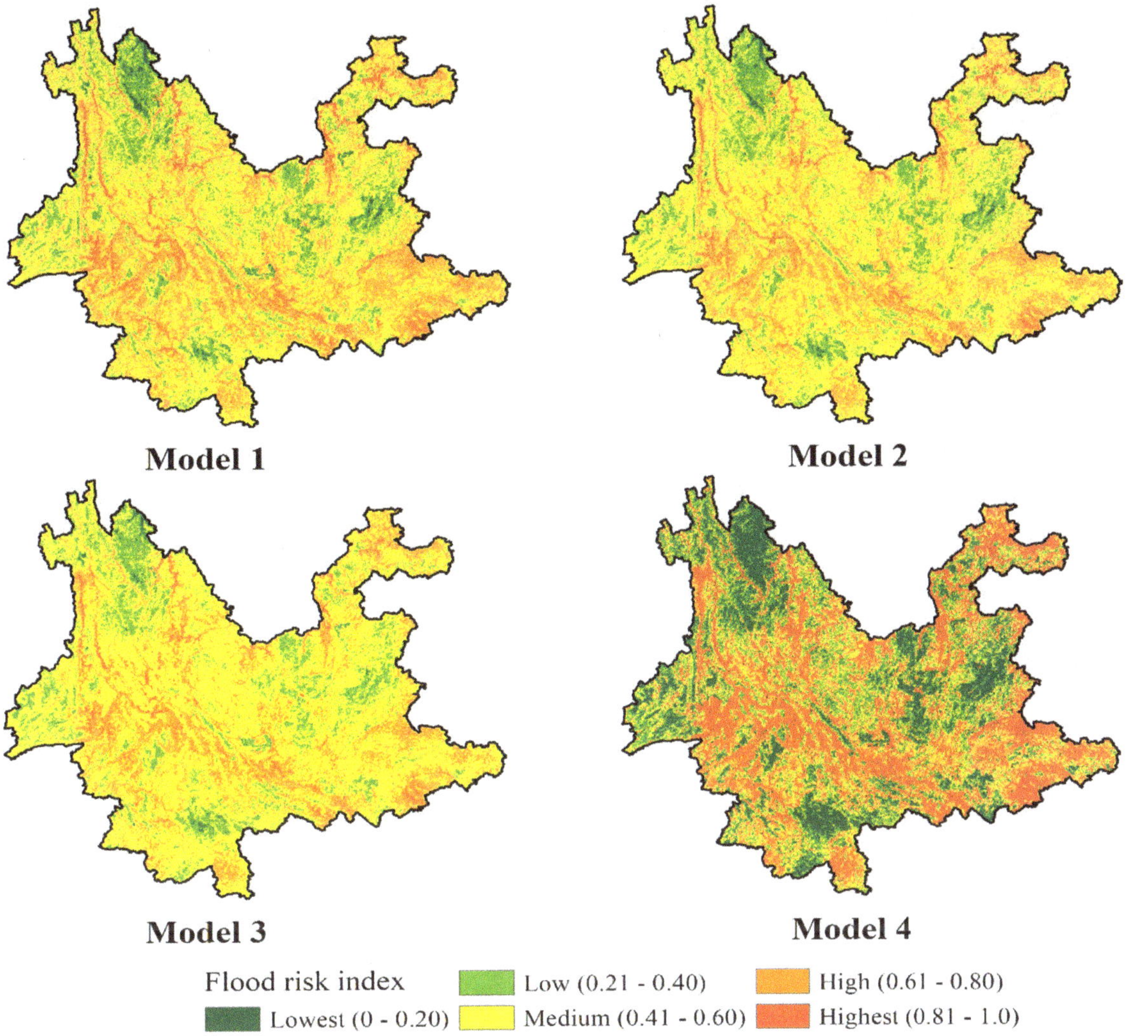

Figure 5. Flood risk index distribution of different models. (Model 1: LSSVM + LN, model 2: LSSVM + RBF, model 3: LSSVM + PL, model 4: LR).

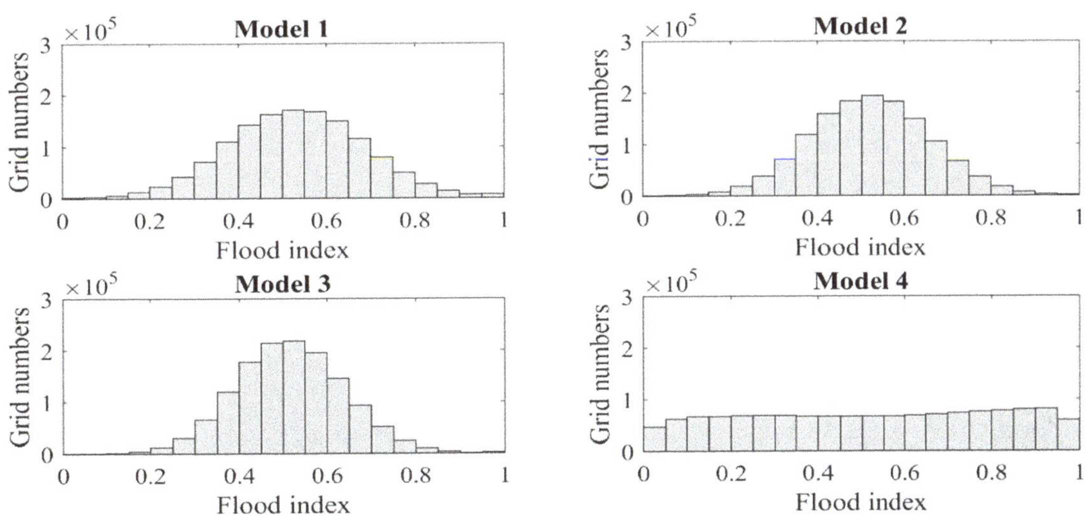

Figure 6. Histogram of Flood index in different models. (Model 1: LSSVM + LN, model 2: LSSVM + RBF, model 3: LSSVM + PL, model 4: LR).

Many studies have utilized some statistical methods to conduct flash flood risk assessments in other areas. For example, Smith (2010) proposed the Flash Flood Potential Index (FFPI) model, considering slope, land use, soil texture and so forth. FFPI values from 1 to 10 correspond to the risk probability from the minimum to the maximum and has been tested in central Iowa, Colorado and upstate New York and Pennsylvania [34,35]. Based on the AHP and information entropy theory, Zeng et al. (2016) selected some relevant indicators (e.g., soil, slope, rainfall and flood control measures), utilized expert scoring method to explore their different weights and finally obtained the risk map of Yunnan Province [18]. In this study, the LSSVM method is firstly used for flash flood risk assessment. LSSVM can directly assess flood risk without setting factor weights. The contribution of each factor to flood risk is assessed by the correlation coefficient between factors and the flood risk, with a more significant advantage.

Figure 7 showed the correlation coefficient of each factor with the flash flood risk from LSSVM-RBF. The greater the correlation coefficient, the greater impact of this indicator on flash floods risk. Obviously, the correlation coefficient of CN is the largest, exceeding 0.5, followed by 7 indicators (DEM, SL, RD, FFP, TWI, 24-H-P, 3-H-P) between 0.1 and 0.5 and the remaining 5 indicators (AP, POP, SM, GDP, VC) are less than 0.1. Combined with the previous analysis, CN identifies the runoff generation capacity. DEM mainly responds to the topography of the study area and SL, RD and TWI all derived from DEM. Therefore, the flash flood risk of Yunnan Province is mainly affected by local runoff capacity, topography. Meanwhile, the correlation coefficient of FFP is 0.3, reflecting that positive man-made measures can largely prevent the occurrence of flash floods. However, compared with topographical factors, we found that the precipitation factor shows a relatively low correlation with the flash floods risk. This mainly because flash floods are caused by intensive rainfall but casualties are usually occurred and reported in low-lying areas. In addition, the effects of short-term precipitation (e.g., 24-H-P, 3-H-P) are greater than the annual precipitation. Our proposed model can concern all flash flood explanatory factors and give an accurate assessment for flash flood risk. In the future, we will further combine water depth and flow as a more reasonable indicator for flood assessment.

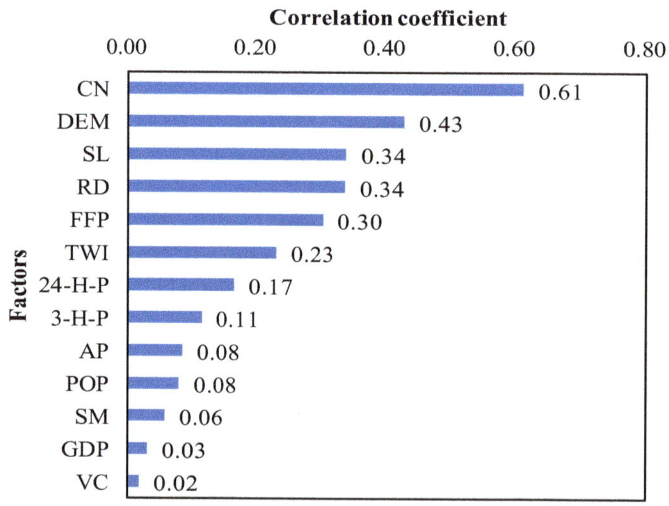

Figure 7. The correlation coefficient between the flash flood risk and 13 indicators.

4. Conclusions

Flash floods have brought huge economic losses and casualties to China. An accurate flash flood risk assessment can identify flood-prone areas and give people enough time to prevent flood disasters in advance. In this study, LSSVM was selected to assess flash flood risk based on 13 explanatory factors. The main conclusions are as follows:

(1) LSSVM can provide a more accurate risk assessment than LR and LSSVM with RBF kernel evaluates best.

(2) The risk of flash flood in Yunnan Province is shown as a normal distribution. The highest risk areas are mainly concentrated in the central and western regions and the lowest risk areas are distributed in the northwest regions.

(3) Flash floods are caused by the combination of various factors and the rank of various factors affecting flash floods is as follows: CN > DEM > SL > RD > FFP > TWI > 24-H-P > 3-H-P > AP > POP > SM > GDP > VC.

In conclusion, the paper utilized the LSSVM method to assess the flash flood risk for the first time and verifies that LSSVM with RBF kernel is suitable for assessing flash floods risk at large or medium scales. Since this method primarily collects explanatory factors and local flood records, where the explanatory factors are mainly derived from public datasets (remote sensing images and statistic bulletin) that can easily get for other areas. Thus, this method is feasible to apply in other regions by collecting local historical flood inventories. This method is highly dependent on data and lacks obvious physical mechanisms. Some problems, such as the shortage and uncertainty of flood inventories, limited the accuracy of model results. In particular, the historical flood record in this study was obtained through investigations by the authority of Yunnan Province, which limited the application of the research results to other regions. With the development of data mining technology, historical flood records from websites or media are desired to use for model development especially for data sparse areas in future works.

Author Contributions: All of the authors contributed to the conception and development of this manuscript. M.M. and G.Z. carried out the analysis and wrote the paper. C.L. designed the system framework and developed the project implementation plan. P.J. collected data and drew the study area map. D.W. participated in the results analysis. H.X., H.W. and Y.H. proposed many useful suggestions to improve its quality.

Funding: This research was funded by the projects of Application of remote sensing on water and soil conservation in Beijing and its demonstration (grant number Z161100001116102), Key technology on dynamic warning of flash flood in Henan Province (China) and its application(grant number HNSW-SHZH-2015-06), Study on infiltration mechanisms of special underlying surface in coalmine goal in Shanxi Province (China) and application of runoff generation and concentration theory(grant number ZNGZ2015-008_2), Research on spatial-temporal variable source runoff model and its mechanism(grant number JZ0145B2017) and National Natural Science Foundation of China (NSFC. General Projects: (grant number. 41471430)).

Acknowledgments: The authors are grateful to the editors and the anonymous reviewers for their insightful comments and suggestions, which helped to improve the manuscript.

References

1. Baker, V.R.; Kochel, R.C.; Patton, P.C. *Flood Geomorphology*; John Wiley and Sons: New York, NY, USA, 1987; ISBN 978-0-12-394846-5.
2. Gruntfest, E.; Handmer, J. *Coping with Flash Floods*; Nato Science: Washington, DC, USA, 2001.
3. Gourley, J.J.; Flamig, Z.L.; Vergara, H.; Kirstetteret, P.E.; Argyle, E.; Terti, G.; Erlingis, J.M.; Hong, Y.; Howard, K.W.; Arthur, A.; et al. The flooded locations and simulated hydrographs (FLASH) project: Improving the tools for flash flood monitoring and prediction across the United States. *Bull. Am. Meteorol. Soc.* **2016**, *98*, 361–372. [CrossRef]
4. Mousavi, M.E.; Irish, J.L.; Frey, A.E.; Olivera, F.; Edge, B.L. Global warming and hurricanes: The potential impact of hurricane intensification and sea level rise on coastal flooding. *Clim. Change* **2011**, *104*, 575–597. [CrossRef]
5. Creutin, J.D.; Borga, M.; Gruntfest, E.; Lutoff, C.; Zoccatelli, D.; Ruin, I. A space and time framework for analyzing human anticipation of flash floods. *J. Hydrol.* **2013**, *482*, 14–24. [CrossRef]
6. Klijn, F.; Kreibich, H.; Moel, H.D.; Penning-Rowsell, E.C. Adaptive flood risk management planning based on a comprehensive flood risk conceptualization. *Mitig. Adapt. Strateg. Glob. Chang.* **2015**, *20*, 845–864. [CrossRef] [PubMed]
7. Dutta, D.; Herath, S.; Musiake, K. A mathematical model for flood loss estimation. *J. Hydrol.* **2003**, *277*, 24–49. [CrossRef]

8. Dottori, F.; Baldassarre, G.D.; Todini, E. Detailed data is welcome, but with a pinch of salt: Accuracy, precision, and uncertainty in flood inundation modeling. *Water Resour. Res.* **2014**, *49*, 6079–6085. [CrossRef]
9. Alfieri, L.; Salamon, P.; Bianchi, A.; Neal, J.C.; Bates, P.; Feyen, L. Advances in pan-European flood hazard mapping. *Hydrol. Process.* **2014**, *28*, 4067–4077. [CrossRef]
10. Sampson, C.C.; Smith, A.M.; Bates, P.D.; Neal, J.C.; Alfieri, L.; Freer, J.E. A high-resolution global flood hazard model. *Water Resour. Res.* **2015**, *51*, 7358–7381. [CrossRef]
11. Mcmillan, H.K.; Brasington, J. Reduced complexity strategies for modelling urban floodplain inundation. *Geomorphology* **2007**, *90*, 226–243. [CrossRef]
12. Bao, H.; Wang, L.; Zhang, K.; Li, Z. Application of a developed distributed hydrological model based on the mixed runoff generation model and 2D kinematic wave flow routing model for better flood forecasting. *Atmos. Sci. Lett.* **2017**, *18*, 284–293. [CrossRef]
13. Huang, P.N.; Li, Z.J.; Li, Q.L.; Zhang, K.; Zhang, H.C. Application and comparison of coaxial correlation diagram and hydrological model for reconstructing flood series under human disturbance. *J. Mt. Sci.* **2016**, *13*, 1245–1264. [CrossRef]
14. Ma, Z.; Shi, Z.; Zhou, Y.; Xu, J.; Yu, W.; Yang, Y. A spatial data mining algorithm for downscaling TMPA 3B43 V7 data over the Qinghai-Tibet Plateau with the effects of systematic anomalies removed. *Remote Sens. Environ.* **2017**, *200*, 378–395. [CrossRef]
15. Tehrany, M.S.; Pradhan, B.; Mansor, S.; Ahmad, N. Flood susceptibility assessment using GIS-based support vector machine model with different kernel types. *Catena* **2015**, *125*, 91–101. [CrossRef]
16. Kalteh, A.M. Improving forecasting accuracy of streamflow time series using least squares support vector machine coupled with data-preprocessing techniques. *Water Resour. Manag.* **2016**, *30*, 747–766. [CrossRef]
17. Pradhan, B. Flood susceptible mapping and risk area delineation using logistic regression, GIS and remote sensing. *J. Sp. Hydrol.* **2010**, *9*, 1–18.
18. Zeng, Z.; Tang, G.; Long, D.; Zeng, C.; Ma, M.; Hong, Y.; Xu, J. A cascading flash flood guidance system: Development and application in Yunnan Province, China. *Nat. Hazards* **2016**, *84*, 2071–2093. [CrossRef]
19. Duan, C.C.; Zhu, Y.; You, W.H. Characteristic and formation cause of drought and flood in Yunnan province rainy season. *Plateau Meteorol.* **2007**, *26*, 402–408.
20. Lindsay, J.B.; Rothwell, J.J.; Davies, H. Mapping outlet points used for watershed delineation onto DEM-derived stream networks. *Water Resour. Res.* **2008**, *44*, 370–380. [CrossRef]
21. Zhao, G.; Pang, B.; Xu, Z.; Yue, J.; Tu, T. Mapping flood susceptibility in mountainous areas on a national scale in China. *Sci. Total Environ.* **2018**, *615*, 1133–1142. [CrossRef]
22. Pettorelli, N.; Ryan, S.; Mueller, T.; Bunnefeld, N.; Jędrzejewska, B.; Lima, M.; Kausrud, K. The normalized difference vegetation index (NDVI): Unforeseen successes in animal ecology. *Clim. Res.* **2011**, *46*, 15–27. [CrossRef]
23. Zeng, Z.; Tang, G.; Hong, Y.; Zeng, C.; Yang, Y. Development of an NRCS curve number global dataset using the latest geospatial remote sensing data for worldwide hydrologic applications. *Remote Sens. Lett.* **2017**, *8*, 528–536. [CrossRef]
24. Sørensen, R.; Zinko, U.; Seibert, J. On the calculation of the topographic wetness index: Evaluation of different methods based on field observations. *Hydrol. Earth Syst. Sci.* **2005**, *10*, 101–112. [CrossRef]
25. Abelen, S.; Seitz, F.; Abarca-del-Rio, R.; Güntner, A. Droughts and floods in the La Plata basin in soil moisture data and GRACE. *Remote Sens.* **2015**, *7*, 7324–7349. [CrossRef]
26. Jongman, B.; Kreibich, H.; Apel, H.; Barredo, J.I.; Bates, P.D.; Feyen, L.; Ward, P.J. Comparative flood damage model assessment: Towards a European approach. *Nat. Hazards Earth Syst. Sci.* **2012**, *12*, 3733–3752. [CrossRef]
27. Guo, L.; He, B.; Ma, M.; Chang, Q.; Li, Q.; Zhang, K.; Hong, Y. A comprehensive flash flood defense system in China: Overview, achievements, and outlook. *Nat. Hazards* **2018**, *92*, 1–14. [CrossRef]
28. He, B.; Huang, X.; Ma, M.; Chang, Q.; Tu, Y.; Li, Q.; Hong, Y. Analysis of flash flood disaster characteristics in China from 2011 to 2015. *Nat. Hazards* **2017**, *90*, 1–14. [CrossRef]
29. Dos Santos, G.S.; Luvizotto, L.G.J.; Mariani, V.C.; Dos Santos Coelho, L. Squares support vector machines with tuning based on chaotic differential evolution approach applied to the identification of a thermal process. *Expert Syst. Appl.* **2012**, *39*, 4805–4812. [CrossRef]
30. Zhu, B.; Wei, Y. Carbon price forecasting with a novel hybrid ARIMA and least squares support vector machines methodology. *Omega Intern. J. Manag. Sci.* **2013**, *41*, 517–524. [CrossRef]

31. Budimir, M.E.A.; Atkinson, P.M.; Lewis, H.G. A systematic review of landslide probability mapping using logistic regression. *Landslides* **2015**, *12*, 419–436. [CrossRef]
32. He, J. *Assessment and Regionalization of Drought Disaster Risk in Yunnan Province[D]*; Yunnan University: Kunming, China, 2016.
33. Wang, L.; Wang, S.; Wang, X.; Wang, F.; Fan, C. Risk zoning of drought disaster based on AHP and GIS in Yunnan province. *Water Sav. Irrig.* **2017**, *10*, 100–103, 106.
34. Smith, G.E. Development of a Flash Flood Potential Index Using Physiographic Data Sets within a Geographic Information System. Ph.D. Thesis, The University of Utah, Salt Lake City, UT, USA, 2010.
35. Minea, G. Assessment of the flash flood potential of Basca River Catchment (Romania) based on physiographic factors. *Cent. Eur. J. Geosci.* **2013**, *5*, 344–353. [CrossRef]

12

Time-Series Evolution Patterns of Land Subsidence in the Eastern Beijing Plain, China

Junjie Zuo [1,2], Huili Gong [1,2], Beibei Chen [1,2,*], Kaisi Liu [1,2], Chaofan Zhou [1,2] and Yinghai Ke [1,2]

1 Base of the Key Laboratory of Urban Environmental Process and Digital Modeling, Capital Normal University, Beijing 100048, China; 2173602001@cnu.edu.cn (J.Z.); gonghl_1956@126.com (H.G.); kaisiliu@yeah.net (K.L.); chaofan0322@126.com (C.Z.); yke@cnu.edu.cn (Y.K.)

2 BCollege of Resources Environment and Tourism, Capital Normal University, Beijing 100048, China

* Correspondence: cnucbb@yeah.net.

Abstract: Land subsidence in the Eastern Beijing Plain has a long history and is always serious. In this paper, we consider the time-series evolution patterns of the eastern of Beijing Plain. First, we use the Persistent Scatterer Interferometric Synthetic Aperture Radar (PSI) technique, with Envisat and Radarsat-2 data, to monitor the deformation of Beijing Plain from 2007 to 2015. Second, we adopt the standard deviation ellipse (SDE) method, combined with hydrogeological data, to analyze the spatial evolution patterns of land subsidence. The results suggest that land subsidence developed mainly in the northwest–southeast direction until 2012 and then expanded in all directions. This process corresponds to the expansion of the groundwater cone of depression range after 2012, although subsidence is restricted by geological conditions. Then, we use the permutation entropy (PE) algorithm to reverse the temporal evolution pattern of land subsidence, and interpret the causes of the phenomenon in combination with groundwater level change data. The results show that the time-series evolution pattern of the land subsidence funnel edge can be divided into three stages. From 2009 to 2010, the land subsidence development was uneven. From 2010 to 2012, the land subsidence development was relatively even. From 2012 to 2013, the development of land subsidence became uneven. However, subsidence within the land subsidence funnel is divided into two stages. From 2009 to 2012, the land subsidence tended to be even, and from 2012 to 2015, the land subsidence was relatively more even. The main reason for the different time-series evolution patterns at these two locations is the annual groundwater level variations. The larger the variation range of groundwater is, the higher the corresponding PE value, which means the development of the land subsidence tends to be uneven.

Keywords: land subsidence; SDE; PE; groundwater level; compressible sediment layer

---

1. Introduction

During the last few decades, interferometric synthetic aperture radar (InSAR) has become an important tool for the mapping and monitoring deformation processes [1–4]. With the InSAR technique, we can measure deformation over a large scale, from millimeters to centimeters. However, this method faces the problems of spatial and temporal decorrelation and atmospheric distribution. Persistent Scatterers InSAR (PSI) was proposed to overcome the limitations of InSAR [5,6]. The PSI techniques has shown its potential for ground deformation monitoring in a number of applications, including land subsidence [7–10], seismic faults [11–13], and landslide-prone slopes [14–16].

Since the 1960s, land subsidence has been found in the Beijing Plain, which has experienced rapid development. Currently, land subsidence is extremely uneven, and it has formed two major settlements centers in the north and south. The north settlement center has become the largest ground settlement funnel group on the Beijing Plain. Many scholars study the spatiotemporal evolution characteristics of the settlement on the Beijing Plain. The studies indicate that the deformation has mainly occurred in the eastern and northern part of the Beijing Plain, and Chaoyang and the northeastern part of Tongzhou have experienced the most severe subsidence. Land subsidence shows an increasing trends in the rate and extent over time [17–21]. The uneven settlement has developed rapidly, and the changes in land subsidence in the north–south direction are more pronounced than those in other directions [22]. The deformation of the Beijing Plain shows seasonal variations, and the spatial location of land subsidence funnels is consistent with the location of the groundwater cone of depression, although not entirely [23]. The spatial distribution of the deformation rate from 2003 to 2010 was similar to that from 2010 to 2016, but the subsidence rate from 2010 to 2016 was higher that from 2003 to 2010 [24].

Most of these studies focus on the spatial distribution of land subsidence, however, fewer studies examine the time-series evolution pattern. In this paper, we pay attention to the time series of the land subsidence pattern. First we use the standard deviation ellipse (SDE) method to reveal the spatial evolution pattern of land subsidence; then we adopt the permutation entropy (PE) algorithm in order to determine how the land subsidence changes over time. PE is a complex parameter based on the comparison of adjacent data in a long time series. It amplifies the imperceptible changes in the signal by quantitatively describing the variations in signal spatial complexity [25,26]. This complexity shows a difference. PE mainly compares the differences between the variations in settlement data in the former time interval and those in the latter time interval. PE monitors the differences in the land subsidence process in a long time series. More specifically, the difference reflects whether land subsidence develops uniformly or not in the long time series. The process of PE value increase is the process of increasing difference in land subsidence, which means land subsidence develops more unevenly over this period of time. The process of PE value decrease is the process of decreasing difference in land subsidence, which means land subsidence develops more evenly during this time. At present, many scholars who have studied the PE method. The main applications of this method include medicine, biology, climate, and image processing [27–30]. Most scholars approximately analyze the overall development of deformation in the whole research period through the time-series settlement. Using the PE method, we can reveal the different development processes of subsidence with time. PE amplifies the settlement details, which are hard to determine using other methods.

This paper is organized as follows. The background of the study area and datasets used are described in Section 2. The data methods are presented in Section 3. In Section 4, we use the SDE method combined with the changes in the land subsidence funnel area, in the center of gravity, and in the spatial distribution and range to reveal the spatial evolution pattern of land subsidence. Meanwhile, we adopt the PE method to respond to the time-series evolution pattern. In Section 5, we discuss the relationship between the spatial evolution pattern obtained by SDE and hydrogeologic data. Moreover, we explore the causes of the different PE results for the land subsidence funnel and funnel edge and find that this phenomenon is mainly due to the difference in the annual variations in the groundwater level.

2. Study Area and Dataset

2.1. Study Area

The study area (Figure 1) is located in the eastern part of the Beijing Plain, which is an area with serious land subsidence and a part of the warm temperate zone, with a semihumid and semiarid continental monsoon climate and an annual average temperature of 11–12 °C. The precipitation distribution in the study area is extremely uneven. The precipitation in summer is approximately 70% of the annual precipitation, which is much higher than that in winter [31].

The groundwater system in the study area is composed of three water systems. They are the Yongding River system, Wenyu River system, and Chaobai River system. According to the groundwater supplementation, diameter, drainage conditions, groundwater exploitation horizon, and genesis type, the Quaternary aquifers in the study area are mainly divided into three main aquifer groups [19,23]. The first group of aquifers is in Holocene and upper Pleistocene strata, which are widely distributed. The depth of the floor is 35–75 m and it is a multilayer structure. The underground water types include upper stagnant water, interlayer water, diving, and shallow confined water. The lithologies includes fine sand, silt, silty sand, and sandy clay. The second aquifer group is the middle Pleistocene strata. The types of groundwater mainly include variations in the shallow groundwater level in the study area, which is mainly affected by the precipitation infiltration of medium and deep confined water. The alluvial-diluvial fan floor of the Yongding River is buried as deep as 150 m and that of the Chaobai River is as deep as 190 m. The lithologies include multiple types of gravel, sand, and clay soil. The last aquifer group is in the lower Pleistocene strata, which are composed of medium coarse and gray sand. The water-bearing group is distributed in the middle and lower parts of the alluvial–diluvial fan with a multilayer structure. The groundwater type is mainly deep confined water, and the roof is buried as deep as 190 m.

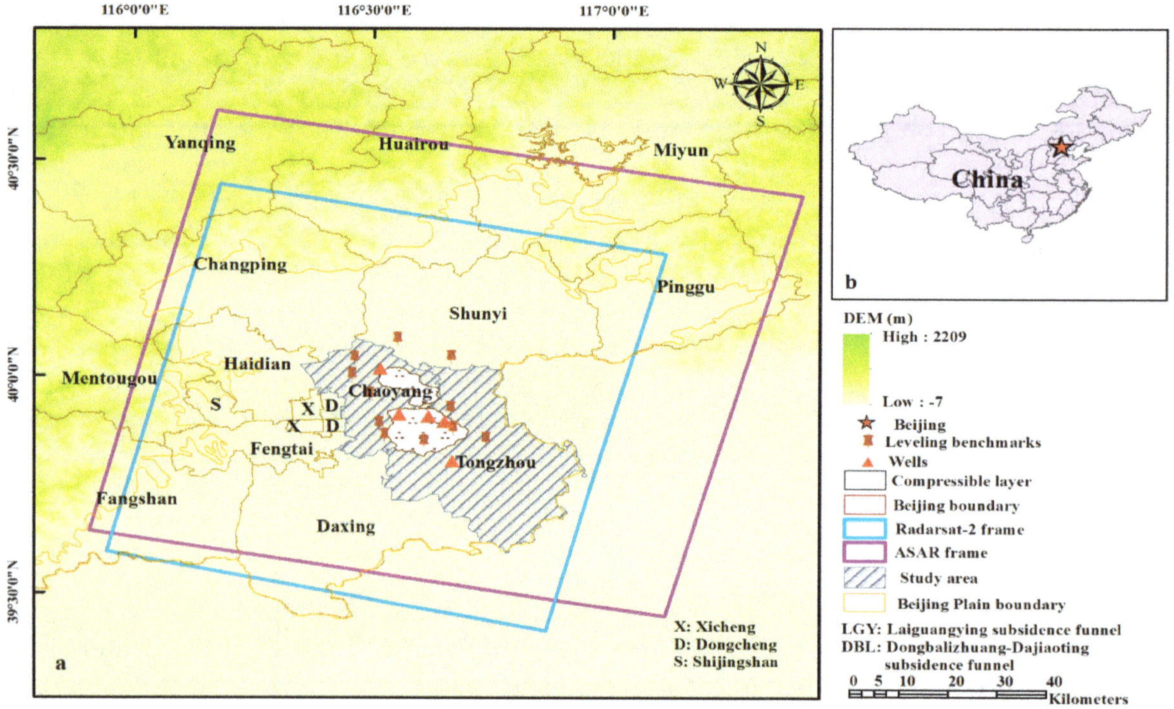

Figure 1. The geographical location of the study area. The violet box and blue box represent the ASAR and Radarsat-2 data spatial coverage, respectively. The colored pushpins indicate the locations of the leveling benchmarks. The red points represent the locations of wells and are named JingshunLu (JSL), Dengfuzhuang (DFZ), Baliqiaocun (BLQ), Luhe middle school (LHZ), and Nanhuofa (NHF) in the study area.

2.2. Dataset

The dataset used in this study comes from two different satellites. One set is the 31 C bands descending track ASAR images acquired from January 2007 to August 2010, with a 35 day revisit cycle, which was provided by the European Space Agency. The other set is the 48 C bands Radarsat-2 images acquired from Oct. 2010 to Nov.2015, with a 24 day revisit cycle, which was provided by the Canadian Space Agency. The spatial resolutions of both Envisat ASAR and Radarsat-2 images were 30 m. The coverage of the SAR images is shown in Figure 1, and the detailed parameters of the SAR images are summarized in Table 1.

We use the SARPROZ software to handle our SAR images. And we use the Shuttle Radar Topography Mission (SRTM) digital elevation model (DEM) with a spatial resolution of 90 m to remove the topographic phase and to geocode interferograms. The groundwater level contours are provided by the Beijing Water Authority and are used for comparing the relationship between the groundwater levels and subsidence, and the well locations are shown in Figure 1a.

Table 1. The parameters of the interferometric synthetic aperture radar (In-SAR) data used.

Parameter.	Envisat ASAR	Radarsat-2
Band	C	C
Wavelength (cm)	5.6	5.5
Polarization	VV	VV
Orbit directions	Descending	Descending
Track no.	2218	60115
Incidence angle (°)	22.9	27.6
Heading (°)	−164	−168.8
Spatial resolution (m)	30	30
No. of images	31	48
Data range	January 2007–August 2010	October 2010–November 2015

3. Methods

3.1. PSI Method

The Persistent Scatterer Interferometric Synthetic Aperture Radar (PSI) was proposed by Ferretti [5]. The technique reduces the incoherence and atmospheric effects in the time and space domains. It is capable of extracting the targets points with a strong and stable radiometric property, and of obtaining surface deformation by separating the topographic phase of the ground targets. In this study, we use the SARPROZ software to acquire the surface deformation information for the study area.

The surface deformation phase can be obtained using the PSI procedure by decomposing the interferometric phase based on Equation (1):

$$\Delta\phi_{\text{int}} = \phi_{flat} + \phi_{topo} + \phi_{def} + \phi_{atmos} + \phi_{noise} \tag{1}$$

where $\Delta\phi_{\text{int}}$ is the interferometric phase, and ϕ_{flat} is the flat earth phase, which can be removed by the precise orbital state vector of the satellite, obtained using the SARPROZ software, when reading SAR images.ϕ_{topo} is the topographic phase contributed by the topographic relief, and it can be removed by the external DEM in the SARPROZ software; ϕ_{def} is the deformation phase caused by the displacement of the ground during the two image acquisitions. What we truly would like to obtain is the ϕ_{def}. ϕ_{atmos} is the atmospheric phase due to the contribution of atmospheric components, and APS processing in the SARPROZ software can eliminate this phase. ϕ_{noise} is the thermal noise and coregistration errors, which can be removed by the linear models in APS processing. However, ϕ_{def} is the deformation phase in the radar line-of-sight (LOS) direction including the horizontal and vertical directions. Hence, the LOS (d_{los}) can be converted into vertical displacement (d_v) by the following Equation (2):

$$d_v = d_{los} / \cos\theta \tag{2}$$

where θ is the incidence angle.

3.2. Standard Deviation Ellipse Method

The standard deviation ellipse (SDE) method was first proposed by Lefever in 1926 to analyze the spatial distribution characteristics of discrete datasets [32]. In this study, we use numerous parameters of SDE, such as the ellipse center, long axis, and short axis, to analyze the spatial characteristics of land subsidence. These parameters can be calculated as follows:

$$SDE_x = \sqrt{\frac{\sum_{i=1}^{n}(x_i - \overline{X})^2}{n}}$$
$$SDE_y = \sqrt{\frac{\sum_{i=1}^{n}(y_i - \overline{Y})^2}{n}} \tag{3}$$

where x_i and y_i are the coordinates of PSI points, $\{\overline{X}, \overline{Y}\}$ is the mean center of PSI points, and n is the total number of PSI points. The rotation angle is calculated as follows:

$$\begin{aligned} \tan\theta &= \frac{A+B}{C} \\ A &= \left(\sum_{i=1}^{n} x_i^2 - \sum_{i=1}^{n} y_i^2\right) \\ B &= \sqrt{\left(\sum_{i=1}^{n} x_i^2 - \sum_{i=1}^{n} y_i^2\right)^2 + 4\left(\sum_{i=1}^{n} x_i y_i\right)^2} \\ C &= 2\sum_{i=1}^{n} x_i y_i \end{aligned} \tag{4}$$

where x_i and y_i are the deviation of xy and the mean center, respectively. The standard deviations of the X and Y axes are given by:

$$\sigma_x = \sqrt{\frac{\sum_{i=1}^{n}(x_i\cos\theta - y_i\sin\theta)^2}{n}}$$
$$\sigma_y = \sqrt{\frac{\sum_{i=1}^{n}(x_i\sin\theta - y_i\cos\theta)^2}{n}} \tag{5}$$

where θ is the azimuth of the ellipse, indicating the direction of the north clockwise rotation angle to the long axis of the ellipse, σ_x is the standard deviation of the X axis and σ_y is the standard deviation of the Y axis.

3.3. Permutation Entropy Method

Christoph Bandt et al. [25], proposed an entropy parameter to measure the complexity of the one-dimensional time series, called the permutation entropy (PE). It is similar to the LyaPullov index in terms of performance reflecting one-dimensional time-series complexity. However, compared with complex parameters such as the LyaPullov index and fractal dimension, PE has the characteristics of simpler calculation and stronger anti-noise interference ability.

Given a land subsidence time series $[x(i), i = 1, 2, \ldots n]$, for various n, n increasing to ∞ [25], any of the elements $x(i)$ is a phase space reconstruction of the elements $x(i)$, and then the following is obtained:

$$X(i) = [x(i), x(i+1), \ldots, x(i+(m-1)l)] \tag{6}$$

In Equation (6), m and l are the embedded dimension and delay time, respectively. The m components of $\{x(i), x(i+1), \ldots x[i+(m-1)l]\}$ are rearranged as:

$$\{x[i+(j_1-1)l]\} \leq \{x[i+(j_2-1)l]\} \leq \ldots \leq \{x[i+(j_m-1)l]\} \tag{7}$$

If $x[i+(j_1+1)l] = x[i+(j_2+1)l]$ exists, these values are sorted by the size of the j value at this time. This step means when $j_{i1} < j_{i2}$, then $x[i+(j_1+1)l] \leq x[i+(j_2+1)l]$. Thus, any vector x_i can produce a sequence of symbols:

$$A(g) = [j_1, j_2, \ldots, j_m] \tag{8}$$

In Equation (8), where $g = 1, 2, \ldots, k$, and $k \leq m!$, there are $m!$ kinds of different arrangements of m different symbols, which indicates that there are $m!$ kinds of symbol sequences. The symbol sequence of $A(g)$ is one example. The occurrence probability of each symbol, such as $p_1, p_2, \ldots, p_k$ is calculated. Hence, the k kinds of symbol sequences of the time series can be defined in the form of the Shannon information entropy:

$$H_p(m) = -\sum_{v=1}^{k} p_v \ln p_v \tag{9}$$

When $p_v = 1/m!$, $H_p(m)$ is up to the maximum value $\ln(m!)$. For convenience, usually $H_p(m)$ can be labeled as:

$$0 \leq H_p = H_p(m)/\ln(m!) \leq 1 \tag{10}$$

The value of H_p indicates the degree of randomness of the time series $([x(i), i = 1, 2, \ldots, n])$. The smaller the value of H_p is, the more regular of the time series; the higher the value of H_p is, the more random of the time series. With the various values of H_p, pamplifies the small variations in the time series $([x(i), i = 1, 2, \ldots, n])$.

In this paper, we use the MATLAB software to realize the PE method. Because the study period is relatively short (only nine years), we choose three embedded dimensions, and the delay time is two years, which indicates that the cumulative land subsidence for the first two years is a training sample. The sliding window step size is one, which ensures that the variations in the PE are due to the changes in the state of settlement at the later time nodes.

4. Results

4.1. Land Subsidence Information Monitoring by PSI Validation

The Eastern Beijing Plain has two major deformation bowls, which are named the Laiguangying (LGY) and Dongbalizhuang-Dajiaoting (DBL) land subsidence funnels. They are the places with the earliest and most serious subsidence in Beijing, and the two subsidence funnels have been developing since they were discovered in 1983 (Figure 1). The two subsidence funnels are located in Chaoyang and Tongzhou Districts, respectively. Chaoyang is the industrial base of Beijing, and Tongzhou is the subcenter of Beijing. Meanwhile, these districts are key areas for future planning and construction in Beijing, including of the Central Business District (CBD), in the CBD east expansion zone. Hence, understanding the spatial and temporal evolution of land subsidence in these two regions is important for the urban development of Beijing.

We selected 11 leveling points in 2009, which were chosen to verify the PSI processing results (Figure 2). In 2009, the maximum error of the two measurements was 19.91 mm, and the minimum error was 1.18 mm. The error is caused by the deviation between the position of the PSI point and the position of the leveling point. However, the correlation coefficient between the leveling point and the PSI points is 0.94, which proves that the two sets of points show good consistency. The PSI monitoring results are reliable.

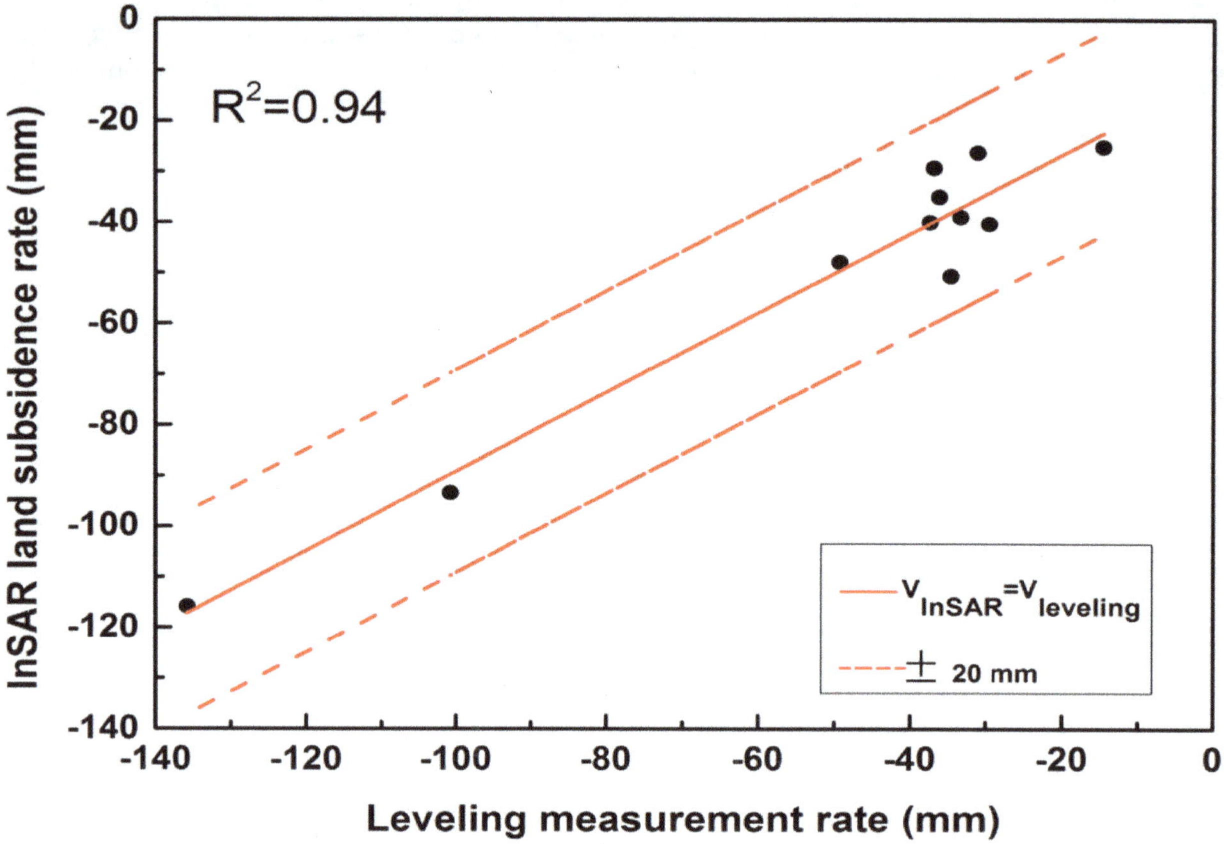

Figure 2. The comparison of land subsidence rates derived from the Persistent Scatterer Interferometric Synthetic Aperture Radar (PSI) technique and from leveling measurements rates. The location of benchmarks are shown in Figure 1.

4.2. Spatial Evolution Pattern by the SDE Method

The SDE is an effective approach, which can accurately reveal the geographical spatial distributions and other characteristics using spatial statistical methods. We use the annual land subsidence rate as the weight and adopt the spatial analysis method to obtain the SDE. Figure 3 shows the cumulative land subsidence information and SDE shapes from 2007 and 2015 in the study area. The maximum accumulated land subsidence reached 1.184 m by 2015. We can observe that land subsidence was serious and the spatial distribution of land subsidence increased with time. The major axis of the SDE is oriented northwest–southeast. This position reflects that the development of land subsidence in the northwest–southeast direction is more obvious than those in other directions. We take the annual subsidence of 60 mm as the dividing line and define the place with more than 60 mm as the subsidence funnel areas. We find that after 2008, the LGY and DBL land subsidence funnels evolves into a single area. We describe the spatial distribution of land subsidence in the eastern part of Beijing Plain with respect to the following four factors: the changes in the land subsidence funnel area; the changes in the center of gravity of land subsidence; the changes in the distribution range of land subsidence and the changes in the distribution of land subsidence.

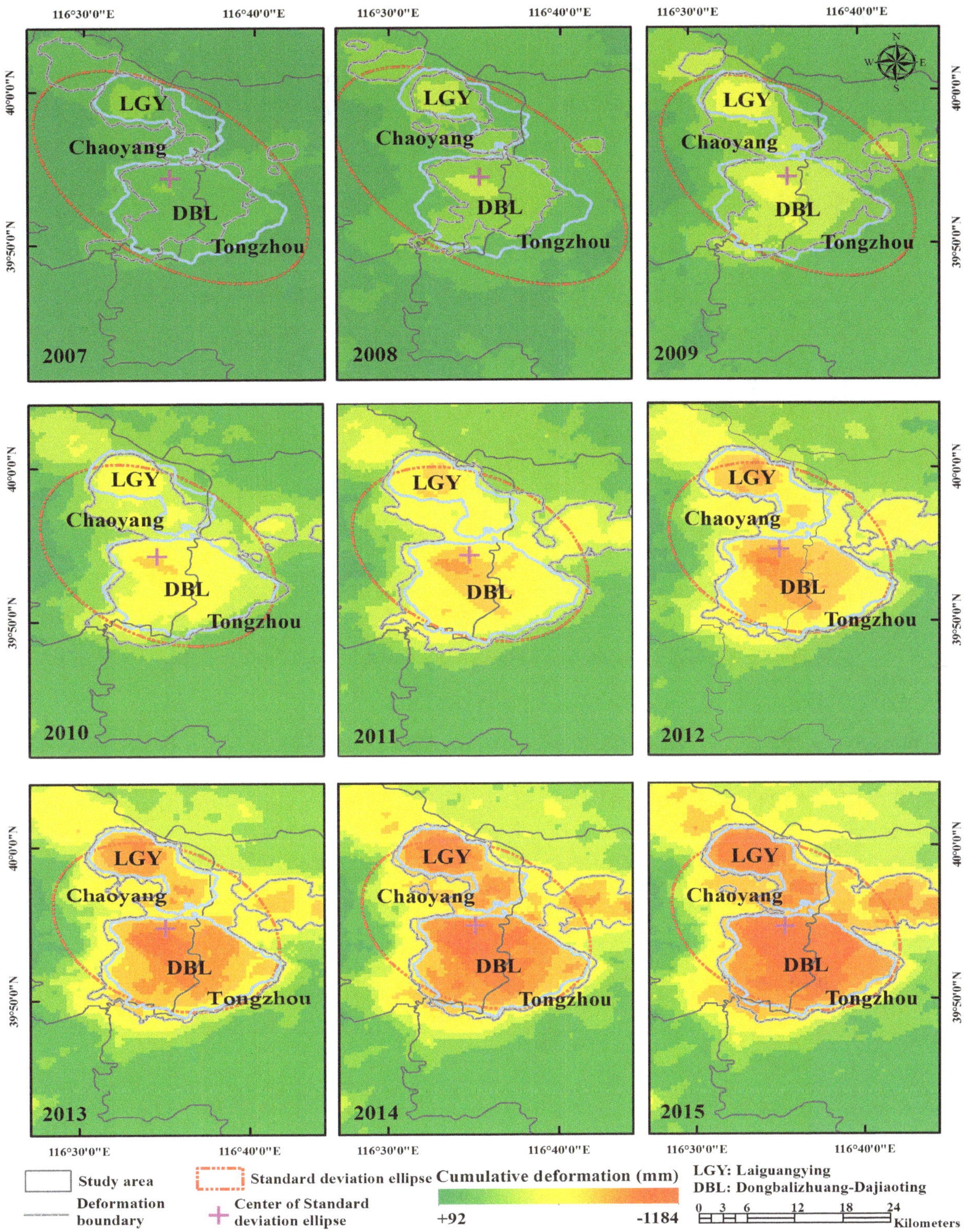

Figure 3. Cumulative deformation in the study area from 2007 to 2015 measured by the PSI technique using ASAR from 2007 to 2010 and Radarsat-2 data from 2010 to 2015. The red ellipse indicates the standard deviation ellipse (SDE) of the study area. The amethyst cross represents the center of the SDE.

4.2.1. Changes in the land subsidence funnel area

We calculated the area of the land subsidence funnels from 2007 to 2015 (Figure 4). From Figure 4, we can see the land subsidence funnel area was 253.28 km^2 in 2008, which was the smallest value and accounted for 10.97% of the total study area. The land subsidence funnel area was 312.04 km^2 in 2011, which was the greatest value and accounted for 21.79% of the total study area. From 2007 to 2008, the land subsidence funnel area decreased; then, it continued to increase rapidly until 2011, especially from 2008 to 2009 and 2010 to 2011. Between 2011 and 2012, the area decreased, and from 2012 to 2015, the area increased slowly. Therefore, we can find that the area of land subsidence in the study area is still increasing, but the growth rate of the area has been lower, especially since 2012. This result means that land subsidence in the study area is still slowly developing.

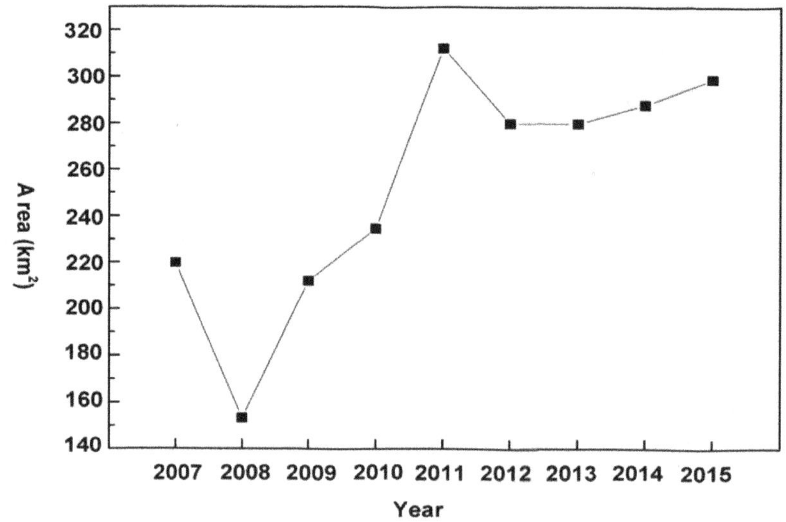

Figure 4. The area of the land subsidence funnels from 2007 to 2015.

4.2.2. Changes in the center of the gravity of land subsidence

The changes in the central migration trajectory of SDE from 2007 to 2015 are shown in Figure 5. We find that the central coordinates of the SDE in 2007 were 116.584° E, 39.908° N and those in 2015 were 116.588° E, 39.913° N. The center of the SDE of the study area moved toward the northeast as a whole. From 2007 to 2008, the center moved to the southeast; in 2008 and 2009, it moved to the northeast; between 2009 and 2010, it moved to the southwest; between 2010 and 2014, it moved to the north; and from 2014 to 2015, it moved to the southeast. Combining this information with that in Figure 2, the SDE center of the study area has been changing toward the direction of the northern part of the DBL land subsidence funnel. We estimate that this land subsidence funnel will still be the land subsidence development center of the Chaoyang and Tongzhou Districts.

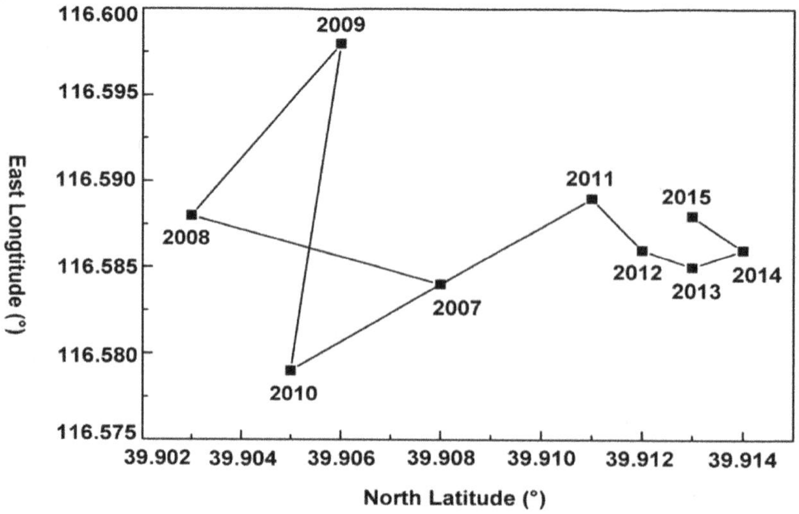

Figure 5. Changes in the SDE central migration trajectory from 2007 to 2015.

4.2.3. Changes in the distribution range of land subsidence

The length of the long axis of the SDE represents the distribution range of land subsidence. The spatial variation distributions of land subsidence in the study area from 2007 to 2015 is shown in Figure 6. The long axis decreased from 154.438 m in 2007 to 119.448 m in 2015. From 2007 to 2008, the long axis increased, suggesting that the spatial distribution range of land subsidence increased. Combining this information with that in Figure 4, we can see that the land subsidence funnel area was decreasing during this period, which means that the distribution of land subsidence was dispersed. From 2008 to 2012, the long axis decreased rapidly, indicating that the distribution range of land subsidence was reduced during this period. However, the land subsidence funnel area expanded in 2008 and 2012, which indicated that the land subsidence of the area was concentrated; meanwhile, land subsidence showed a tendency to merge into one region during this time. From 2012 to 2015, the long axis was increasing slowly, which indicated that the distribution range of land subsidence was increasing. The area of the land subsidence funnel expanded, indicating that land subsidence increased during this time.

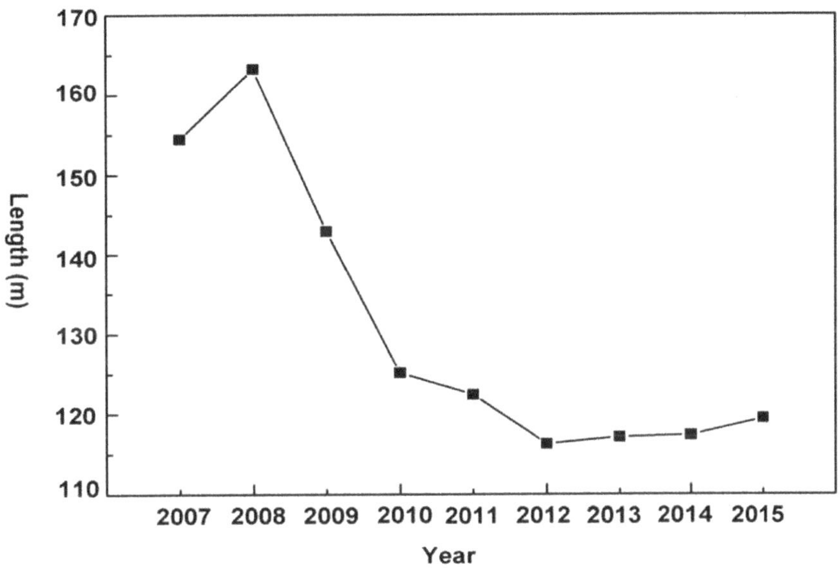

Figure 6. Temporal changes in the length of the SDE long axis.

4.2.4. Changes in the distribution of land subsidence

The ratio of the short axis to the long axis of the SDE represents the spatial distribution shape of land subsidence. When the ratio is close to 1, the spatial distribution shape of land subsidence is close to a circle. That is, land subsidence evolves more uniformly in all directions. The spatial distribution shape changes in the study area are shown in Figure 7. From 2007 to 2015, the spatial distribution shape changed significantly. Between 2007 and 2008, the ratio of the short axis to the long axis decreased, which means that land subsidence mainly changed toward the long axis in the northwest–southeast direction. From 2008 to 2012, the ratio of the two axes increased rapidly, suggesting that during this period, land subsidence intensified in the short axis direction (northeast–southwest direction). Between 2012 and 2015, the ratio of the two axes decreased slowly, indicating that land subsidence was relatively uniform in all directions during this time. In short, the main direction of land subsidence development was in the northwest–southeast direction before 2008. After 2008, land subsidence intensified in the northeast–southwest direction and showed a trend of expansion.

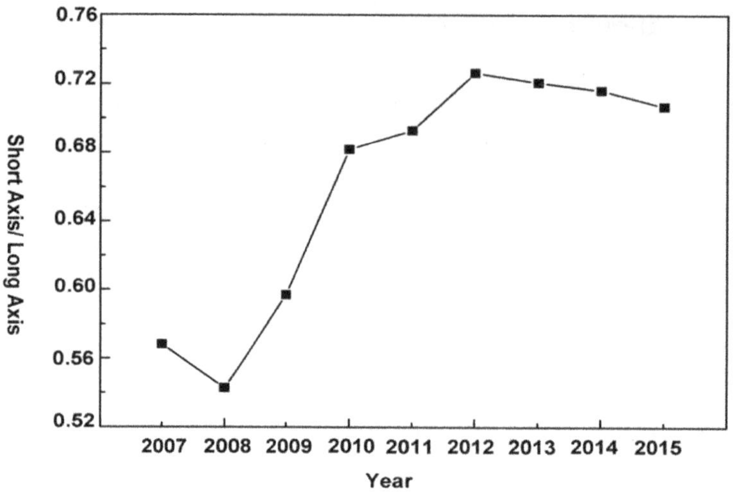

Figure 7. Changes in the SDE short axis/long axis ratio from 2007 to 2015.

4.3. Temporal Evolution Pattern by the PE Method

According to the LGY and DBL land subsidence funnel cumulative deformation rate, we selected five wells and analyzed the time-series evolution pattern of land subsidence in the deformation funnels and funnel edges using the PE method. The results are shown in Figure 8.

The PE results for well JSL are shown in Figure 8a. JSL lies at the edge of the LGY subsidence funnel area. The cumulative land subsidence ranged from −50–450 mm between 2007 and 2015. It showed a linear increasing trend from 2009 to 2010. After 2012, it showed an increasing trend with volatility. In contrast to the PE result, since 2009, the PE value of JSL was increasing and reached a maximum in April 2010. This phenomenon indicates that the land subsidence development was relatively uneven during the period from 2009 to April 2010. The value continued to decrease from April 2010 to October 2011, and it reached a minimum value in October 2011. This phenomenon indicates that the state of the land subsidence development developed from a nonuniform state to a uniform state during this period. After October 2011, the PE value increased, which means that land subsidence developed toward a nonuniform state.

The PE results for well DFZ are shown in Figure 8b. DFZ belongs to the DBL subsidence funnel. The cumulative land subsidence varied from 0 to −740 mm from 2007 to 2015. The cumulative land subsidence showed a downward trend. However, the PE value continued decreasing from 2009 to November 2013. This phenomenon indicates that land subsidence tended to develop toward a uniform state after 2009. The value returned to zero during the period from November 2013 to 2015. This result means that after this time, the cumulative land subsidence continued to increase, with no decrease. This result indicates that the land subsidence in DFZ has been in a more uniform state since 2013.

The PE results for well BLQ are shown in Figure 8c. BLQ is located in the DBL subsidence funnel. The cumulative land subsidence ranged from 0 to −740 mm between 2007 and 2015. The cumulative land subsidence showed a linear rise in the study period. The PE value decreased from 2009, and the minimum value appeared in August 2012. Then, in the following years, the value was zero. In other words, land subsidence was moving toward a uniform trend from 2009 to 2012. In addition, after 2012, the cumulative land subsidence continued to increase, which indicates that the land subsidence of this well will was relatively more uniform.

The PE results for well LHZ are shown in Figure 8d. LHZ lies in the DBL subsidence funnel. During the research period, the cumulative settlement was increasing linearly. The PE value, decreased form 2009, until August 2012, when it reached its minimum. This result indicates that land subsidence moved from a nonuniform state to a uniform state. After August 2012, the PE value became zero, which showed that the state of land subsidence development was more uniform than that in August 2012. This result reflected that the value of cumulative land subsidence was progressively increasing.

The PE results for well NHF are shown in Figure 8e. NHF is located at the edge of the DBL subsidence funnel. The cumulative land subsidence ranged from 20 to −90 mm during the period from 2007 to 2015, with fluctuations. The value of PE increased from the beginning of January 2009 to April 2009. This phenomenon reflected that in this period, the development of land subsidence was uneven. From April 2009 to July 2011, the value decreased significantly, which means that the development of land subsidence in during this period trended to be even. The value increased between July 2011 and 2015, which indicates that land subsidence developed unevenly.

According to the value of the PE, the development of land subsidence at the edge of the funnel is different from that within the land subsidence funnel. The PE results at the edge of the funnel increased first, then decreased, and finally increased, indicating that land subsidence developed nonuniformly first, then uniformly, and nonuniform again during the research period. However, since 2009, the value of PE for land subsidence decreased during the research period; even after 2012, the value was even zero. This phenomenon indicates that, since 2009, land subsidence at these three wells trended to be uniform. It became relatively more uniform after 2012.

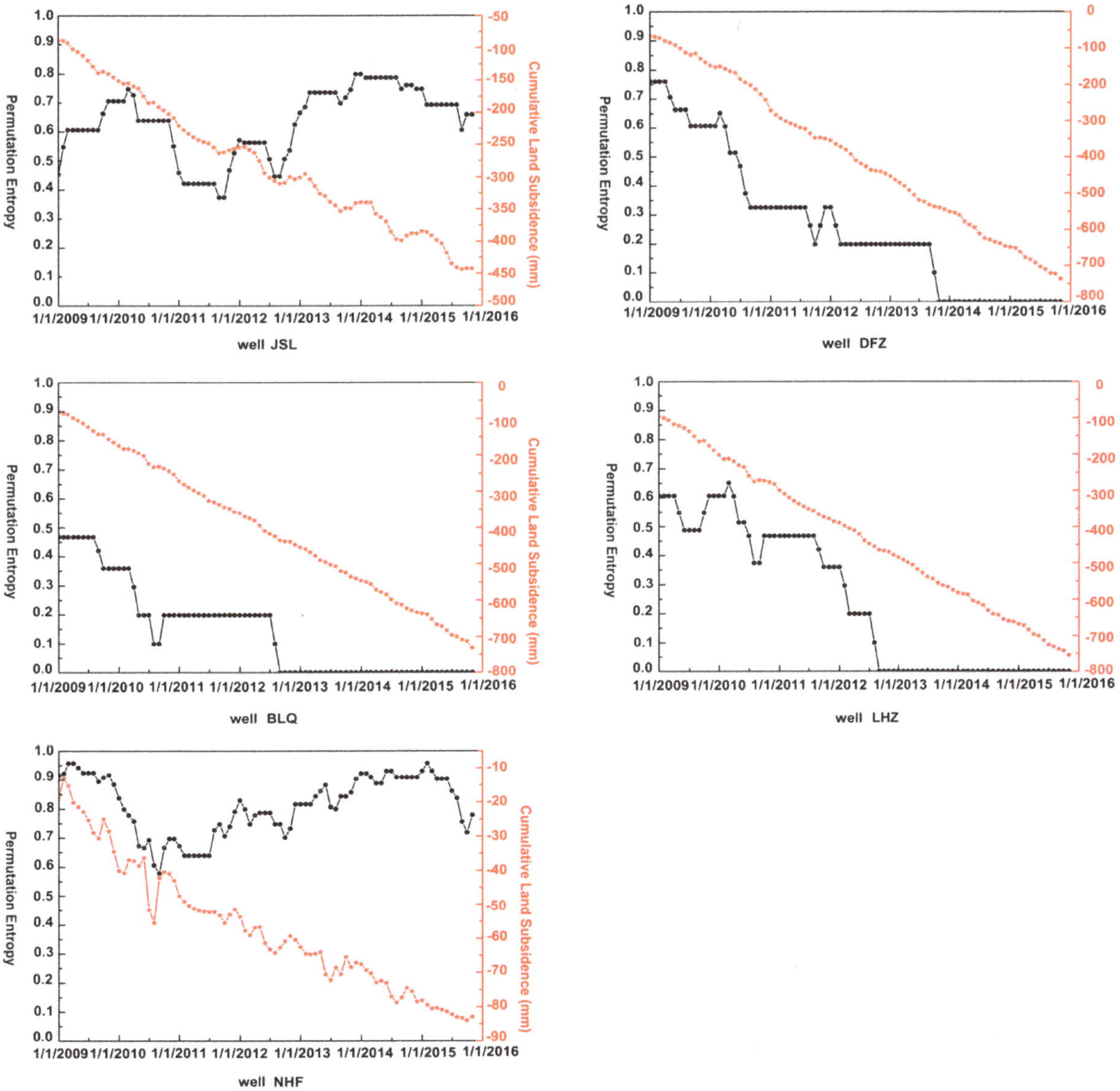

Figure 8. The PE results for the five wells.The black lines indicate the PE values, and the red lines represent the cumulative land subsidence from 2009 to 2015.

Thus, based on the PE results, the state of land subsidence development at the edge of the land subsidence funnel can be divided into three stages. The first stage was from 2009 to 2010, when the PE value increased, which indicated that the land subsidence tended to be nonuniform. The second stage was from 2010 to 2012, when the value of PE decreased, implying that land subsidence developed from a nonuniform to a uniform state. In the third stage, from 2012 to 2015, the PE value increased, indicating that land subsidence developed in a uniform state. The state of land subsidence development in the land subsidence funnel can be arranged in two phases. During the first stage, the value of PE continually decreased between 2009 and 2012, which means that land subsidence developed uniformly in this period. The second stage was from 2012 to 2015, when the PE value returned to zero. This state indicates that the land subsidence during this period developed more uniformly than it had during the previous period.

5. Discussion

5.1. Relationship Between the Spatial Evolution Pattern by the SDE Method and Hydrogeologic Data

Combining the groundwater level contours and compressible sediment layer of the study area from 2007 to 2015 with the SDE obtained by PSI during the period, the correlation between the subsidence response patterns and the phreatic groundwater flow field was analyzed comprehensively (Figure 9). Comparing the groundwater levels of 2015 with the 2007 levels shows that the groundwater level declined by 15–25 m throughout the eastern part of the Beijing Plain.

When referring to the SDE, we find that it outlined two subsidence funnels. As times passed, the extent of land subsidence in the northwest–southeast direction expanded after 2012; then, the SDE started to develop in this direction. The expansion of land subsidence distribution was due to the groundwater cone of depression. In 2007, the groundwater elevations varied from 10 to 20 m, and there was a groundwater cone of depression in the LGY subsidence funnel. As time passed, the groundwater contours in this area became more compact, as in 2015. This result means that the groundwater consumption increased greatly. The LGY groundwater cone of depression expanded after 2012, and correspondingly, the rate and range of land subsidence in this area increased. The DBL subsidence funnel was not in the zone of the groundwater cone of depression bowl. However, land subsidence in this area achieved a maximum value, and the deformation increased from 2007 to 2015. This effect may have occurred due to soil consolidation, which causes hysteresis in the groundwater.

The stratum structure of the Beijing Plain is characterized by a transformation from a single structure zone to a multilayered structure zone from the northwest to the southeast. The sediment particles changed from coarse to fine, the thickness gradually increases, and the proportion of clay soil increases gradually [33]. Note that the thickness of the loose Quaternary sediments in the study area ranges from 80 to 210 m. The thickness of the compressible clay layer in LGY varies from 130 to 220 m and the thickness changed from 110 to 180 m in DBL. Meanwhile, land subsidence in these two areas achieved the maximum values. This result means that the spatial distribution of land subsidence and the thickness of the compressible sediment layer are significantly correlated. The greater the cumulative thickness of the sediment layer is, the greater the total land subsidence.

From comprehensive groundwater and compressible soil data, we find that the spatial evolution characteristics of land subsidence in the Eastern Beijing Plain are basically consistent with those of groundwater, although subsidence is restricted by geological conditions. The compressible soil layer provides the environment for land subsidence and the overexploitation of groundwater is the main cause of land subsidence.

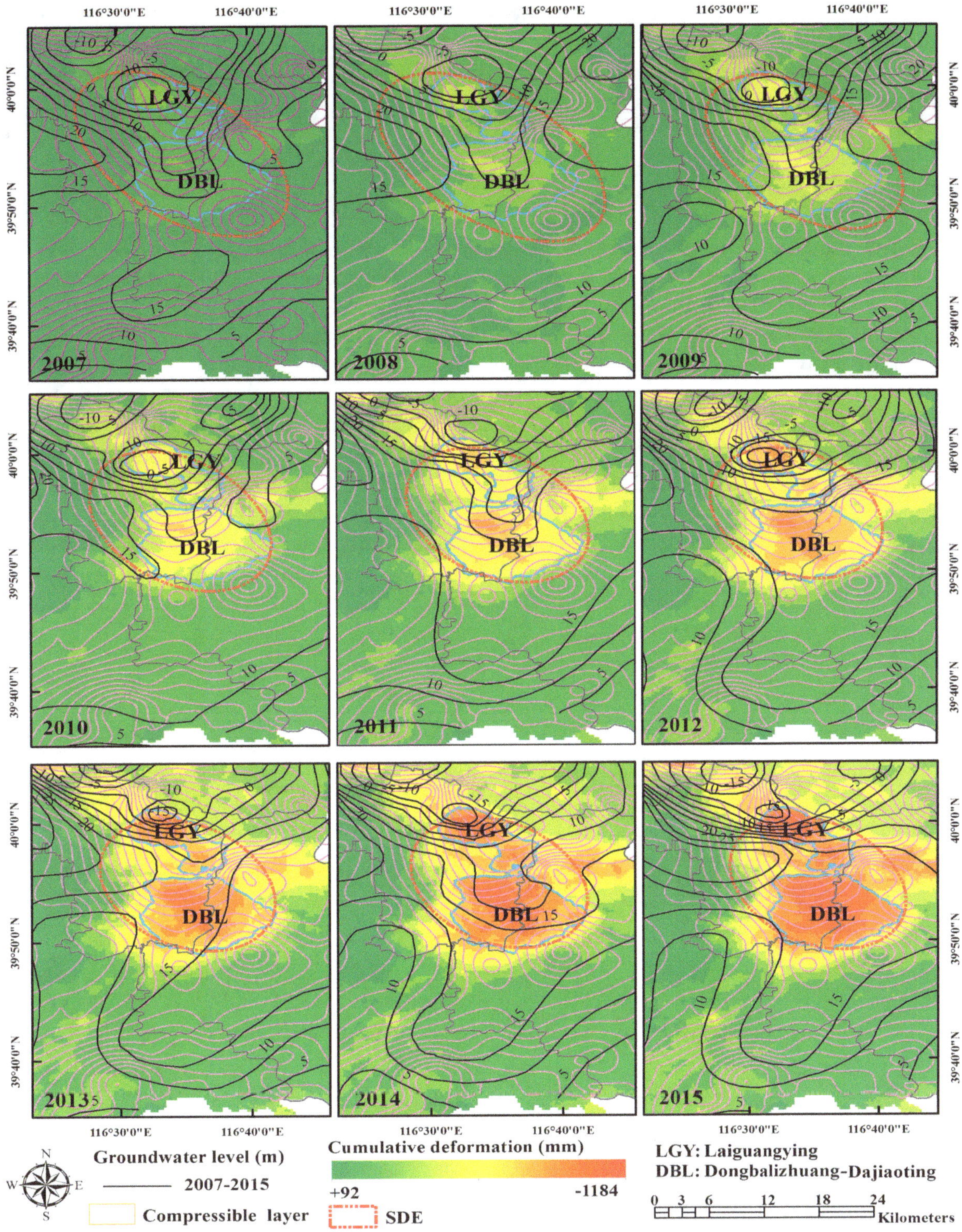

Figure 9. The spatial evolution patterns of land subsidence obtained by the SDE method. The black countours indicate the changing groundwater level data from 2007 to 2015. The heliotrope lines represent the compressible sediment layer data. The red ellipses indicate the SDEs for every year. In addition, the groundwater level is referenced to sea level.

5.2. Relationship Between the Temporal Evolution Pattern by the PE Method and Groundwater

From the permutation entropy results, we show that the results differed between the land subsidence funnel and the edge of the funnel. Combining this information with the land subsidence rates, we can discover that the cumulative land subsidence rate at the edge of the land subsidence funnel fluctuated after 2012, while the cumulative land subsidence rate in the land subsidence funnel was increasing during this time. The PE describes the differences in the cumulative settlements. Thus, these differences show that the PE is reliable.

As we know, the overextraction of groundwater is the main cause of settlement in the Beijing Plain. Hence, comprehensive groundwater level variation data are used to discover the relationship between the PE variations and groundwater level changes.

Only well JSL obtained groundwater level variation data from 2011 to 2015. Therefore, JSL was selected to analyze the relationship between the changes in the groundwater level and the PE results (Figure 10). The average annual variations in the groundwater level in well JSL were 6.71 m, 3.46 m and 3.6m between 2009 and 2010, from 2010 to 2012, and from 2012 to 2015, respectively. Combining the groundwater data with the PE results, indicates that between 2009 and 2010, the average annual variation in groundwater was greater than that between 2010 and 2012, while the average annual variation in the groundwater was smaller than that between 2012 and 2015. These results lead to the conclusion that the average annual settlement fluctuation was more pronounced between 2009 and 2010 than between 2010 and 2012. Meanwhile, the average annual settlement fluctuation was less pronounced between 2010 and 2012 than between 2012 and 2015. Therefore, the development of the accumulative settlement between 2009 and 2010 was more uneven than that between 2010 and 2012. In addition, the development of the cumulative settlement between 2010 and 2012 was more uniform than that between 2012 and 2015. This phenomenon corresponds to the process of the PE value increasing between 2009 and 2010 and decreasing between 2010 and 2012. This feature also explains, why PE increased between 2012 and 2015.

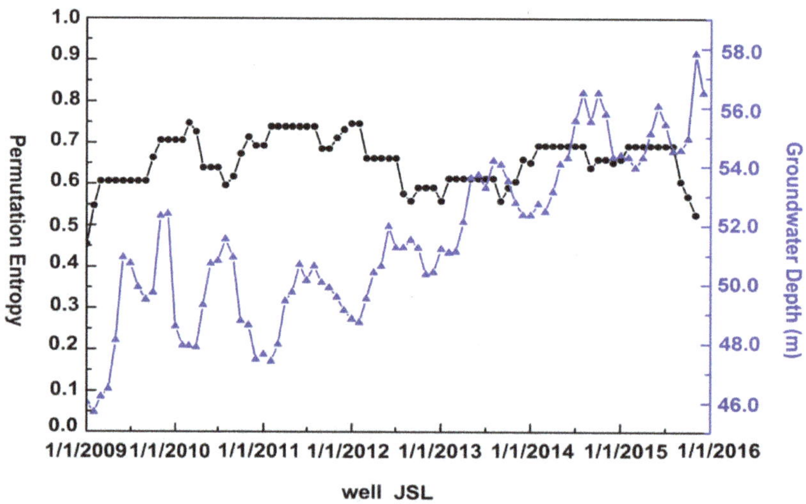

Figure 10. Comparison of the PE results for well JSL and the variations in the groundwater level. The blue line indicates the groundwater level variations. The black line represents the PE results. In addition, the groundwater level is referenced to sea level.

Wells DFZ, BLQ, and LHZ lie within the settlement funnel area of the research area, but BLQ acquired the groundwater level variation data only from 2009 to 2013. Therefore, DFZ and LHZ are selected to analyze the relationship between the variation in the groundwater level and the PE results (Figure 11, Figure 12).

Figure 11 reveals the groundwater level variations in well BLQ. The average annual variation in the groundwater level in well BLQ was 1.51 m between 2009 and 2012, and the groundwater level average annual variation was 0.81 m between 2013 and 2015. Thus, the fluctuation in the groundwater level between 2009 and 2012 was almost twice that between 2013 and 2015. This result indicates that the settlement changes at well BLQ between 2009 and 2012 were faster than those between 2013 and 2015. Therefore, the PE value decreased. However, the groundwater fluctuation between 2013 and

2015 was less than 1 m per year. Thus, during this period, the land subsidence rate fluctuated little, and the accumulative land subsidence showed a slow increase. This result is expressed by the PE value returning to zero.

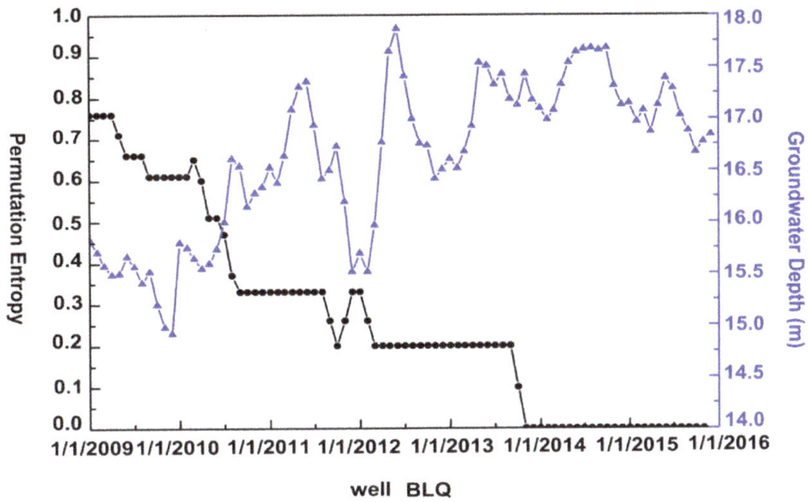

Figure 11. Comparison of the PE results for well BLQ and the variations in the groundwater level. The blue line indicates the groundwater level variations. The black line represents the PE results. The groundwater data reference datum is the same as that in Figure 10.

Figure 12 shows the groundwater level variations in well LHZ. During the period from 2011 to 2012, the groundwater level fluctuated greatly, and the average annual variation in groundwater in this period was 3.83 m. The average annual change in the groundwater level was 1.51 m between 2013 and 2015. From the above results, the groundwater level variation between 2011 and 2012 was more than twice that between 2013 and 2015. This result proves that the fluctuation in the land subsidence rate changed more quickly between 2011 and 2012 than that between 2013 and 2015, which means that between 2011 and 2012, the PE value is greater than that between 2013 and 2015. Thus, this process reflected the PE value decreasing. After 2013, the changes in groundwater were much smaller. Then, the land subsidence rate showed little fluctuation. Hence, the cumulative settlement continued slowly decreasing, which is expressed by the PE value becoming zero.

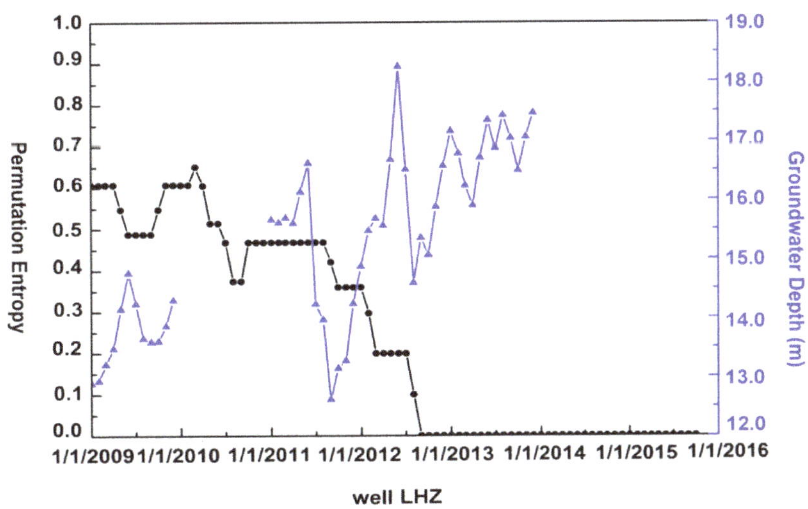

Figure 12. Comparison of the permutation entropy results for well LHZ and the variations in the groundwater level. The blue line indicates the groundwater level variations. The black line represents the PE results. The groundwater data reference datum is the same as that in Figure 10.

The groundwater level variation data for the three wells clearly show that the PE value for JSL was higher than those for DFZ and LHZ during the entire research time. This effect occurred because the average annual change of groundwater in JSL was greater than 3 m, while the average annual groundwater variation levels in DFZ and LHZ were approximately 1 m during the research period.

In summary, the groundwater level change is the main reason for why PE value for the five typical wells differ and the groundwater level changes correlate with the variations in PE values. The larger the variation range of groundwater is, the higher the corresponding PE value.

6. Conclusions

In this work, based on Envisat and Radarsat-2 data from 2007 to 2015, we detect the spatial and temporal evolution patterns of the two most serious land subsidence funnels in the Beijing Plain based on the SDE and PE methods. Integrating hydrogeological data, we conclude that the expansion of the groundwater cone of depression led to the development of the SDE in all directions after 2012. Moreover, the differences in the annual variations in the groundwater levels are the main reason for the different PE values of the land subsidence funnel and the funnel edge.

First, we utilize the PSI technique to obtain the long-term displacement in the Eastern Beijing Plain, China. The vertical displacement rates agree well with the measurements from ground leveling surveys: the correlation coefficient is 0.94, which indicates that our PSI results are reliable.

Then, we adopt the SDE to analyze the spatial evolution pattern of land subsidence. The SDE results suggest that the development of land subsidence in the southeast-northwest direction is more obvious than those in other directions; however, land subsidence develops obviously in all directions after 2012. Land subsidence is serious, and the spatial distribution of land subsidence increases over time. However, after 2012, the rate of increase of land subsidence decreases. Comparing the spatial evolution pattern using the SDE methods with the groundwater level changes and compressible sediment thicknesses, we find that the groundwater cone of depression expands and the range increases after 2012. At the same time, the compressible sediment thickness in the SDE range is large, providing an environment for land subsidence.

Finally, we use the PE method to reveal the time-series evolution pattern of land subsidence, and the results show that the time-series evolution pattern at the edge of the land subsidence funnel is different from that within the land subsidence funnel. The settlement development process at the edge of the land subsidence funnel is divided into three stages. From 2009 to 2010, the land subsidence development is uneven. From 2010 to 2012, the land subsidence development is relatively uniform. From 2012 to 2015, the development of land subsidence becomes nonuniform. We divide the land subsidence development of the land subsidence funnel into two stages. From 2009 to 2012, land subsidence tends to be uniform. From 2012 to 2015, land subsidence is more even than that in the previous period. Comparing the PE results with the groundwater level change data, the value of the PE increase corresponds to the process of increase in the groundwater level variations. This result indicates that the process whereby land subsidence tends to be nonuniform. Conversely, the value of PE decreases, corresponding to the process of decrease in the groundwater level variation. This result shows that the process whereby land subsidence tends to be uniform. The larger the variation range of the groundwater is, the higher the corresponding PE value.

In this paper, we use the PE method to discover the temporal evolution process of deformation which can help us to better understand land subsidence. In the future, we will decompose the effects of the groundwater on land subsidence to determine more details.

Author Contributions: J.Z. performed the experiments, analyzed the data and wrote the paper. H.G. and B.C. provided crucial guidance and support through the research. K.L. and C.Z. processed the data. Y.K. made important suggestions on writing the paper.

Acknowledgments: This work was funded by the National Natural Science Foundation of China (number 41771455/D010702; 41401493), the Beijing Outstanding Young Scientist Project, the Beijing Natural Science Foundation (8182013), the China Postdoctoral Science Foundation (2018M641407), Beijing Youth Top Talent Project, the program of Beijing Scholars, and the National "Double-Class" Construction of University Projects, Capacity Building for Sci-Tech Innovation-Fundamental Scientific Research Funds(025185305000/194). Thanks for the excellent software package SARPROZ.

References

1. Wright, T.J.; Parsons, B.E.; Lu, Z. Toward mapping surface deformation in three dimensions using InSAR. *Geophys. Res. Lett.* **2004**, *31*, 169–178. [CrossRef]
2. Lohman, R.B.; Simons, M. Some thoughts on the use of InSAR data to constrain models of surface deformation: Noise structure and data downsampling. *Geochem. Geophys. Geosyst.* **2013**, 6. [CrossRef]
3. Chaussard, E.; Wdowinski, S.; Cabral-Cano, E.; Amelung, F. Land subsidence in central Mexico detected by ALOS InSAR time-series. *Remote Sens. Environ.* **2014**, *140*, 94–106. [CrossRef]
4. Eriksen, H.Ø.; Lauknes, T.R.; Larsen, Y.; Corner, G.D.; Bergh, S.G.; Dehls, J.; Kierulf, H.P. Visualizing and interpreting surface displacement patterns on unstable slopes using multi-geometry satellite SAR interferometry (2D InSAR). *Remote Sens. Environ.* **2017**, *191*, 297–312. [CrossRef]
5. Ferretti, A.; Prati, C.; Rocca, F. Nonlinear subsidence rate estimation using permanent scatterers in differential SAR interferometry. *IEEE Trans. Geosci. Remote Sens.* **2000**, *38*, 2202–2212. [CrossRef]
6. Ferretti, A.; Colesanti, C.; Prati, C.; Rocca, F. Radar permanent scatterers identification in urban areas: Target characterization and sub-pixel analysis. In Proceedings of the IEEE Remote Sensing and Data Fusion Over Urban Areas, IEEE/isprs Joint Workshop, Rome, Italy, 8–9 November 2001; p. 52.
7. Gong, H.; Zhang, Y.; Li, X.; Lu, X.; Chen, B.; Gu, Z. Land subsidence research in Beijing based on the Permanent Scatterers InSAR technology. *Prog. Nat. Sci.* **2009**, *19*, 1261–1266.
8. Ng, A.H.; Ge, L.; Li, X.; Abidin, H.Z.; Andreas, H.; Zhang, K. Mapping land subsidence in Jakarta, Indonesia using persistent scatterer interferometry (PSI) technique with ALOS PALSAR. *Int. J. Appl. Earth Obs. Geoinf.* **2012**, *18*, 232–242. [CrossRef]
9. Chaussard, E.; Amelung, F.; Abidin, H.; Hong, S.H. Sinking cities in Indonesia: ALOS PALSAR detects rapid subsidence due to groundwater and gas extraction. *Remote Sens. Environ.* **2013**, *128*, 150–161. [CrossRef]
10. Raspini, F.; Loupasakis, C.; Rozos, D.; Adam, N.; Moretti, S. Ground subsidence phenomena in the Delta municipality region (Northern Greece): Geotechnical modeling and validation with Persistent Scatterer Interferometry. *Int. J. Appl. Earth Obs. Geoinf.* **2014**, *28*, 78–89. [CrossRef]
11. Salvi, S.; Atzori, S.; Tolomei, C.; Antonioli, A.; Trasatti, E.; Merryman Boncori, J.P.; Pezzo, G.; Coletta, A.; Zoffoli, S. Results from INSAR monitoring of the 2010–2011 New Zealand seismic sequence: EA detection and earthquake triggering. In Proceedings of the IEEE International Geoscience and Remote Sensing Symposium, Munich, Germany, 22–27 July 2012; pp. 3544–3547.
12. Champenois, J.; Fruneau, B.; Pathier, E.; Deffontaines, B.; Lin, K.-C.; Hu, J.-C. Monitoring of active tectonic deformations in the Longitudinal Valley (eastern Taiwan) using Persistent Scatterer InSAR method with ALOS PALSAR data. *Earth Planet. Sci. Lett.* **2012**, *337*, 144–155. [CrossRef]
13. Krishnan, S.P.V.; Kim, D.; Jung, J. Subsidence in the Kathmandu Basin, before and after the 2015 Mw 7.8 Gorkha Earthquake, Nepal Revealed from Small Baseline Subset-DInSAR Analysis. *GISci. Remote Sens.* **2018**, *55*, 604–621.
14. Farina, P.; Colombo, D.; Fumagalli, A.; Marks, F.; Moretti, S. Permanent Scatterers for landslide investigations: Outcomes from the ESASLAM project. *Eng. Geol.* **2006**, *88*, 200–217. [CrossRef]
15. Mikhailov, V.; Kiseleva, E.; Smolyaninova, E.; Golubev, V.; Dmitriev, P.; Isaev, Y.; Dorokhin, K.; Hooper, A.; Esfahany, S.; Hanssen, R.; Khairetdinov, S. PS-InSAR Monitoring of Landslides in The Great Caucasus, Russia, Using Envisat, ALOS And TerraSAR-X SAR Images. In Proceedings of the ESA Living Planet Symposium, Edinburgh, United Kingdom, 9–13 September 2013; p. 283.
16. Kiseleva, E.; Mikhailov, V.; Smolyaninova, E.; Dmitriev, P.; Golubev, V.; Timoshkina, E.; Hooper, A.; Samiei-Esfahany, S.; Hanssen, R. PS-InSAR Monitoring of Landslide Activity in the Black Sea Coast of the Caucasus. *Procedia Technol.* **2014**, *16*, 404–413. [CrossRef]
17. Zhu, L.; Gong, H.; Li, X.; Wang, R.; Chen, B.; Dai, Z.; Teatini, P. Land subsidence due to groundwater withdrawal in the northern Beijing plain, China. *Eng. Geol.* **2015**, *193*, 243–255. [CrossRef]

18. Chen, B.; Gong, H.; Li, X.; Lei, K.; Gao, M.; Zhou, C.; Ke, Y. Spatial–temporal evolution patterns of land subsidence with different situation of space utilization. *Nat. Hazards* **2015**, *77*, 1765–1783. [CrossRef]
19. Chen, M.; Tomás, R.; Li, Z.; Motagh, M.; Li, T.; Hu, L.; Gong, H.; Li, X.; Yu, J.; Gong, X. Imaging Land Subsidence Induced by Groundwater Extraction in Beijing (China) Using Satellite Radar Interferometry. *Remote Sens.* **2016**, *8*, 468. [CrossRef]
20. Zhang, Y.; Wu, H.A.; Kang, Y.; Zhu, C. Ground Subsidence in the Beijing-Tianjin-Hebei Region from 1992 to 2014 Revealed by Multiple SAR Stacks. *Remote Sens.* **2016**, *8*, 675. [CrossRef]
21. Zhou, C.; Gong, H.; Zhang, Y.; Warner, T.; Wang, C. Spatiotemporal Evolution of Land Subsidence in the Beijing Plain 2003–2015 Using Persistent Scatterer Interferometry (PSI) with Multi-Source SAR Data. *Remote Sens.* **2018**, *10*, 552. [CrossRef]
22. Zhou, C.; Gong, H.; Chen, B.; Guo, L.; Gao, M.; Chen, W.; Liang, Y.; Si, Y.; Wang, J.; Zhang, X. Spatiotemporal characteristics of land subsidence in Beijing from small baseline subset interferometric synthetic aperture radar and standard deviational ellipse. In Proceedings of the International Workshop on Earth Observation and Remote Sensing Applications, Guangzhou, China, 4–6 July 2016; pp. 78–82.
23. Chen, B.; Gong, H.; Li, X.; Lei, K.; Zhu, L.; Gao, M.; Zhou, C. Characterization and causes of land subsidence in Beijing, China. *Int. J. Remote Sens.* **2017**, *38*, 808–826. [CrossRef]
24. Yang, Q.; Ke, Y.; Zhang, D.; Chen, B.; Gong, H.; Lv, M.; Zhu, L.; Li, X. Multi-Scale Analysis of the Relationship between Land Subsidence and Buildings: A Case Study in an Eastern Beijing Urban Area Using the PS-InSAR Technique. *Remote Sens.* **2018**, *10*, 1006. [CrossRef]
25. Bandt, C.; Pompe, B. Permutation entropy: A natural complexity measure for time series. *Phys. Rev. Lett.* **2002**, *88*, 1–4. [CrossRef] [PubMed]
26. Feng, F.Z.; Rao, G.Q.; Si, A.W.; Sun, Y. Application and Development of Permutation Entropy Algorithm. *J. Acad. Armored Force Eng.* **2012**, *02*, 34–38. [CrossRef]
27. Liu, J.; Zhang, C.; Zheng, C.; Yu, X. Mental fatigue analysis based on complexity measure of multichannel electroencephalogram. *J. Xian Jiaotong Univ.* **2008**, *42*, 1555–1559.
28. Soriano, M.C.; Zunino, L.; Larger, L.; Fischer, I.; Mirasso, C.R. Distinguishing fingerprints of hyperchaotic and stochastic dynamics in optical chaos from a delayed opto-electronic oscillator. *Opt. Lett.* **2011**, *36*, 2212–2214. [CrossRef] [PubMed]
29. Veisi, I.; Pariz, N.; Karimpour, A. Fast and Robust Detection of Epilepsy in Noisy EEG Signals Using Permutation Entropy. In Proceedings of the IEEE International Conference on Bioinformatics and Bioengineering, Boston, MA, USA, 14–17 October 2007; pp. 200–203.
30. Wang, L. A universal algorithm to generate pseudo-random numbers based on uniform mapping as homeomorphism. *Chin. Phys. B* **2010**, *19*, 090505.
31. *Beijing Water Resources Bureau*; Beijing Water Resources Bulletin: Beijing, China, 2017.
32. Lefever, D.W. Measuring geographic concentration by means of the stand deviational ellipse. *Am. J. Sociol.* **1926**, *31*, 88–94. [CrossRef]
33. Lei, K.; Luo, Y.; Chen, B.; Guo, G.; Zhou, Y. Distribution characteristics and influence factors of land subsidence in Beijing area. *Geol. China* **2016**, *43*, 2216–2228.

Permissions

All chapters in this book were first published in MDPI; hereby published with permission under the Creative Commons Attribution License or equivalent. Every chapter published in this book has been scrutinized by our experts. Their significance has been extensively debated. The topics covered herein carry significant findings which will fuel the growth of the discipline. They may even be implemented as practical applications or may be referred to as a beginning point for another development.

The contributors of this book come from diverse backgrounds, making this book a truly international effort. This book will bring forth new frontiers with its revolutionizing research information and detailed analysis of the nascent developments around the world.

We would like to thank all the contributing authors for lending their expertise to make the book truly unique. They have played a crucial role in the development of this book. Without their invaluable contributions this book wouldn't have been possible. They have made vital efforts to compile up to date information on the varied aspects of this subject to make this book a valuable addition to the collection of many professionals and students.

This book was conceptualized with the vision of imparting up-to-date information and advanced data in this field. To ensure the same, a matchless editorial board was set up. Every individual on the board went through rigorous rounds of assessment to prove their worth. After which they invested a large part of their time researching and compiling the most relevant data for our readers.

The editorial board has been involved in producing this book since its inception. They have spent rigorous hours researching and exploring the diverse topics which have resulted in the successful publishing of this book. They have passed on their knowledge of decades through this book. To expedite this challenging task, the publisher supported the team at every step. A small team of assistant editors was also appointed to further simplify the editing procedure and attain best results for the readers.

Apart from the editorial board, the designing team has also invested a significant amount of their time in understanding the subject and creating the most relevant covers. They scrutinized every image to scout for the most suitable representation of the subject and create an appropriate cover for the book.

The publishing team has been an ardent support to the editorial, designing and production team. Their endless efforts to recruit the best for this project, has resulted in the accomplishment of this book. They are a veteran in the field of academics and their pool of knowledge is as vast as their experience in printing. Their expertise and guidance has proved useful at every step. Their uncompromising quality standards have made this book an exceptional effort. Their encouragement from time to time has been an inspiration for everyone.

The publisher and the editorial board hope that this book will prove to be a valuable piece of knowledge for researchers, students, practitioners and scholars across the globe.

List of Contributors

Miguel Laverde-Barajas
IHE-Delft Institute for Water Education, Westvest 7, 2611 AX Delft, The Netherlands
Department of Water Management, Faculty of Civil Engineering and Geosciences, Delft University of Technology, 2628 CN Delft, The Netherlands
SERVIR-Mekong, SM Tower, 24th Floor, 979/69 Paholyothin Road, Samsen Nai Phayathai, Bangkok 10400, Thailand
Asian Disaster Preparedness Center, SM Tower, 24th Floor, 979/69 Paholyothin Road, Samsen Nai Phayathai, Bangkok 10400, Thailand

Dimitri P. Solomatine
IHE-Delft Institute for Water Education, Westvest 7, 2611 AX Delft, The Netherlands
Department of Water Management, Faculty of Civil Engineering and Geosciences, Delft University of Technology, 2628 CN Delft, The Netherlands

Gerald A. Corzo
IHE-Delft Institute for Water Education, Westvest 7, 2611 AX Delft, The Netherlands

Chinaporn Meechaiya, Susantha Jayasinghe and Peeranan Towashiraporn
SERVIR-Mekong, SM Tower, 24th Floor, 979/69 Paholyothin Road, Samsen Nai Phayathai, Bangkok 10400, Thailand
Asian Disaster Preparedness Center, SM Tower, 24th Floor, 979/69 Paholyothin Road, Samsen Nai Phayathai, Bangkok 10400, Thailand

Ate Poortinga and Farrukh Chishtie
SERVIR-Mekong, SM Tower, 24th Floor, 979/69 Paholyothin Road, Samsen Nai Phayathai, Bangkok 10400, Thailand
Spatial Informatics Group, LLC, 2529 Yolanda Ct., Pleasanton, CA 94566, USA

Amanda Markert
Earth System Science Center, The University of Alabama in Huntsville, 320 Sparkman Drive, Huntsville, AL 35805, USA
SERVIR Science Coordination Office, NASA Marshall Space Flight Center, 320 Sparkman Drive, Huntsville, AL 35805, USA

David Saah
SERVIR-Mekong, SM Tower, 24th Floor, 979/69 Paholyothin Road, Samsen Nai Phayathai, Bangkok 10400, Thailand
Asian Disaster Preparedness Center, SM Tower, 24th Floor, 979/69 Paholyothin Road, Samsen Nai Phayathai, Bangkok 10400, Thailand
Geospatial Analysis Lab, University of San Francisco, 2130 Fulton Street, San Francisco, CA 94117, USA

Lam Hung Son and Sothea Khem
Mekong River Commission, Flood Management and Mitigation Programme, 576 National Road No. 2, Sangkat Chak Angre Krom, Khan Menachey, Phnom Penh 12353, Cambodia

Surajate Boonya-Aroonnet and Winai Chaowiwat
Hydroinformatic Institute, 901 Ngam Wong Wan Road, Lat Yao, Chatuchak, Bangkok 10900, Thailand

Remko Uijlenhoet
Hydrology and Quantitative Water Management Group, Department of Environmental Sciences, Wageningen University, 6700 AA Wageningen, The Netherlands

Augusto Getirana and Kristi Arsenault
Hydrological Sciences Laboratory, NASA Goddard Space Flight Center, Greenbelt, MD 20771, USA
Science Applications International Corporation, Greenbelt, MD 20771, USA

Dalia Kirschbaum
Hydrological Sciences Laboratory, NASA Goddard Space Flight Center, Greenbelt, MD 20771, USA

Felipe Mandarino
Instituto Pereira Passos, Rio de Janeiro 22221-070, Brazil

Marta Ottoni
Geological Survey of Brazil, Rio de Janeiro 22290-240, Brazil

Sana Khan
Hydrological Sciences Laboratory, NASA Goddard Space Flight Center, Greenbelt, MD 20771, USA
Earth System Science Interdisciplinary Center, University of Maryland, College Park, MD 20740, USA

Qin Yang
College of Resources Environment and Tourism, Capital Normal University, 105 North Road of the 3rd Ringroad, Haidian District, Beijing 100048, China
Laboratory Cultivation Base of Environment Process and Digital Simulation, Beijing 100048, China

Dongyi Zhang, Mingyuan Lv, Lin Zhu and Xiaojuan Li
College of Resources Environment and Tourism, Capital Normal University, 105 North Road of the 3rd Ringroad, Haidian District, Beijing 100048, China
Laboratory Cultivation Base of Environment Process and Digital Simulation, Beijing 100048, China
Beijing Laboratory of Water Resources Security, Beijing 100048, China

Huili Gong, Yinghai Ke and Beibei Chen
Base of the Key Laboratory of Urban Environmental Process and Digital Modeling, Capital Normal University, Beijing 100048, China
College of Resources Environment and Tourism, Capital Normal University, 105 North Road of the 3rd Ringroad, Haidian District, Beijing 100048, China
Laboratory Cultivation Base of Environment Process and Digital Simulation, Beijing 100048, China
Beijing Laboratory of Water Resources Security, Beijing 100048, China

Eunna Jang, Yoojin Kang and Jungho Im
School of Urban and Environmental Engineering, Ulsan National Institute of Science and Technology (UNIST), Ulsan 44919, Korea

Dong-Won Lee, Jongmin Yoon and Sang-Kyun Kim
Environmental Satellite Centre, Climate and Air Quality Research Department, National Institute of Environmental Research, Incheon 22689, Korea

Minsang Kim
School of Urban and Environmental Engineering, Ulsan National Institute of Science and Technology, Ulsan 44919, Korea
Satellite Analysis Division, National Meteorological Satellite Center, Jincheon-gun, Chungcheongbuk-do 27803, Korea

Myong-In Lee
School of Urban and Environmental Engineering, Ulsan National Institute of Science and Technology, Ulsan 44919, Korea

Myung-Sook Park
Korea Ocean Satellite Center, Korea Institute of Ocean Science and Technology, Busan 49111, Korea

Seonyoung Park
Satellite Operation and Application Center, Korea Aerospace Research Institute, Daejeon 34133, Korea

Boksoon Myoung
APEC Climate Center, 12 Centum 7-ro, Haeundae-gu, Busan 48058, Korea

Seung Hee Kim, Shenyue Jia, Kristen Whitney and Menas C. Kafatos
APEC Climate Center, 12 Centum 7-ro, Haeundae-gu, Busan 48058, Korea
Center of Excellence in Earth Systems Modeling and Observations, Chapman University, Orange, CA 92866, USA

Son V. Nghiem
Jet Propulsion Laboratory, California Institute of Technology, Pasadena, CA 91109, USA

Anming Bao
State Key Laboratory of Desert and Oasis Ecology, Xinjiang Institute of Ecology and Geography, Chinese Academy of Sciences, Urumqi 830011, China

Jieyun Zhang
State Key Laboratory of Desert and Oasis Ecology, Xinjiang Institute of Ecology and Geography, Chinese Academy of Sciences, Urumqi 830011, China
University of Chinese Academy of Sciences, Beijing 100039, China

Qingling Zhang
State Key Laboratory of Desert and Oasis Ecology, Xinjiang Institute of Ecology and Geography, Chinese Academy of Sciences, Urumqi 830011, China
School of Aeronautics and Astronautics, Sun Yat-Sen University, Guangzhou 510275, China

Yujuan Wang
China-Asean Environment Cooperation Center, Beijing 100875, China

Jiaxing Ye and Hiroshi Tsuda
National Metrology Institute of Japan (NMIJ), The National Institute of Advanced Industrial Science and Technology (AIST), Tsukuba 305-0045, Japan

Yuichi Kurashima and Takeshi Kobayashi
Department of Electronics and Manufacturing, The National Institute of Advanced Industrial Science and Technology (AIST), Tsukuba 305-8564, Japan

Teruyoshi Takahara and Wataru Sakurai
Sabo Department, National Institute for Land and Infrastructure Management (NILIM), Tsukuba 305-0804, Japan

Jae-Hyun Ryu and Jaeil Cho
Department of Applied Plant Science, Chonnam National University, 77 Yongbong-ro, Gwangju 61186, Korea

Kyung-Soo Han and Yang-Won Lee
Department of Spatial Information Engineering, Pukyong National University, 45 Yongsoro, Namgu, Busan 48513, Korea

Sungwook Hong
Department of Environment, Energy, and Geoinfomatics, Sejong University, 209 Neungdong-ro, Gwangjin-gu, Seoul 05006, Korea

No-Wook Park
Department of Geoinformatic Engineering, Inha University, 100 Inha-ro, Nam-gu, Incheon 22212, Korea

Yeonjin Lee, Myoung-Hwan Ahn and Su Jeong Lee
Department of Atmospheric Sciences and Engineering, Ewha Womans University, 52 Ewhayeodae-gil, Seodaemun-gu, Seoul 03760, Korea

Daehyeon Han
School of Urban and Environmental Engineering, Ulsan National Institute of Science and Technology, Ulsan 44919, Korea

Meihong Ma
China Institute of Water Resources and Hydropower Research, Beijing 100038, China
College of water sciences, Beijing Normal University, Beijing 100875, China

Changjun Liu
China Institute of Water Resources and Hydropower Research, Beijing 100038, China

Huixiao Wang
College of water sciences, Beijing Normal University, Beijing 100875, China

Gang Zhao
School of Geographical Sciences, University of Bristol, Bristol BS8 1SS, UK

Hongjie Xie
Department of Geological Sciences University of Texas at San Antonio, San Antonio, TX 78249, USA

Pengfei Jia
CITIC Construction Co., Ltd., Beijing 100027, China

Dacheng Wang
Lab of Spatial Information Integration, Institute of Remote Sensing and Digital Earth, Chinese Academy of Sciences, Beijing 100101, China

Yang Hong
School of Earth and Space Sciences, Peking University, Beijing 100871, China

Junjie Zuo, Kaisi Liu and Chaofan Zhou
Base of the Key Laboratory of Urban Environmental Process and Digital Modeling, Capital Normal University, Beijing 100048, China
College of Resources Environment and Tourism, Capital Normal University, Beijing 100048, China

Index

www.ingramcontent.com/pod-product-compliance
Lightning Source LLC
LaVergne TN
LVHW081030210926
848485LV00004B/991